Canada and the Global Economy
The Geography of Structural and Technological Change

A collection of essays by twenty-three of Canada's leading economic geographers, *Canada and the Global Economy* is a comprehensive study of the evolving economic and geographic patterns of Canadian development. It provides a benchmark for research on the spatial development of the Canadian economy.

The contributors explore four central themes: the locational impacts of the openness of the Canadian economy, Canada's relatively simple conomic geography in terms of regional variations in resources and urban development, the problems of keeping pace with rapid advances in technology, and the role of government in maintaining a national market and assisting economic development. They outline the essential elements of Canada's contemporary economic geography and highlight the origins and spatial imprint of change in the Canadian economy; in particular they provide an assessment of Canada's participation in significant international patterns of economic change.

Canada and the Global Economy is concerned not only with the economic size and location of consumption and production but also with institutional changes and shifts in employment, the sectoral composition of economic activity, and the organizational structure and locational behaviour of particular industries and firms. Special attention is given to the technological development of both established industries and new service and manufacturing activities. A timely addition to the field, it provides a geographic perspective on significant changes in jobs and types of work that result from the transformation of economic activities.

JOHN N.H. BRITTON is professor of geography, University of Toronto.

Canada and the Global Economy: The Geography of Structural and Technological Change

EDITED BY

JOHN N.H. BRITTON

McGill-Queen's University Press
Montreal & Kingston • London • Buffalo

© McGill-Queen's University Press 1996
ISBN 0-7735-0927-5 (cloth)
ISBN 0-7735-1356-6 (paper)

Legal deposit second quarter 1996
Bibliothèque nationale du Québec

Printed in Canada on acid-free paper

McGill-Queen's University Press acknowledges the
financial support of the Government of Canada through
the Canadian Studies and Special Projects Directorate of the
Department of Canadian Heritage.

Canadian Cataloguing in Publication Data

Main entry under title:
  Canada and the global economy: the geography of structural
  and technological change
  (Canadian Association of Geographers series in Canadian
  geography)
  Includes bibliographical references and index.
  ISBN 0-7735-0927-5 (bound) –
  ISBN 0-7735-1356-6 (pbk.)
  1. Canada – Economic conditions – 1991–
  2. Technological innovations – Canada. 3. Canada –
  Foreign economic relations. 4. Economic geography.
  I. Britton, John N.H. (John Nigel Haskings), 1939–
  II. Series.
  HC115.C1865 1996   330.971′0648   C96-900023-5

This book was typeset by Typo Litho Composition Inc.
in 10/12 Times.

Contents

Contributors

TREVOR J. BARNES   Department of Geography, University of British Columbia

JOHN N.H. BRITTON   Department of Geography, University of Toronto

JAMES CANNON   Department of Geography, Queen's University

WILLIAM J. COFFEY   Département de géographie, Université de Montréal

J. TAIT DAVIS   Department of Geography, York University

GEOFFREY P. DOBILAS   Department of Geography, University of Toronto

WILLIAM C. FOUND   Faculty of Environmental Studies, York University

MERIC S. GERTLER   Department of Geography, University of Toronto

JAMES M. GILMOUR   Consultant, Ottawa, formerly at Science Council of Canada

ROGER HAYTER   Department of Geography, Simon Fraser University

JOHN HOLMES   Department of Geography, Queen's University

A.C. LEA   Compusearch, Toronto

IAN MACLACHLAN   Department of Geography, University of Lethbridge

VIRGINIA W. MACLAREN   Department of Geography, University of Toronto

ALAN D. MACPHERSON   Department of Geography, State University of New York at Buffalo

GLEN NORCLIFFE   Department of Geography, York University

D. MICHAEL RAY   formerly at Department of Geography, Carleton University

TOD RUTHERFORD   Department of Geography, University of Waterloo

R. KEITH SEMPLE   Department of Geography, University of Saskatchewan

JAMES W. SIMMONS   Department of Geography, University of Toronto

WILLIAM SMITH   Department of Geography, University of Auckland, formerly at Science Council of Canada

GUY P.F. STEED   formerly at Science Council of Canada

IAIN WALLACE   Department of Geography, Carleton University

NIGEL M. WATERS   Department of Geography, University of Calgary

Preface

In this book 24 geographers seek to capture the essential elements of Canada's contemporary economic geography, especially the dynamics of change in the economy and its spatial imprint. Inevitably we have ranged widely, from the impact of international competition on regional resource economies, with their boom-and-bust cycles, and the locational repercussions of technological change, to the industrial implications of North American free trade. Throughout, we have been alert to the more general changes occurring in regional economic development at the international scale, to generic changes in work and technology that affect the competitiveness of Canadian industries and regions, and to the necessity for effective design of public initiatives to increase the rate of industrial innovation.

It will be quickly apparent that this book is motivated by the same concerns that stimulate many writers in policy and professional fields and in other social sciences. But as geographers we are interested ultimately in the changing spatial distribution of economic activity. Even so, it is important to recognize that the geographic scale at which we engage particular problems may vary. This book ranges over issues such as Canada's international economic geography, the urban or regional effects of innovations in various industries, and the significance of investment decisions made by exemplary businesses.

From the outset it was determined that individual chapters would draw on the current research of authors and would build on these blocks of expertise. The book organizes that research around several important themes in the economic geography of Canada. Once these common elements were established, I commissioned chapters, bearing in mind the size constraint for the finished work. The resultant product is true to its title and demonstrates much of the best current research on Canadian economic geography.

Given the number of prior commitments by the authors, this volume could never be an "instant book." Nor was it designed to be. These chapters, however, are considered reflections of the patterns of location and processes of change that have been long-term features of the Canadian economy or that have begun to emerge in the 1990s.

Inevitably, some powerful current influences on Canada's economic geography have eluded analysis and explanation. We still do not have empirical appraisals of the immediate effects of the North American Free Trade Agreement (NAFTA) or of the Canada–United States Free Trade Agreement (FTA), which preceded it. Moreover, the shaky transition of the Canadian economy from recession to recovery obscures and complicates these inquiries. The effects of the renegotiations of world subsidy and trading arrangements, under the Uruguay Round of the General Agreement on Tariffs and Trade (GATT), are still of unknown scale. Nonetheless, this book provides a benchmark for future analysis and a sense of the significant processes of change in economic activities. We hope that it, above all, provides cogent and robust guides to the pattern of locational development of the Canadian economy and provides points of departure for ensuing research.

John N.H. Britton
Toronto

Canada and the Global Economy

CHAPTER ONE

Introduction

JOHN N.H. BRITTON

This is a singularly appropriate time to examine the structure, performance, and pattern of changes in the Canadian economy, to assess what appear to be current trends, and to account for the sources of uncertainty that it faces. In addition to the worldwide recession, a number of powerful forces have been causing the reorganization of economic activities at the global scale. The application of advanced information technology in all areas of production and consumption, for example, is permitting new forms of industrial organization, and a wide variety of companies and participants in many markets now experience greatly reduced geographic constraints on their functions. Changes in the regulation of international trade similarly affect the organization of economic activity. The impact of these changes is shown by the locational shifts in manufacturing and services occurring across North America and elsewhere. In effect, old arrangements of production and service activities are being challenged as new locations threaten to replace them, and skills in use of new machines are being introduced as advanced technologies are developed and adopted in the workplace.

As a small industrial economy, Canada has initiated few of these changes, though its contributions, especially in telecommunications equipment, are recorded in its pattern of international trade. Like its industrial competitors, the Canadian economy is grappling with the need to encourage technological change, but it lags behind many others in speed of adoption of new technology, in training of labour, and in increasing productivity. Though there is no clear explanation of this slow adaptation, perhaps Canada's long period of extended job growth reduced the sense of urgency to pursue technological change.

Canada, exceptional among industrial economies, generated over 1 million jobs from the late 1970s through the mid-1980s through employment growth that reached 3.1 per cent per annum during the 1970s. Since then, the rate has halved. Despite that, Canadian employment growth (1.6 per cent annual average), along with u.s., has led other G7 countries by a wide margin. Its rate of annual growth in the labour force (1.7 per cent), however, has since the 1950s remained higher than its rate of job generation (Foot and Milne 1994). While this situation reflects the influence of immigration and the size of the baby-boom generation, part of the difference was the result of social changes reflected in significant increases in the female participation rate from the 1960s onward;[1] women claimed more than 60 per cent of new jobs in the 1970s and 1980s. Though employment for women grew disproportionately in clerical, sales, and service jobs, many of which are part-time, other gains were made. In particular, women enjoyed a greater increase in managerial and professional jobs than did men – a pattern consistent across the country (Simmons 1993).

Unfortunately, the increase of labour supply since the 1960s has not been matched by a comparable expansion in demand for Canadian products: while growth in employment and the labour force peaked in the 1970s, the rate of growth of real GDP reached its zenith in the 1960s. As a result, the rate of unemployment encountered even at the peak of business cycles is now more than double its level of two decades ago. The inability of the production sector to use available labour reflects an amalgam of factors that constitute the Canadian variant of the economic and employment changes sweeping the industrial world. In particular, slow adoption of new technology in the workplace and in new products has impeded the response of Canadian firms to international market trends for highly manufactured products. Simultaneously, employment contraction has also derived from productivity increases in manufacturing and the primary industries as a result of technological change. Industries producing traditional manufactured products, for which there is price competition, have proved vulnerable.

An indicator of the depth of these structural changes in work is the longer time the unemployed remain so, verifying the extent of the mismatch between skills and new jobs (Department of Finance 1994). There are age factors built into changes in Canada's employment structure, too, with less-educated young men and men over 45 years old having diminished shares of jobs available to them. As noted above, this is really an educational-training factor: low-skilled and blue-collar jobs have declined, while better job retention and job expansion are associated with professional, managerial, and technical work and with younger people "who are likely to be more educated, flexible, innovative and receptive to the introduction of new ideas and practices in the workplace" (Foot and Milne 1994).

Until now, Canada had not undergone the massive regional dislocation that has shaken the U.S. economy since the 1970s. Nevertheless, Canada is no stranger to employment downturns, though, its joblessness has traditionally been concentrated strongly in certain regions and economic sectors and, over the long term, the decline in jobs in the production of resource commodities has usually been compensated by renewed markets and employment in other activities in the same (or different) regions.

For a long time, however, this relationship has been at its weakest in Atlantic Canada, while the largest metropolitan region, Toronto, had proved the most resilient. Toronto, however, now faces different circumstances. The slump in jobs and incomes induced by the recession of the 1990s has coincided with the stresses of structural unemployment, and the unemployment rate for Toronto exceeded the national rate for the first time during 1992 (Figure 1). In the past, employment diversity buffered the

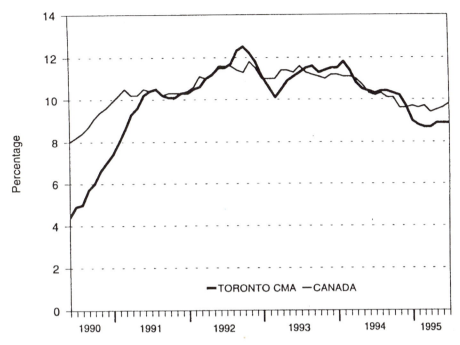

Figure 1
Seasonally adjusted rates of unemployment, Toronto CMA and Canada, 1990–95. *Sources*: *Metro Toronto Facts* (Toronto: Municipality of Metropolitan Toronto, various years).

region against downturns; now structural changes, especially in secondary manufacturing, have reduced that capacity, and the core of the Canadian industrial economy contributed much of the nation's job contraction. With 15 per cent of Canadian employment, Toronto experienced 35 per cent of the job losses, and the rest of Ontario, 25 per cent (Little 1994); furthermore, the anticipated rebound has proved less rapid in the Toronto region than in other parts of southern Ontario.

Significant structural economic change in Canada's trading performance has been expected, since many conventions and agreements that govern the rules of North American trade have been rewritten by the FTA and then NAFTA, which Canada sought, so as to reduce constraints on its North American exports. Despite attempts to assess the effects, the time elapsed is too short to test competing experiences and hypotheses. Consequently, the positive and negative results (for Canadian imports, exports, and output) are difficult to distinguish from several other factors – cumulative effects of the rise in the costs of manufacturing in Canada in the late 1980s (slower increase in productivity and faster increase in wages); growth in global trading, investment, and production systems; and dynamic effects of technological change, recession, and movements in the exchange value of the Canadian dollar. Some of these influences may reinforce each other. Limited economic modelling estimates put the net employment losses in manufacturing since 1989 attributable to freer North American trade at about 15 per cent of the total (Waverman 1993), while political rhetoric has produced a different account; research probing strategic choices made by business is only in its early stages.

Within the context of economic changes, many of them global in nature, this book is concerned directly with institutional, sectoral, and regional influences on competitiveness, technology, and markets, but the view that it describes is obtained through a wide window, overlooking the shifting economic patterns and locational sequences of Canadian development. It excludes the turmoil of quarterly and other short-run indices of change because only longer periods allow any sound judgment about the structure and geography of the Canadian economy. Our task has been to capture Canada's deeper and abiding economic characteristics, to assess the extent to which Canada is responsive to world-wide forces of change, and to identify problems inhibiting spatial and structural evolution of the economy.

Though the balance shifts between chapters, the geographic character of the economy and its patterns of change are of equal importance to questions about sectoral development and those about transformation of industries and types of work. All authors have joint concern with the economic size and relative location of places of consumption and production, as well as with the sectoral composition of GDP and the organizational structure and locational behaviour of particular industries. Economic activity is viewed here at various scales of resolution (sectors, industries, and firms), while in the spatial dimension complex processes are examined by means of case studies of particular regions or through systematic assessment of the similarities and differences in the locational dynamics of particular economic sectors or of activities common to the organization of many or all sectors. By placing their work simultaneously within both the economic and spatial domains, economic geographers have defined economic interdependence through the economic links that bind Canadian cities and their surrounding regions into urban systems. These connections, effected by transportation and telecommunications facilities, result in financial flows and many other types of traffic between these places. In other words, input-output relationships are always conceived as forming connections between places as well as activities.

SECTORS, SCALE, AND SPECIALIZATION IN CANADIAN DEVELOPMENT

Canada is a member of G7, the group of seven industrial nations, by virtue of its level of industrial output, its volume of trade, and its standard of living. In nominal per-capita terms, Canadian output ($748.6 billion in 1994) is only 7 per cent less than American, and it is this similarity that often generates a false sense of the parity between these two economies. The Canadian economy, with 29 million people, is less than 11 per cent of the size of the U.S. economy and generates just under 10 per cent as much GDP. Thus a fairer scale comparison might be with California, though Canada

Table 1
Sectors of the Canadian Economy

	Output (% GDP) 1992	Employment Canada (%) 1992	Output/ worker ($) 1989	Exports/ output (%) 1989	Average annual growth output 1979–89	Output variability (%) 1979–89
Agriculture	2.3	3.5	25,140	58	0.4	8.6
Fish, Forest, Mine	5.2	2.1	90,556	74	0.4	10.1
Manufacturing	18.2	14.6	50,859	34*	2.2	11.8
Construction	6.2	5.6	53,228	negl	–5.9	11.8
Transportation, Communications and Utilities	12.9	7.5	59,524	"	3.2	12.7
FIRE	18.1	6.2	127,967	"	3.5	13.1
Services	24.4	36.0	32,017	"	2.7	8.7
Trade	12.7	17.6	30,236	"	4.1	16.5
Public Admin.	7.3	6.8	43,029	"	1.4	4.8
CANADA	$688.5 billion	12.24 million	$45,825	24†	2.9	10.8

Sources: Statistics Canada, various publications.

 * By 1993 this proportion had risen to 48percent.

 † 1993 data show this proportion to have resumed its upward trajectory to about 29 per cent.

has more complex public-sector functions, since it has three levels of government. Geographic differences reduce the value of other apparent similarities: for example, both Canada and California have two major metropolitan centres, though the geographic forms of urbanization are vastly different, with Toronto and Montreal being much smaller and more compact. The opposite is true for Canada's urban system, which is strongly linear and embraces vast distances, reflecting extensive use of land within much of the Canadian ecumene.

Canada has long viewed itself as a resource economy, though only 9 per cent of output is derived from the primary sector (or 10 per cent when primary processing is included). But resource industries account for an even smaller proportion of employment (Table 1), and when resource, manufacturing, and construction activities are taken together, the resultant goods (production) sector accounts for only 26 per cent of employment and 32 per cent of output. Nevertheless, primary commodities and primary manufactures dominate Canada's exports.

Moreover, the economic base and thus the fortunes of much of the Canadian economy (Figure 2) are specialized in the production and processing for export of one or two resource-based commodities associated with particular regions. The economy of the regions of the west, for example, reflects the strength of the forest products industry, solo or in combination with oil and gas in Alberta or mineral extraction, which stretches from British Columbia across the northern reaches of the country to Labrador. Agriculture, especially wheat, dominates the economy of a large part of the southern prairies in Saskatchewan, but elsewhere agriculture has this significance

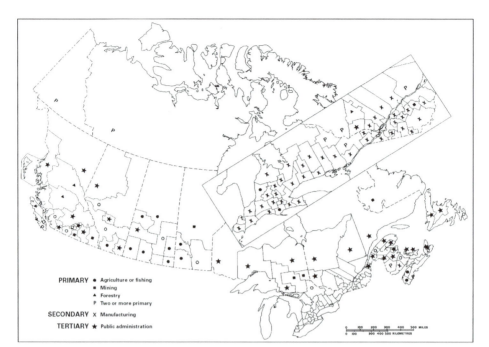

Figure 2
Regional specialization of production, Canada, 1981. *Source*: Simmons (1991). This map combines urban centres with the counties comprising their hinterlands, includes primary processing with associated primary production, and excludes tertiary activities serving the local population.

only in pockets. An equally striking conclusion to be drawn from Figure 2 is the locational concentration of secondary manufacturing in the Windsor–Quebec corridor. We thus complete the identification of the heartland–periphery dichotomy that has often been used to describe the spatial structure of the Canadian economy.

In secondary manufacturing, only the auto (assembly and parts) towns of southern Ontario and Quebec, such as Oakville, Oshawa, and St-Thérèse, have been exposed to the dynamics of foreign markets to an extent comparable with that of the resource regions and their urban centres.

The high export ratios and production scale and wages of these resource and manufacturing activities signify the economic benefits that derive from success in external markets, but the temporal variability of these export markets presents basic difficulties for the management of the Canadian economy as a whole and of regions based on a single export product. Bad weather (particularly severe for agriculture) and upswings and downturns in international markets caused by cyclical, competitive, and technological factors disadvantage these highly specialized regions, which lack other activities to buffer the loss of jobs. Possibly most significant, these activities have declining or modest rates of growth of output (Table 1).

The relative growth in output of tertiary activities (Table 1) over the 1980s was particularly pronounced in employment. While the largest cities benefited the most, there has been a shift down the urban hierarchy. The resource-dependent regions in particular have acquired more of the headquarters of firms in local primary production, and employment structures of the largest cities in these regions reflect the emergence of technical, financial, and other firms that serve the industry. Thus Vancouver, whose economy depends on forest products, has broadened its economic base through the growth of producer services, beginning with those linked to the forest sector. This trend is supported by the quality of the labour supply across the country, giving rise to software and engineering firms, head office administration, and employment in the insurance industry in many non-industrial cities.

Cities in the second tier of the urban system have discovered that some new tertiary activities – tourism and some office activities – can generate high income – multiplier effects. Thus more centres and regions now have a larger share of some of the functions that were once more exclusively associated with Toronto or Montreal because of the traditional locational pattern of manufacturing, banks, and corporate head offices. At the same time, however, the concentration of employment growth in tertiary activities has permitted the very largest cities to grow more rapidly than the rest during the 1980s. Despite expansion of services, Canadian tertiary industries as a group are not successful exporters, except for some consulting engineering firms, and as a whole the sector is a source of trade deficits.

The crude relationship between production- and tertiary-sector jobs is a 1:2 ratio, but the structure and dynamics of the economy preclude any simple basic/non-basic relationship. The sectoral structure of the Canadian economy has to be interpreted in the light of complex interdependencies of different activities within a more broadly defined set of production relationships. A growing proportion of services responds directly to the producer's demands for inputs. The growth of these technical, commercial, and managerial services reflects the rate of technological change generally and the expansion of information technology in particular; firms in all economic sectors now acquire a much larger variety of inputs from other businesses.

Apart from agriculture, public administration has experienced apparently the slowest growth of output, except in personal and community services, such as education, health, and social services, that are funded directly by government and have expanded. Strongest growth occurs in private-sector services, trade, finance, and transportation-communications. This expansion has lower rates of variability, too, reflecting the more diversified producer and consumer markets served, though they are almost exclusively domestic markets.

Once evaluation of Canada's economic geography moves to the economic specializations underlying the broad (one-digit) sectors of Table 1, the shallowness of Canadian economic development becomes clear. Only a few of the 871 four-digit industries are substantial. Employment by industry follows a log-normal distribution, with a few large industries and many small activities. The transportation-equipment group of industries (two-digit), for example, has 18 four-digit industrial divisions, two of which produce half of the output. Similarly, among Canada's 700,000 corporations the largest 3,000 generate 60 per cent of the revenue. Thus, in terms of industries and firms, the economy is comprised of a small number of relatively easily identified parts; within each activity a small number of oligopolistic firms compete within particular product markets, while most firms are "niche" producers. While this pattern is common throughout the industrial world, Canada's large indigenous firms are of modest scale by world standards, and many of its biggest industrial producers, such as General Motors and Imperial Oil, are subsidiaries or branches of foreign firms.

The log-normal distribution also describes the spatial dispersion and fragmentation of the production system, as reflected in the scale of urban markets and production, and this subject is a particular concern of geographers. Toronto and Montreal, between them, account for one-quarter of the market and production activity, the next eight places add another quarter, and the 100 largest places make up more than three-quarters of the total. The relationships between this market distribution and the size of firms are intriguing, since they jointly generate much of the economic geography of Canada. In particular, the high spatial concentration of firms generates the industrial heartland of Canada and, within it, the strategic importance of Toronto and Montreal. Certainly, in most urban centres outside the Windsor–Quebec corridor the number of major producers is very small, producing heavy locational specialization and great vulnerability.

Each economic activity and the operations of each corporation are typically concentrated in a limited set of communities; for example, integrated steel production is localized in Hamilton, where both leading producers have facilities. Various production locations – economic nodes – can be linked together in a number of ways. Multi-locational firms manage their corporate spatial system by producing/selling common products at different locations, while smaller firms within an industry tend to operate within regional sets of locations. Flows of goods and services between firms and within multi-activity corporations are particularly important because they tie together producers of different goods and services within the same places and with other locations in Canada or abroad.

Thus production and distribution chains – for example, those that ultimately link energy sources and raw materials, such as coal and iron-ore mines, with parts producers, assembly plants, and auto dealerships – determine the strength of both Canadian industrial clusters and interregional and international impulses of growth and stagnation. While the auto industry provides the most highly developed Canadian example of such flows, the production-chain concept applies to many processing and assembly industries, and Toronto and Montreal have grown as major articulation points in Canada's economic geography because of their key positions in generating the flows that sustain industries such as aircraft, machinery, computers, and other electrical and electronic goods.

The spatial concentration of the production and distribution systems of the Canadian economy has proved remarkably stable. Given the recent significant regional

Table 2
Provincial shares of capital stock (%), Canada, selected years, 1961–90

Province	1961	1966	1971	1976	1981	1986	1990
Newfoundland	1.8	2.0	2.4	2.4	2.2	2.3	2.1
PEI	0.4	0.4	0.4	0.3	0.3	0.3	0.3
Nova Scotia	2.7	2.6	2.9	2.8	2.7	2.8	2.8
New Brunswick	2.4	2.4	2.5	2.6	2.6	2.5	2.5
Quebec	23.3	23.9	22.8	23.0	22.2	21.6	22.0
Ontario	33.7	33.6	33.8	33.6	32.0	32.2	34.4
Manitoba	5.3	5.1	5.0	4.9	4.4	4.0	3.8
Saskatchewan	5.9	5.8	5.5	5.2	5.0	4.7	4.2
Alberta	11.9	11.8	11.9	12.2	15.1	16.0	14.7
British Columbia	12.6	12.4	12.9	13.0	13.5	13.7	13.3

Source: Bougrine (1993).

shifts in the U.S. economy, it may seem surprising that Canada has resisted the trend: 56 per cent of Canadian capital stock continues to be located in Ontario and Quebec, these regions having lost less than 1 percentage point of their share over the past 30 years (Table 2). Even more striking are the shares of output generated by this capital stock in the two central provinces (1991: 40.1 per cent in Ontario and 23.1 per cent in Quebec) and the income earned (41.5 per cent and 23.2 per cent, respectively), while within Central Canada, Ontario has increased its share only very slightly over the long run. While stability of the pattern of spatial concentration at the centre and within particular regions of the country is the strongest characteristic of the space economy, there have been interesting changes in the distribution of capital stock. Over the long term, for example, the prairies have lost relative to the rest of the country, but Alberta and British Columbia have increased their share (3.5 percentage points) as a result of investments in resource-based industries.

The oil crises beginning in 1976 initiated substantial growth in Alberta, and there were also high rates of capital accumulation in British Columbia in the 1970s and early 1980s. Since the late 1980s, however, there has been a shift in the direction of rates of new investment (Bougrine 1993), reversing the pattern of new growth. In the period 1986–90, Ontario moved back to first rank (from fourth place) in its rate of total capital growth, eclipsing the rate of capital accumulation in Alberta and British Columbia; but the rate of manufacturing investment rose to first rank in British Columbia because of resource investments, while it sagged in Ontario, pointing to some reorientation in investment in Ontario favouring, on the margin, new activities such as producer-service industries.

The share of output contributed by most regions has moved in the same direction as their share of capital stock. Most notably, there is a substantial increase in production associated with oil and gas developments in Alberta. From 1976 that province began generating higher GDP per capita than Ontario, the former leader, but Alberta's fortunes peaked in 1981 (Table 3). Elsewhere, changes in production have been more

Table 3
Provincial GDP per capita (Canada = 100), selected years, 1961–94

Province	1961*	1966*	1971*	1976*	1981*	1981	1986	1991	1994
Newfoundland	50.0	52.1	56.2	53.6	52.0	57.9	59.9	65.1	65.3
PEI	49.4	48.4	52.3	52.2	50.5	56.5	59.1	63.5	70.8
Nova Scotia	65.3	63.0	67.9	66.0	61.3	61.2	74.8	78.2	76.8
New Brunswick	60.5	61.3	63.7	63.8	63.8	60.5	71.1	75.5	77.2
Quebec	91.1	89.9	88.9	88.1	86.0	85.8	90.1	91.2	89.7
Ontario	119.9	117.4	117.3	109.4	106.5	102.5	111.5	109.2	108.0
Manitoba	90.4	87.1	90.7	91.4	88.1	87.2	86.2	85.5	86.7
Saskatchewan	77.3	99.6	86.9	101.2	108.8	102.8	85.1	80.5	89.2
Alberta	108.8	109.3	110.8	137.1	146.0	156.9	121.0	114.6	118.3
British Columbia	111.1	109.2	106.8	108.6	109.4	112.5	99.5	104.6	106.4
Disparity ratio[†]	2.4	2.4	2.2	2.6	2.8	2.8	2.6	2.4	1.8

* Data for these years use definitions that were revised in 1981.

† Disparity ratio = value highest-ranked province/value lowest-ranked province.

Sources: Statistics Canada (various years to 1994); Savoie (1986).

modest, though in Atlantic Canada cumulative, small increases in efficiency have moved these disadvantaged provinces toward the national GDP per capita average. Nova Scotia and New Brunswick have seen larger gains, especially during the 1980s (Table 3). This late move of convergence with the national average by Atlantic Canada is reflected even more strongly by data on earned income per capita (Table 4) – a measure less sensitive than GDP per capita to provincial variations in the aggregate capital:labour ratio. Regardless, Atlantic Canada has an outmoded urban structure associated with a small, residual industrialization and a declining resource base, and recent shifts in earned income and GDP, though encouraging, do not establish whether it has reached the bottom of a long-term regional cycle of structural unemployment and stagnation. In the prairies the economy has not significantly diversified, and recent data for Saskatchewan show that a cyclical swing placed it behind Nova Scotia.

Regional economic differences have proved remarkably enduring and are still a major feature of the Canadian economic landscape. Across the country, there have been no significant changes in the rank order of economic performance of provinces in 1994 compared with 1961: for example, the three top- and three bottom-ranked provinces are unchanged. This pattern, the position of Atlantic Canada, and the relative income levels of Quebec and Ontario all point to limited results from more than three decades of regional development programs mounted by federal and provincial governments. Economic growth in the west, however, demonstrates the power of richer natural resources and the impact of market-driven investment.

Regardless of the impact of regional development programs, Canada's spatial economy continues to be influenced heavily by governments, though their actions are fragmented by responsibilities divided unevenly among the three levels and by a

Table 4
Provincial earned income per capita (Canada = 100), selected years, 1961–94

Province	1961	1966	1971	1976	1981	1986	1991	1994
Newfoundland	57.5	57.2	65.1	61.0	58.0	57.4	63.4	64.1
PEI	55.1	59.3	68.2	66.1	65.0	66.3	67.1	71.2
Nova Scotia	68.2	72.7	62.5	75.5	74.6	79.6	79.4	80.3
New Brunswick	64.3	68.9	78.8	72.9	68.5	70.4	73.4	75.0
Quebec	104.4	104.2	112.2	105.5	89.9	90.8	91.6	92.3
Ontario	116.4	114.3	125.8	103.5	106.6	114.8	113.0	109.4
Manitoba	90.7	89.2	100.5	92.0	90.9	89.9	85.3	89.8
Saskatchewan	70.3	90.9	84.6	97.8	97.2	86.8	78.6	79.9
Alberta	112.8	105.7	115.1	119.9	121.9	106.7	104.8	105.9
British Columbia	123.2	116.7	125.8	115.0	115.2	99.8	105.3	107.5
Yukon and NWT	na.	84.8	100.1	93.8	105.1	122.8	106.4	107.9
Disparity ratio	2.1	2.0	1.9	1.9	2.1	2.1	1.8	1.7

Sources: Statistics Canada (various years to 1994).

Note: New methods of measuring employment income adopted in 1981 have yielded higher indices of income per capita for the Atlantic provinces, Alberta, and British Columbia and lower indices for Ontario, Manitoba, and Saskatchewan. In this table the old data series has been linked to the new to provide continuity; there was some change in rank order of the leading provinces 1961–76.

proliferation of departments, programs, and service and regulatory agencies. The collective expenditures of the provinces are of the same order of magnitude as Ottawa's, but the public economic environments of the provinces (Table 5) vary enormously in terms of government employment, provincial expenditures, and fiscal transfers by the federal government. The differences in scale between provinces – Ontario has financial resources 70 times larger than those of Prince Edward Island – demonstrate that provinces are in no sense equivalent in their abilities to provide services, to redistribute income, or to negotiate with each other, the federal government, or private enterprise. These discrepancies of size and economic significance are powerful characteristics of Canada's economic geography and constrain how it functions and how it may be managed through public policy. One important area where provincial power has prevailed over the federal (and perhaps general) interest is the slow rate at which trade barriers within the country have been reduced: restrictions prevail on labour mobility in some areas of work and, on interprovincial sale of certain types of goods, while public-sector purchases are often arranged on a preferential basis.

ORGANIZING THEMES OF THIS VOLUME

As economic geographers have accumulated knowledge about Canada's economic geography, using both generic and Canadian theories, four organizing themes have emerged:

Table 5
The size of government
Part A

	Revenues 1993 (% GDP)	Transfers into revenues from other governments 1993 (% GDP)	Expenditures 1993 (% GDP)	Transfers to other governments 1993 (% of expenditures)
Federal	19.0	negl	23.4	18.6
Provincial	20.7	4.2	22.9	31.0
Local	8.6	4.1	7.8	0.2
Hospitals	3.2	3.2	3.3	–

Part B

Province	Canadian GDP 1994 (%)	Canadian population 1994 (%)	Government employment 1994 (%)	Total employment 1994 (%)	Federal transfers 1993 (%)	Provincial expenditures 1993 (%)
Newfoundland	1.3	2.0	1.7	1.5	4.9	2.1
PEI	0.4	0.5	0.6	0.4	1.0	0.5
Nova Scotia	2.5	3.2	4.0	2.9	5.1	2.9
New Brunswick	2.0	2.6	2.8	2.3	5.1	2.7
Quebec	22.3	24.9	24.9	23.7	26.6	27.1
Ontario	40.3	37.4	36.1	38.8	26.0	34.0
Manitoba	3.4	3.9	4.4	3.8	5.9	4.0
Saskatchewan	3.1	3.5	3.7	3.4	5.0	3.6
Alberta	11.0	9.3	9.5	10.1	7.5	10.0
British Columbia	13.4	12.5	11.3	13.0	8.6	12.3

Source: Statistics Canada (various years to 1994; 1994).

- the continuing importance of the openness of Canada's economy to external influences that have shaped its sectoral and spatial development,
- the regional variation in the resource base and the pattern of urbanization, which yield a relatively simple economic geography,
- the rapidity of changes in technology and markets to which Canada must continually adjust, and
- the efforts of government to organize and maintain a national market based on such policies as tariffs, transportation investments, and equalization payments.

These core issues connect the chapters of this book. Individual authors use research methods ranging from micro to macro economic analyses of data, which vary in source and nature from public files to original case studies. They make reference to the influence of the openness of the economy, regional specialization, the rate of innovation, and/or the influence of public policy.

Openness

The strong external economic connections of the Canadian economy have developed because of the openness of this economic system. They have long been viewed as the basis for a continuously high level of Canadian per-capita income. Openness strongly influences the economic structure of regional economies, too, since the long-term strength of Canada's resource exports represents the comparative advantage which has enabled this relatively small economy to achieve a standard of living that the domestic market could not have supported. At the same time regional specializations lead to vulnerability, because international competition in markets served by Canadian resource firms means that the health of Canada's staples (like that of its secondary industry) depends on technology and investment.

Openness of the Canadian economy is also reflected in competition for capital investment. Massive inflows of foreign direct investment (FDI) have been encouraged at various times, and their legacy is a substantial stream of reinvested profits, which has increased the importance of foreign corporations in Canada. The deficit on the services account within the current account of international balance of payments is a complementary result, reflecting the way large corporations, especially multinationals, create their own internal financial and service environments. These imports of managerial and professional services outstrip comparable exports, and dividends and royalties (related to FDI) flow out of Canada.

As a counterpoint to the analysis of foreign investment and the imbalance of associated international flows, it is useful to assess the influence of predominantly indigenous firms of medium and small size, especially on the development of regional economies. Firms with fewer than 100 workers employ only 49 per cent of the total (1986), but throughout the 1980s they generated jobs when larger firms cut back. More and more, modern technology suitable for small firms has become available, increasing their flexibility and their potential to serve niche markets, including those for technological and marketing services, industrial services, and sub-contracting capacity to firms of all scales. From the perspective of industrial growth, small, innovative firms are of particular interest as a source of new competitive strength in both traditional industrial regions and in locations without strong industrial histories.

Geographic Patterns

Though the Canadian economy is sectorally diverse, its geographic patterns, from regions specializing in primary production to locations of metropolitan development, prove relatively simple to generalize, even though they reflect considerable spatial differences in resource base, influence of export trade, role of tariffs, and positive externalities of urban development. This situation is illustrated by the remarkably robust regional patterns of earned income per capita above and below the national average, especially the persistence of the ranking of the Atlantic provinces. Likewise, the high level of urbanization in the Windsor–Quebec corridor and the enormous concentration of economic activity in Toronto and Montreal superimpose an equally important pattern.

Both Montreal and Toronto and their smaller industrial neighbours have risen to prominence as loci of manufacturing (see Figure 1, part III) and distribution, though the cities of southern Ontario have benefited from their position on the obvious route of entry of manufactured components and capital from the United States. Ironically, tariffs on many manufactured imports and other restrictions to entry on services – that is, closure of the Canadian market – have promoted this pattern. Within Canada, returns to scale have been gained as transportation costs have been traded off against the advantages of centralized production, and Toronto and Montreal have long dominated the national urban system as an outcome of the agglomeration of higher-order retailing, business services, office functions, and manufacturing. Similarly, information and financial flows are centralized through the functions of the banks, stock markets, and other financial institutions.

In many ways this description will remain valid under the new trade regime, though increased openness to U.S. financial institutions is likely to cause flows to and from U.S. regions increasingly to replace interaction between Canadian cities. Industrial and service activities in central Canada must compete more and more with American firms, since FTA and, NAFTA have changed the rules of trade and investment and have defined a time scale over which Canadian markets and production points will be rationalized within a North American set of regions and cities.

Technological Change

There is inconsistency in Canada's technological profile. Canada is a leader in certain areas of telecommunications technology and an effective competitor in some types of aircraft and space technology. In some other areas, such as nuclear generation of electricity, it is very highly advanced. However, compared with other industrial nations, Canada lags in its rate of industrial innovation. Its failure to adopt modern production equipment – such as robots and computer numerically controlled (CNC) machines – as rapidly as other G7 countries has caused serious lags in the quality, costs, and potential exports of traditional, modern, and research-intensive products. It has been slow to generate and adopt new technology to complement its endowment in natural resources and extend its capacity to expand economic development. While Canada has been exceedingly well served by the scale and quality of the natural resources it has exploited, this basis of development is being challenged at an increasing rate by technology-intensive substitutes (such as fibre optics for copper cable) and by lower-priced competitors in commodity markets.

Canada has only one viable strategy – to increase the level of technological input into production and to be a product innovator. The aims are to increase production efficiency and lower the costs of output, with the goal of the latter being to meet the needs of performance-maximizing markets, which implies adding more value to resource output so that the product embodies a greater input of human capital. The introduction or adoption of techniques such as "clean technologies" that contribute to sustainable development are an increasingly important specification of both product and production technology. There is, then, common interest in the need for resource and primary manufacturing industries to innovate if they are to remain significant sources of employment and GDP. A high-wage country whose economy is as open to international trade as Canada's cannot afford to remain disadvantaged. Since changes in the international division of labour rest on the application of new industrial technology by advanced as well as industrializing countries, the comparatively low level of commitment of Canadian firms to industrial innovation limits their ability both to challenge penetration of imports into the Canadian market and to generate competitive advantage in secondary manufacturing. It should be stressed this is a dynamic problem, and the technology gap is not narrowing.

Government Policy

By implication, Canada needs to increase the range and effectiveness of its policy initiatives to respond to the evolving trade environment and to accelerate the rate of technological change. From the beginning of Canada, development policies ranging from infrastructure investment to subsidies to encourage the increased technical capability of firms and workers have influenced the development of the Canadian market and the emergence of industrial concentrations. Stimulation of industrial development in regions of high unemployment has also been a politically significant concern. Though policies designed to attract employers to these regions have failed, consumer's purchasing power has been maintained by means of family income support, and substantial income equalization among provinces has ensured that essential services such as health and education can be provided throughout the country.

Economic development policies of various levels of government have never been perfectly coherent, despite periodic federal-provincial agreements. While many policies have had the goal of defining a Canadian market, provincial governments, as noted above, have persisted in balkanizing the national market with subsidies, public-sector purchasing policies, and controls on consumer markets that have favoured local producers (as with beer). Similarly, they have imposed geographic constraints on contractors and labour.

Regulation at the federal level has limited the pricing power of monopolies (such as utilities), and Canadians have obtained the price advantages of large-scale production. As a result of more open international trade, however, governments are under severe pressure from external and some domestic business interests to deregulate. Similarly, bilateral and multilateral trade agreements are inducing movement toward an unfettered domestic market, while industrial subsidies that were compatible with tariffs are under constant threat from U.S. regional industrial interests. Canada will have to minimize its exposure to subsidy disputes with the United States in such areas as raw materials (for example, stumpage fees under dispute in the forest products trade) and energy (as with costs of electricity used in mineral refining).

Canada is at a sensitive juncture in its political-economic evolution. It is increasingly apparent that many established economic policies and programs are misfits in a world where liberalization of trade is defining new markets and new expectations and setting new policy constraints. The incomes of Canadian farmers, for example, have long been supported by government policies, but the provisions of the Uruguay Round of GATT will change the organization of agricultural production. Canada now finds itself a participant in a program of agricultural trade liberalization. Ultimately, the power of marketing boards, developed to manage supply, will wane, but for another decade Canadians will continue to pay prices for chicken and dairy products that in North American terms are uncompetitive.

STRUCTURE OF THIS STUDY

Given the thematic foundations of this book, there were two alternatives for its organization – a systematic regional arrangement of chapters or the integrated sectoral-issues format that has been chosen. The intellectual latitude provided by this format allows the regional texture of the economy to emerge. Some of the chapters focus on one region to constrain the range of data they examine for a particular industry or process. They seek to reveal the forces that have affected firms in the one region, but their additional goal – and this is true of all chapters – is to explain the generic characteristics of spatial economic processes. At the opposite end of the geographic scale of inquiry, other authors have perceived Canada as an economic region in a dynamic international environment of changing opportunities, competitors, and rules. Some chapters use an explicitly spatial approach to the development of industries or sectors, particularly for those activities that are widely, though not uniformly, distributed. Analysis of the urban system thus allows a unique insight into the Canadian economy based on both variations in sectoral dynamics and a focus on corporate decisions. To complete the range of approaches employed here, other chapters reveal geographic detail as it is structured by the economic processes that cause spatial concentration and agglomeration.

The five sections provide flexibility in the organization of individual chapters. Different dimensions of the international openness of the economy are examined in the first section in order to analyse the macro-economic, institutional, and regional dimensions of openness. The next three sections explore the resource-based industries that have provided the traditional export strengths of Canada, the technologically distressed manufacturing sector, and the burgeoning service industries, respectively. Though the geographic dimension of political-economic issues pervades the volume, the last section explicitly considers Canadian governments as an important economic sector, which undertakes many tasks of production, distribution, and redistribution. The influence of governments on the operation of the private sector is the focus for the concluding chapter, which returns to the pressing issues of the rate of technological change and the ability to innovate, which still function as sheet-anchors on the pace of Canada's economic development.

NOTES

1 While this increased the size of the labour force, it also showed population growth, though here other factors were also significant.

REFERENCES

Bougrine, Hassan. 1993. "The Role of Capital Formation in Economic Disparities among Canadian Regions: 1961–1990." *Canadian Journal of Regional Science*, 15, 21–33.

Department of Finance, Canada. 1994. *Agenda: Jobs and Growth – A New Framework for Economic Policy.* Ottawa.

Foot, David K., and Milne, William J. 1994. "Population-Output Linkages: A National and Regional Perspective." Policy and Economic Analysis Program, Policy Study 94–10, Institute for Policy Analysis, University of Toronto.

Little, Bruce. 1994. "Toronto Struggles to Recover." *Globe and Mail*, 14 Nov., B1–2.

Savoie, Donald J. 1986. *Regional Economic Development: Canada's Search for Solutions.* Toronto: University of Toronto Press.

Simmons, James W. 1991. "The Urban System." In Trudi Bunting and Pierre Filion, eds., *Canadian Cities in Transition*, 100–24. Toronto: Oxford University Press.

– 1993. "Canada – Employment Growth, 1971–1986." Map 32.2. In Energy, Mines ands Resources Canada, *National Atlas of Canada*, 5th Edition.

Statistics Canada. Various years to 1994. *Provincial Economic Accounts, Annual Estimates.* Cat. No. 13–213.

– 1994. *Sector Employment and Wages and Salaries.* Cat. No. 72–209.

Waverman, L. 1993. "The NAFTA Agreement: A Canadian Perspective." In Steven Globerman and Michael Walker, eds., *Assessing NAFTA: A Trinational Analysis*, 32–59. Vancouver: Fraser Institute.

PART ONE

The Open Economy

The institutional conditions that have governed Canadian trade and thence investment, particularly adoption and maintenance of tariffs on manufactured imports, have
been pivotal in its economic history. The National Policy origins of this approach to
assisting development are now over 100 years old, and significant reductions to the
barriers against open trading have been made in a series of steps since the Second
World War. By signing the Canada-U.S. Auto Pact, by participating in various tariff-
reducing agreements under the General Agreement on Tariffs and Trade (GATT), and
by seeking the Canada–United States Free Trade Agreement (FTA) and the North
American Free Trade Agreement (NAFTA), Canada has reduced progressively the extent to which the central government can channel the evolution of Canadian trade and
of economic development.

Historically, exports have been seen as the way in which the Canadian economy
could be developed, and the involvement of the Canadian economy in international
trade has reached the level where exports comprise nearly 27 per cent of gross domestic product (GDP). Norcliffe (this volume, chap. 2) demonstrates that this is an impressive indication of the openness of the economy by evaluating the exports of a broad
spectrum of comparable countries. Over the past 40–50 years, despite major changes
in the world's economy, Canada has continued to rely for its export earnings on the
primary resource base, particularly on forest products and minerals.

Thus Canada begins the new era of North American free trade with a dependence
(57 per cent of exports) on the export of primary and semi-processed commodities. In
addition, there are enormous two-way flows of autos, trucks and parts (18 per cent of
exports and 22 per cent of imports), though Canada has minimal design capacity and
an ever-increasing technology gap, reflected by an expanding deficit in the trade balance on end products. Despite the rather limited industrial depth of Canadian exports,
and the slow pace of change to higher-value products, Canada is a prime example of
the successful application of a program of protection by a small economy seeking to
build an urbanized, industrial structure. The core of this position, as espoused by Paul
Krugman (1991), for example, is that the scale and sophistication of the economy are

now sufficiently developed to allow Canada to compete internationally in services and secondary manufactures in addition to maintaining its traditional exports.

We should still wonder, however, how well trade has diversified from traditional market connections. This is an especially important question, since Canada's exports have been strongly North American in their geography and have been unresponsive to the markets in Asia and the Pacific Rim that have been growing the fastest. Norcliffe, in reflecting on Canada's dominant trade connection with the United States contrasts the merits of NAFTA with the gains that would be made from better-developed connections with a complementary – rather than a regionally competitive – trade partner, especially Japan, or a more balanced portfolio of trading partners, which could produce a profile of international economic signals that has less variance. MacPherson (this volume, chap. 4) also points to the faster-rising incomes of these trading areas, which Canada has yet to develop fully as commercial partners.

At the regional level, as explained by Barnes (this volume, chap. 3) using a contemporary interpretation of Harold Innis's ideas on Canadian economic development, export staples are both friend and foe. They are responsible, directly or indirectly, for the economic development of most regions of the country, but they are prone to substantially greater cyclical instability than manufacturing or the services, and this situation has fed directly into great variations in regional patterns of unemployment and (consumer) income. Regional specialization of production and the volatility in incomes derived from resource markets have unfortunate consequences. This is demonstrated by Barnes's analysis of the BC economy and by the temporal variability in Saskatchewan farm incomes (Simmons 1986).

Canada has been a major capital importer especially over the past 50 years and has acquired the branches of international manufacturing, service, and retail corporations, which have been direct investors of this capital. Overseas companies have long displayed similar interest in Canadian resources, and the history of Canadian resource exploitation is strongly associated with foreign direct investment. While the consequences of this pattern of foreign investment for Canadian development were viewed very positively for a long time, since the 1970s the opportunity costs in development that derive from a very high level of foreign ownership have been a recurring theme among some geographers. This is revealed in trade in a variety of ways. Norcliffe points out, for example, that the strongly negative trade position for invisibles that had developed by the mid-1970s has deteriorated further. Also, the deficit on investment income is now nearly $24 billion (1993): this is the trade category that includes not only interest payments to foreign bond-holders but dividends and royalties generated by foreign direct investments.

The accessibility of the economy to foreign direct investment (FDI) – reflected, partially at least, in the capital account of the international balance of payments – continues to be an exceedingly important dimension of the open identity of the Canadian economy. With a substantial proportion of the largest European and U.S. firms already operating in Canada, direct investment inflows faltered during the last recession (early 1980s), but two factors have caused a more recent increase

(MacPherson, this volume, chap. 4). Company profits not distributed from Canadian operations in the form of dividends have been a major source of funds for increasing FDI. Furthermore, the FTA signalled greater U.S. interest in foreign acquisition of Canadian resource companies, and particular attention has been paid by other new investors to Canadian high-technology enterprises, such as Mitel, Connaught Laboratories, and Lumonics, which have provided them with access to high-tech design and production. What acquisition of these Canadian sources of technology by foreign companies does in a positive sense for the Canadian economy has never been clear; strategic alliances are thought to have advantages, but they, too, often lead to take-overs (Ahern 1993).

Simultaneously, Canadian companies have been making substantial foreign investments in order to secure markets (MacPherson, this volume, chap. 4) Northern Telecom, Noranda, and Alcan, which have been doing so for a number of years, are prime examples of manufacturing and resource companies highly involved in this process. The United States has been the long-term focus, but more and more investments are being made in western Europe and the Pacific Rim. A second area of investment activity has involved Canadian banks, financiers, real estate developers, and retail chains, which seem to be at an early stage of learning how to cope with the stresses of international portfolio management.

It is also evident that these new patterns of Canadian investment overseas preceded the advent of FTA. In one sense they were an outgrowth of domestic corporate success and the realization that the Canadian economy had become too small an environment to sustain progressive, or at least aggressive Canadian firms. The technical feasibility of the global organization of economic activity (through information technology) is of at least equal importance, because it has allowed or induced more resource, manufacturing, and service firms to extend their operations abroad, following the earlier example of truly multinational enterprises.

The number of foreign-owned financial corporations in Canada is rising, and interconnection is growing between the "four pillars" of the Canadian financial establishment – banks, insurance companies, brokerage houses, and trust companies – and with foreign affiliates. As explained by Dobilas (this volume, chap. 5) these developments derive directly from the considerable deregulation of the operation of financial institutions in Canada, which followed similar changes elsewhere, especially in the United Kingdom. The trade agreements with the United States have eliminated most operating restrictions on U.S. financial firms. Though the domestic financial system is now institutionally more flexible, there have been difficulties in eliminating the influence of the old regulatory framework, and Dobilas points to its constraining effects on the size of the capital bases of Canadian financial firms, which inhibits development of their international operations.

International financial markets are of growing importance, and the London–New York–Tokyo "axis of control" increasingly determines the activities of national financial systems. The result is loss of national control; as well, the bulk of Canadian overseas (public and corporate) debt is managed by foreign financial institutions, and

Dobilas indicates that the Canadian financial system, as a second-tier node, has much-diminished autonomy.

The conventional vehicles of international economic connection – trade in merchandise and services, portfolio investment, and FDI – are still important but provide an incomplete account of the openness of the Canadian economy, and the increased ability of Canadian banks to form consortia with foreign correspondents to finance overseas acquisitions by Canadian development companies should join this list. While the trend toward a more open economy reflects the greater interest of financial and non-financial firms in Canada, the increased diversity of Canadian firms investing overseas may provide some confirmation of Krugman's (1991) thesis.

REFERENCES

Ahern, Rowena. 1993. "Implications of Strategic Alliances for Small R&D Intensive Firms." *Environment and Planning A* 25, 1511–26.

Krugman, Paul R. 1991. *Geography and Trade*. Leuven: Leuven University Press.

Simmons, J.W. 1986. "The Impact of the Public Sector on the Canadian Urban System." In Gilbert A. Stelter and Alan F.J. Artibise, eds, *Power and Place: Canadian Urban Development in the North American Context*. Vancouver: University of British Columbia Press.

CHAPTER TWO

Foreign Trade in Goods and Services

GLEN NORCLIFFE

Since the time that Bristol's adventurous sailors began fishing on the Grand Banks of Newfoundland in the 1480s – several years before John Cabot's voyage to North America[1] – the history of Canada has been an almost seamless record of resource exploitation for export, much of it initiated by foreign direct investment. For almost four centuries French and then British trading companies were the main agents exporting resources from their North American colonies to European markets, shipping back manufactured goods in exchange. The next phase, marked by Canada's emergence as a self-governing dominion a little over a century ago, was followed by a slow redirection of trade from a transatlantic to a continental axis. During this period sporadic attempts by the Conservatives, particularly Sir John A. Macdonald, to promote a National Policy held out the possibility of a model of economic development that placed greater emphasis on nurturing indigenous industry. Laurier's reciprocity agreements, which cost him the election of 1911, reflected the countervailing pressures of continentalism. The sale of many British assets in Canada to Americans in order to finance the First World War strengthened continental links. And C.D. Howe's promotion of foreign (especially American) investment in Canada after the Second World War solidified this continental trading arrangement. Since the 1960s several Pacific states have become important trading partners, and Japan and Hong Kong major sources of capital investment. But in its bare essentials, the mechanism of Canadian development has not changed in 500 years. FDI in Canadian resources, often with Canadian participation, has facilitated export of staples to foreign markets. Much of the income thus generated has been used to build and modernize national infrastructure and to import manufactured goods.

This foreign trade has altered the geography of Canada, conditioned the reproduction of Canadian society, and even affected the Canadian psyche. These effects have, in turn, tended to reinforce Canada's particular trade orientation. Successive investments in fur trading posts, fish-drying flakes, fishing boats, sawmills, wheat elevators, railways, port infrastructure, oil and gas pipelines, and the like, coupled with the complementary development of skills and technology and the growth of supportive corporate, financial, and governmental institutions, have made Canada a major exporter of staples. In its turn, cyclical success in this trade has promoted further rounds of investment in the same staple-producing and -processing industries and associated infrastructure. From the beginning, foreign capital has been a substantial component of investment flows into the resource sector. Indeed, as Harold Innis made clear in much of his writing, the staple export trade and complementary investment in resource industries and infrastructure have been the most enduring feature of Canadian economic history (see also chapters 2 and 6, below).

The main theme to be developed in this chapter is this: though the new staple economy – particularly as reinforced by Conservative governments 1984–93 – had some short-run successes, in the long run it is placing critical limitations on Canada's economic development. Such an argument is not new; it was spelled out in detail by Britton and Gilmour (1978). However, during the 1980s and early 1990s federal and provincial economic policies put into sharper focus the reasons for not being sanguine about Canada's long-term development prospects as a staple exporter.

I examine this theme in several stages. First, I outline the trade policy framework developed during the 1980s. The second section identifies Canada's main trading partners and explains their importance, particularly to certain regions. This analysis introduces the theme of trade domination. Third, I dissect the composition of the trade, both merchandise and non-merchandise, and recent trends in that trade and then discuss the associated technological gap. The fourth section examines the openness of the economy. The last section highlights Canadian commerce with two very different trading partners, the United States and Japan.

TRADE POLICY IN THE 1980s AND 1990s

The Foreign Investment Review Agency (FIRA) was created in 1974 during a fairly nationalistic phase of the Trudeau government. Its role was not ostensibly to deter FDI, but to make sure that it met certain criteria that protected Canadian interests. Though FIRA was subjected to a great deal of criticism by the investment community and U.S. financial interests, it was something of a toothless dragon from the beginning. Attempts by Industry, Trade and Commerce Minister Herb Gray in 1981 to develop a technology-based industrial export policy that included broadened powers for FIRA were effectively quashed (Williams 1983, 165–6). To the contrary, FIRA's wings were progressively clipped – the phrase of Tom Kierans (1990, 37). Indeed, the main criticism that could be levelled at the agency concerns the glacial pace at which it processed applications; during these delays, it could bargain to increase Canadian spinoffs from FDI. However, FIRA's fundamental problem was that it operated in a policy vacuum (Britton and Gilmour 1978, 186).

Having identified FIRA as a barrier to Canadian development in the 1984 election campaign, following their victory the Conservatives moved swiftly to restructure the agency. They transformed it into Investment Canada, with the mandate to encourage FDI in Canada. There was no longer a policy vacuum: both trade and investment were to be deregulated and guided by laissez-faire principles as Canada joined fully in the globalization of capital flows. The government set about negotiating the other major component of this policy – a free-trade agreement with the United States. After lengthy negotiation, the Canada–United States Free Trade Agreement (FTA) was signed in October 1987.[2] It became the central issue of the 1988 election campaign, which the Conservatives won. Its subsequent enactment (1 January 1989), coupled with privatization, deregulation, and tax reform, set the stage for Canadian trade and investment policy under the second Conservative government. In 1993 the FTA was enlarged to include Mexico through the North American Free Trade Agreement (NAFTA), creating a trading bloc comparable in size to those in Western Europe and East Asia. NAFTA therefore forms one of the three pivots of what Ohmae (1985) has labelled the "triad". It is quite posssible that NAFTA will be enlarged to include Chile and other Latin American countries during the 1990s. Even though the Liberal party opposed the FTA in the 1988 election, Jean Chrétien's Liberal government, elected in 1993, seems committed to NAFTA and has even pressed the now more protectionist U.S. government to expand the area (presumably to further Canadian trading interests, as well as to dilute the powerful American voice).

FDI in Canada has been actively encouraged by Investment Canada, with the two main countries taking advantage of the deregulation of capital flows being the United States and Japan, though British, German, and French corporations have also been active. The main goal for American corporations has been continental corporate restructuring, in practice taking one step further the integration of the North American space economy as a free-trade area. Take-overs and mergers in 1988 totalled $23.7 billion and in 1989 came close to $40 billion, as foreign companies moved quickly to take over Canadian resource, resource-processing, and high-tech companies (Campbell, 1990). By 1990, recession was causing a sharp drop in FDI, though the main trends remained intact. As is shown in the next section, these new investments reinforced existing trade patterns, since the multinationals involved were anxious to guarantee either their supply lines to Canadian resources or access for their manufactured products to the Canadian market.

These domestic policies promoting exports rather than broader industrial development achieve a special poignancy in light of the changes that were simultaneously under way in the external trading environment. The Tokyo Round of GATT, negotiated 1973–79, made major progress in liberalizing world trade. Average tariffs dropped below 10 per cent, though agricultural products, certain processed resources, and some labour-intensive industries remained substantially protected (Stone 1988). Had Canada vigorously pursued an open development strategy (Cohen 1991) aimed at realizing dynamic efficiencies through economic restructuring, including promotion of technology-intensive industries in certain market niches (along the lines suggested in the early days of FIRA), the Tokyo Round would have provided the opportunity to reap its benefits,[3] especially in view of the growing weakness of the U.S. economy in many technology-intensive activities that was later revealed by Cohen and Zysman (1987). The subsequent growth of GATT membership and the further liberalization of world trade that will follow implementation of the Uruguay Round (1986–94) and creation of GATT's successor – the World Trade Organization (WTO) – serve merely to underline the dimension of the lost opportunity.[4]

CANADA'S FOREIGN TRADING PARTNERS

Merchandise Trade

The United States dominates Canada's merchandise trade to an extraordinary degree. Table 1 reports Canada's merchandise trade for 1978 and 1993[5] by continent and for countries with merchandise trade exceeding $1 billion in 1993. Since the signing of the FTA and NAFTA, this domination has increased.[6] Though the proportion of Canadian imports coming from the United States dropped by nearly 4 per cent during the fifteen years up to 1993, the proportion of Canadian exports going to the United States rose by 10 per cent, so the net change was an overall increase of nearly 5 per cent. In aggregate, trade with western Europe declined slightly, from 11 per cent to around 9 per cent, while the trading partners involved changed little. This aggregate picture conceals the fact that exports to western Europe declined (relatively), while imports remained high, creating a serious trade imbalance (these imports are overwhelmingly high-value-added manufactured goods). The United Kingdom is Canada's leading European trading partner, closely followed by Germany and France. Trade with eastern Europe, the Middle East, Africa, Oceania, South America, and Central America is very small (mostly around 1 per cent for each region) and has tended to shrink in recent years. That leaves Asia, where imports have been increasing steadily while exports have declined a little, so that trade with this region is now somewhat more sizeable than trade with western Europe. Japan has become Canada's second trading partner and now accounts for around 6 per cent of Canadian merchandise trade.

Non-merchandise Trade

Published statistics on non-merchandise trade with specific countries are limited, but the essentials are reported in Table 2. This trade amounts to only one-third of the value of merchandise trade, and the distribution by country is somewhat less concentrated. The proportion of the total accounted for by the United States dropped slightly, from 58 per cent to 55 per cent during the 13 years. The European Community (EC) and Japan combined accounted for another quarter, with a very large increase in interest payments to the latter as Japan became a major capital exporter. Significantly, Canada had a large deficit in its trade of invisibles with all the industrialized countries.

The major exception to payments exceeding receipts is the residual category in Table 2, composed mostly of Third World countries, in which Canada has a surplus on non-merchandise trade despite "exports" of tourism to these countries. Canadian

Table 1
Canadian merchandise exports and imports by continent and selected countries, 1978 and 1993

	Exports		Imports	
	1978 %	1993 %	1978 %	1993 %
United States	70.3	80.4	70.7	67.0
Western Europe	10.7	7.0	11.3	10.6
United Kingdom	3.8	1.6	3.2	2.6
(West) Germany	1.5	1.4	2.5	2.1
Netherlands	1.2	0.7	0.5	0.4
France	0.9	0.7	1.4	1.3
Belgium and Luxembourg	0.9	0.6	0.4	0.3
Italy	0.9	0.5	1.0	1.1
Rest of Western Europe	1.5	1.2	2.3	2.8
Eastern Europe	1.7	0.8	0.5	0.4
Former USSR	1.1	0.3	0.1	0.3
Rest of Eastern Europe	0.7	0.2	0.4	0.1
Middle East	1.5	0.9	3.2	0.7
Africa (excluding Middle East)	1.1	0.4	0.7	0.8
Asia (excluding Middle East and USSR)	9.1	8.4	7.6	14.0
Japan	5.8	4.5	4.5	6.3
People's Republic of China	0.9	0.9	0.2	1.8
South Korea	0.4	0.9	0.7	1.3
Hong Kong	0.2	0.4	0.7	0.7
Taiwan	0.2	0.5	0.8	1.5
Rest of Asia	1.6	1.2	0.7	2.4
Oceania	0.9	0.5	0.9	0.8
Australia	0.8	0.4	0.7	0.6
Rest of Oceania	0.1	0.1	0.2	0.2
South America	2.8	1.2	3.8	1.2
Brazil	0.8	0.4	0.5	0.5
Rest of South America	2.0	0.8	3.3	0.7
Mexico, Central America, and Antilles	1.9	0.8	1.2	2.6
Mexico	0.4	0.4	0.4	2.2
Central America and Antilles	1.4	0.4	0.8	0.4
Total	100.0	100.0	100.0	100.0
Total value in $million (current)	53,183	187,347	50,108	169,951

Sources: Statistics Canada, *Exports, Merchandise Trade* (1978 and 1993), Cat. No. 65-202, Tables 1, 2; Statistics Canada, *Imports, Merchandise Trade* (1978 and 1993), Cat. No. 65-203, Tables 1, 2.

receipts from these poorer nations now account for about one-fifth of all Canadian invisible exports. This net inflow on invisibles reflects the indebtedness of the Third World to Canada, in the same way that Canada's net outflow on invisibles to industrialized countries reflects, in turn, Canada's indebtedness to the United States, the United Kingdom, Japan, and other industrialized countries.

Table 2
Canadian non-merchandise exports and imports for selected countries, 1978 and 1993

	Receipts (exports) ($million)		Payments (imports) ($million)	
	1978 %	1993 %	1978 %	1993 %
United States	49.6	56.3	62.4	54.7
United Kingdom	9.6	7.1	6.4	8.6
Rest of EC *	10.6	8.3	9.7	10.0
Japan	3.7	4.4	1.5	7.7
Rest of OECD †	4.1	3.3	5.4	5.8
All other countries	22.5	20.6	14.5	13.1
Total	100.0	100.0	100.0	100.0
Total $million (current)	11,738	40,537	20,953	77,466

Sources: Statistics Canada, *Canada's Balance of International Payments* (1978 and 1993), Cat. No. 67-001, 1978 and 1991, Tables 20, 21, 22 A-E.

* Belgium, Denmark, Federal Republic of Germany, France, Greece, Ireland, Italy, Luxembourg, Netherlands, and (after 1 January 1986) Portugal and Spain.

† Australia, Austria, Finland, Iceland, New Zealand, Norway, Sweden, Switzerland, and Turkey.

Regional Patterns

The national patterns reported above conceal some very large regional variations in patterns of merchandise trade.[7] For instance, in 1993 Japan was proportionately at least ten times more important to British Columbia's trade than it was to Quebec's, while the EC was more than three times more important to the Atlantic provinces than it was to Ontario. The salient trends in regional merchandise trade, which are summarized in Table 3, are as follows.

The Atlantic provinces imported an exceptionally low figure of 29 per cent from the United States, though exports were more typical, at 70 per cent. Also exceptional in Atlantic Canada was the high volume of trade with the "rest of the world" – the major items being crude oil imported mainly from Venezuela and vehicle components from Sweden for the Volvo assembly plant in Halifax. Quebec's ties to the United States were stronger than those of Atlantic Canada, but there were strong import ties to the EC and to the rest of the world. Ontario, in contrast, was particularly heavily tied to the U.S. economy, which accounted for 76 per cent of imports and nearly 90 per cent of exports. A two-way trade in vehicles and parts was a major component of this trade.

The trade pattern of the prairies was also distinctive. This region had the highest level of imports from the United States, but on the export side the rest of the world was more important than in any other Canadian region, due to exports of grain, fertilizer, and minerals. BC trade was exceptionally closely tied to Japan, which accounted for 25 per cent of the total (in no other region did Japan account for more than 7 per cent of the total trade). In sum, these regional differences contain few surprises: the

Table 3
Merchandise trade by region, 1993 (percentages unless otherwise stated)

Imports from	Atlantic provinces	Quebec	Ontario	Prairie provinces	BC, Yukon, and NWT	Canada total
United States	29.4	44.1	76.2	84.5	48.9	67.0
Japan	0.7	2.8	5.1	1.8	25.4	6.3
EC (1986 members)	19.9	19.6	6.0	7.0	4.4	8.7
Rest of world	50.0	33.5	12.7	6.7	21.3	18.0
Total	100.0	100.0	100.0	100.0	100.0	100.0
Total ($billion)	6.37	28.04	106.39	12.12	16.54	169.46

Exports to						
United States	69.6	78.9	89.5	75.3	53.6	80.7
Japan	5.7	1.4	0.6	6.9	24.8	4.5
EC (1986 members)	12.2	11.5	3.5	2.8	8.5	5.7
Rest of world	12.5	8.2	6.4	15.0	13.1	9.1
Total	100.0	100.0	100.0	100.0	100.0	100.0
Total ($billion)	8.05	32.94	95.92	29.84	19.93	186.68

Source: Statistics Canada, Exports by Commodity (1993), Cat. No. 65-004, Table 2; Statistics Canada, Imports by Commodity (1993), Cat. No. 65-007, Table 2.

trade orientation of specific provinces is related both to the economic structure of their economies and to their geographical location.

A Dominated Economy

The preceding discussion highlights a crucial aspect of Canadian trade. Whereas an open economy is typically engaged in multilateral trade among several partners, Canada is very strongly tied to one trading partner: in recent years, the United States has consistently accounted for over 70 per cent of Canada's merchandise trade. If NAFTA, in which the United States is the dominant partner, has its intended effect, this proportion should rise even more. No other high-income country comes anywhere close to this level of trade domination. The closest parallel is Mexico, which traded 65 per cent of its imports and 58 per cent of its exports with the United States in 1986. Australia has a staple economy similar to Canada's, yet in recent years, Japan has accounted for around 24 per cent of Australian trade and the United States around 15 per cent, and several other partners follow not far behind. New Zealand, seen by some as being dominated by Australia in a way that is similar to Canada's relationship with the United States, has divided its trade in recent years roughly as follows: Japan 18 per cent, the United States 17 per cent, Australia 16 per cent, and the United Kingdom 9 per cent. In another conceivable parallel, Portugal is contiguous only with Spain, yet West Germany was its largest trading partner (around 14 per cent of the total). In short, the extreme domination of Canadian trade by one trading partner – the United States – is without parallel.

Two problems result from such a situation. First, where a trading nation has many trading partners, its economy receives a range of economic signals that, ceteris paribus, help to smooth out cyclical economic fluctuations. However, Canada's wagon is hitched to one horse, this being especially true for Ontario (Raynauld 1987). Thus U.S. booms and slumps are inevitably visited on the Canadian economy, often magnified. Second, Canada's long-term economic fortunes are now inextricably linked (under NAFTA) to United States. If the latter were to be the world's leading economic force in the twenty first century, fine. But if, as this author speculates, East Asia and Europe become the pacesetters, then Canada will have hitched itself to the wrong horse.

THE COMPOSITION OF FOREIGN TRADE

Merchandise Trade

The structure of Canadian trade, by commodity, is summarized in Table 4. The totals point to a feature that has generally held true throughout its history: Canada has a surplus on its trade of crude and fabricated materials that is roughly balanced by a deficit in end products, in services, and on investment income (Canada remits a large amount in dividends and interest to foreign countries).[8]

A key feature of Canada's commodity trade, the visible part of the trade balance, is the traditional surplus ($39.7 billion in 1993) in the trade of edible, crude, and fabricated materials. These staples in 1993 still accounted for 52 per cent of Canada's total merchandise exports. Table 4 reports trade for the major "sections" and certain groups of "chapters" (as they are known by Statistics Canada). In 1993, Canada's main trade surpluses included wheat ($2.8 billion), metal ores and concentrates ($1.4 billion), hydrocarbons ($9.6 billion), wood and paper ($22.4 billion), non-ferrous metals ($6.8 billion), passenger autos ($12.2 billion), and trucks ($7.2 billion).

These surpluses on merchandise account have generally been offset by some big deficits, particularly in the trade of end products, where the deficit totalled $30 billion in 1993. The major items in deficit included: industrial machinery ($6.3 billion), motor vehicle engines and parts ($11.5 billion), computers and office equipment ($5.4 billion), and assorted other end products ($7.3 billion). Trends since 1975 for some of the main components of merchandise trade are plotted in Figure 1. The graph shows that the broad patterns have held over the last 15 years. A fluctuating but generally large deficit in the trade of end products can be set against a positive balance in the trade of food, feed, beverage, and tobacco, of crude materials, and of fabricated materials. More sensitive was the trade in automotive products, with the balance swinging from negative to positive several times.

The picture painted thus far might seem to be a healthy case of a state pursuing its comparative advantage within the new staple economy. Canada has either a comparative advantage or, in some cases, an absolute advantage in the export of the various staple resources. Moreover, many of these staples are processed into fabricated ma-

Table 4
Canadian merchandise exports and imports by commodity, 1978 and 1993

	Domestic exports*		Imports	
	1978 %	1993 %	1978 %	1993 %
Fish	2.1	1.4	0.5	0.6
Wheat	3.7	1.6	–	–
Other animals, food, beverages, tobacco	4.4	4.7	7.1	5.5
Sub-total	10.2	7.7	7.6	6.1
Metal ores, concentrates, etc.	4.6	1.9	1.4	1.2
Crude petroleum	3.1	3.9	6.9	2.6
Natural gas	4.2	3.3	–	–
Coal, etc.	1.5	1.1	1.3	0.3
Other crude materials, inedible	3.7	2.4	2.2	1.2
Sub-total	17.0	12.2	11.8	5.3
Lumber	6.2	5.3	0.4	0.3
Wood pulp, etc.	4.2	2.6	0.1	0.1
Paper and paperboard	6.7	5.3	0.6	0.9
Other wood and paper	1.1	1.1	0.4	0.4
Textiles	0.3	0.6	2.2	1.6
Chemicals	4.6	5.1	5.2	6.5
Iron and steel	2.8	1.9	2.2	1.7
Aluminum	2.2	2.1	0.5	0.7
Precious metals	1.1	1.6	0.5	0.5
Other non-ferrous metals	3.1	1.7	0.6	0.5
Other fabricated materials, inedible	4.3	4.8	4.8	5.6
Sub-total	36.5	32.1	17.6	18.8
Industrial machinery	3.0	3.2	8.6	7.1
Agricultural machinery, tractors	1.2	0.4	3.0	1.1
Passenger autos and chassis	9.5	13.7	7.6	7.0
Trucks	5.2	5.9	2.5	2.0
Motor vehicle engines, parts	8.6	6.7	15.4	13.8
Aircraft engines and parts	1.3	2.3	1.5	1.7
Other transportation equipment	1.6	1.9	2.2	2.1
Telephone, television, telecommunication	1.1	4.1	3.2	6.6
Computers, office machines	0.9	2.2	2.1	5.5
Other end products	3.6	6.5	15.8	27.3
Sub-total	35.9	46.9	62.0	67.3
Special transactions	0.3	1.2	1.0	2.6
Grand total	100.0	100.0	100.0	100.0
Total value ($million current)	51,919	176,757	49,938	169,460

Sources: Statistics Canada, *Summary of Canadian International Trade* (1978 and 1993), Cat. No. 65-001, Tables X-3 and M-3.

* Excludes re-exports; hence totals are not comparable with those in Tables 1 and 3.

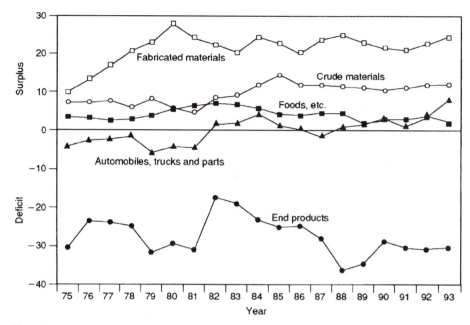

Figure 1

Balance of trade on merchandise account, in billions of constant (1991) dollars, 1975–91. *Source*: Statistics Canada, *Summary of Canadian International Trade 1978 to 1991*, Cat. No. 65-001, Tables X-3 and M-3, with values calculated using GDP deglators in Statistics Canada, *National Income and Expenditure Accounts, Cat. No. 13-201, Table 7.*

terials before being exported, so that value is added in Canada. This, so the argument goes, allows Canada to import manufactured end products, in the production of which we do not have any special advantage. The danger of such an argument becomes clear when the balance in non-merchandise trade (that is, invisibles) is taken into account. As Figure 2 shows, Canada's surplus in merchandise trade has been matched since 1978 by a steadily deteriorating situation in the trade of invisibles (all the figures plotted are in 1991 constant dollars). Since 1975 Canada has run up an annual deficit in the trade of invisibles that rose from $10 billion in 1975 to a staggering $36 billion in 1993. For non-merchandise trade the situation has inexorably deteriorated. Meanwhile, the surplus on merchandise trade grew up to the early 1980s but since 1984 has tumbled, so that the total annual trade balance has been in deficit since 1985 and by 1991 had plummeted to a deficit of $29 billion, improving slightly to −$25 billion in 1993. This deficit has been "balanced" by borrowing, mainly from abroad (which is one reason why Canadian interest rates have had to remain damagingly high). In other words, since 1985 Canada has fallen seriously short on the level of merchandise trade surplus needed to compensate the chronic, and now massive, deficit in non-merchandise trade.

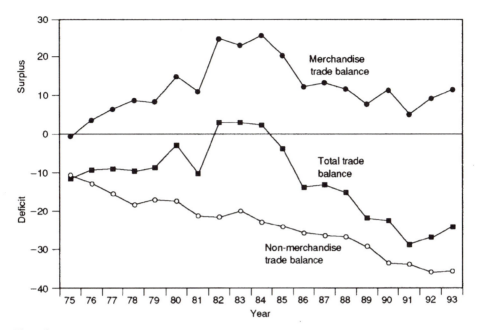

Figure 2
Total balance of trade, merchandise and non-merchandise, in billions of constant (1991) dollars, 1975–91.
Source: Statistics Canada, *Canada's Balance of International Payments, 1991*, Cat. No. 67-001, Table 20.
Values are calculated as in Fig. 1.

Non-merchandise Trade

The main elements of Canadian non-merchandise trade include the travel and vacation account, in which Canada ran a steady annual deficit in the $2–3 billion range from the mid-1970s to the late 1980s (Figure 3; once again, values are 1991 constant dollars). This was partly explained by retired and other Canadians avoiding the rigours of winter. However, in the early 1990s the travel deficit ballooned to over $7 billion. The deficit in business services has followed a similar pattern, but it has not grown in the most recent years as fast as the travel deficit and remains at about $4 billion. Thus the surplus in the trade of such a key staple as wheat, long a cornerstone of Canadian trade, is now matched by the deficit in business services. These services include payments to foreign head offices for administrative services, for patent rights, and for research and development performed elsewhere for subsidiaries located in Canada. Harrington (1989) shows that the main source of Canada's substantial deficit in the trade of business services is the United States.

The truly troubling part of Canada's deficit on invisible trade is the astounding growth of the deficit in investment income. From 1975 to 1981 this deficit (in 1991

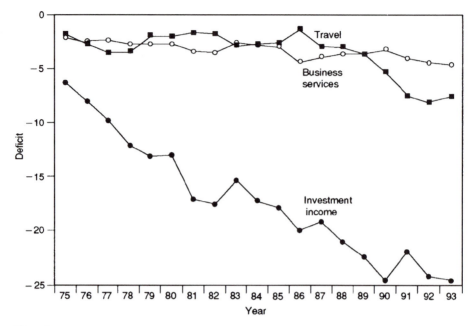

Figure 3
Balance of trade on non-merchandise account, in billions of constant (1991) dollars, 1975–91. *Source*: Statistics Canada, *Canada's Balance of International Payments, 1991*, Cat. No. 67-001, Table 20. Values are calculated as in Fig. 1.

constant dollars) increased nearly threefold, to $17 billion. Since 1981 the situation has worsened, with this deficit exceeding $24 billion in 1990 and remaining close to that level up to 1993. The deficit on investment income cancels out the trade surplus of Canada's single biggest export staple – that of forest products. It is the combined result of the long-established and, under the Conservative governments, renewed sell-off of Canadian resources to foreign interests, coupled with both a federal government budgetary deficit and, since 1985, a serious national trade deficit. Both of these deficits have required borrowing from abroad to balance the books, as is demonstrated in MacPherson's chapter on capital flows (this volume, chap. 4).

The broader picture that emerges is as follows. Through the late 1970s, Canada began to build up a healthy surplus in merchandise trade which peaked in 1984 and slumped thereafter despite a five year boom in world trade. This surplus was achieved during a period of foreign take-overs of Canadian-owned corporations, sometimes associated with foreign direct investment in new projects. For a short while this influx of foreign capital, coupled with a Canadian dollar worth about 82 cents U.S. seems to have helped, in the sense that Canada recorded an overall trade surplus from 1982 to 1984. But since then the merchandise trade surplus has shrunk, the deficit on non-merchandise trade has grown, and the deficit in the total trade balance grew from

$13 billion in 1986 to $29 billion in 1991, with a slight improvement in 1992–93, as world interest rates dropped. What lies behind this troubling dénouement? Three factors have played a key role.

First, productivity in an array of Canadian industries remains stubbornly below the U.S. level. In the era of free trade, Canadian producers that are not competitive will inevitably remain at the margin, the first to close down in periods of deficient demand, and thus a contributor to the cyclical behaviour of the Canadian economy. However, raising productivity will not in itself solve this problem. Efficient Canadian producers have for years been favourite take-over targets for foreign capital, sometimes creating a subsidiary with no export mandate, and often resulting in an increase in the net outflow of interest and dividends.[9] Thus productivity and innovation policies that make Canadian industries competitive with U.S. producers need to be coupled with some mechanism to retain the benefits within Canada. New technologies generate technology rents. The recent record of technological innovation in Canada is less than impressive, but Ottawa has allowed non-Canadian interests to buy out technologies that should have generated economic rents at home.

Second, Canada can and does partly compensate for this productivity difference by maintaining the Canadian dollar below par and by accepting a standard of living lower than that of the United States. However, during recent years, when international capital flows have been minimally regulated, the Canadian dollar's exchange rate has fluctuated over a much wider range than has the U.S.-Canadian productivity differential, which suggests that, all things being equal, exchange rates will have had a bigger influence on international trade and investment patterns than have productivity differences. For example, the rise in the value of the Canadian dollar to over 86 cents U.S. by early 1991 (compared to 76 cents in October 1987) was far more important than high wage settlements in making Canadian merchandise exports expensive in the main export market – the United States.[10]

Third, several Canadian resources are suffering from severe depletion. Major cutbacks in the Atlantic fishing industry in 1990 as a result of overfishing, worrying levels of both inshore and offshore pollution, and serious stock depletion were followed in 1992 by the complete closure of the east coast offshore cod fishery. A range of Canadian resource industries show similar problems. Many forests suffer from poor management, overcutting (particularly in the west), looper and budworm infestation, late adoption of forest management programs, and the lingering impact (in the east) of acid precipitation. Hydrocarbon resources in the Beaufort Sea, in northern sands and shales, and off the Atlantic coast are proving very costly to exploit, pose serious environmental risks, and will need high world oil prices to make them viable economic propositions. Mining resources, by definition, are subject to depletion, though periodic discoveries will trigger flurries of investment. Sudbury is no longer dominant in world nickel production, the iron mines in Schefferville have been closed (Bradbury and St Martin 1983), and several other mines (such as Nanisivik in northern Baffin Island) have proved to have smaller reserves than was once hoped. Meanwhile, Canadian agricultural exports have been squeezed by the subsidies offered to farmers in the United States and particularly in the EC (which has now become an exporter of several agricultural products previously imported). In a nut-shell, Canada's resource base, the very foundation of trade policy for several centuries, shows signs of exhaustion due both to natural depletion and to mismanagement.

Finally, there are the disadvantages associated with extensive foreign ownership. Foreign-owned firms are more likely to import their technology and capital goods than are domestic corporations (Cordell 1971). They are more likely to export their products in a crude or semi-processed state, whereas domestic firms tend to add more value within Canada. And foreign-owned manufacturing firms are more likely to source intermediate imports and producer services abroad. In addition, by permitting the extensive sell-off of Canadian resources to foreign interests, Canada loses the associated resource rents in perpetuity. In Canada's staple economy these resource rents are large, hence the long-term consequences are very serious.[11] This theme is examined more fully below.

The Technology Gap and Canadian Trade

The concerns expressed above for the long-term prospects of Canada's resource base take on their true significance when we bring the sad state of Canada's high-technology industries into the picture. The most successful trading nations since 1945 – France, Hong Kong, Japan, Sweden, South Korea, and West Germany – have developed progressively more sophisticated export industries, including motor vehicles, micro-electronics, scientific instruments, pharmaceuticals, light engineering, electrical equipment, and (alas) armaments. The process has been cumulative, reaching into many aspects of society and economy. Not only are these products typically associated with high income elasticities of demand, but the demand patterns are less cyclical than is demand for Canada's staples. Moreover, because of their scarcity they earn technology rents – a surplus that consumers are prepared to pay to have either the latest technology of an established product or a new product. Canada's burgeoning deficit in its trade of end products reflects the debilitating long-term consequences of this technological gap.

If Canada's technology gap has a single explanation it must be back of interest in the problem and the consequent policy vacuum. Over the years Canada has developed and marketed a number of important technological innovations, but rarely have they been built up into major export industries and retained in Canadian ownership. From Alexander Graham Bell's telephone to De Havilland's Dash 8 aircraft, Canadians have seemingly been unconcerned when foreign interests exploit or take over these innovations. Many have argued that the economy is too small to pursue independently a technology-oriented development policy, but this argument is negated by such success stories as Finland, Sweden, and Switzerland, all with economies far smaller than Canada's, yet with many technology-based exports. The technology gap shows with startling clarity in Canada's trade with Japan, which is reviewed below. Indeed only in Canadian-U.S. trade do Canadian manufactured end products figure prominently, and that is attributable to the Canada-U.S. Auto Pact, which is an example not of free but, of managed trade.

Behind this dismal picture of Canadian trade lies the flow of capital into and out of Canada, which is discussed by MacPherson (this volume, chap. 4). Canada's recent trade deficit has been accompanied by increased foreign borrowing. In 1988 and 1989 extraordinarily large net inflows from abroad of $15.3 and $16.9 billion, respectively, took place into Canadian bonds, largely as a result of the federal government's need to balance its enormous deficit.[12] For foreign direct investment, the net flow into Canada in 1988 was $5.1 billion. This flow subsumes a net inflow of foreign capital buying existing Canadian interests (in effect, take-overs) exceeding $4 billion in 1987 and 1988 and $3 billion in 1989 (see Semple, this volume, chap. 19).

THE OPEN ECONOMY

It might appear that the Canadian economy is not exceptionally open. In 1991 exports of goods and services amounted to 24.7 per cent of GDP which figure was in the same range as that for the larger EC countries such as Italy (19.5 per cent), France (22.6 per cent), West Germany (34.1 per cent) and the United Kingdom (23.4 per cent).[13] But Canada's economy does appear unusually open. The EC countries are supposedly moving toward political union; their territories are comparatively small and, except for the British Isles, form a single, compact land mass; and they had consciously been integrating their economies for 20 years in preparation for economic union in 1992. Thus one would be surprised if they did not trade a great deal with each other. Canada, in contrast, covers a very large area, which, one might suppose, presents greater possibilites for self-sufficiency. And, excepting the United States, it has no contiguous trading partners. Indeed, the contrast is startling, since the United States exported only 10.5 per cent of GDP in 1991, which is less than one-half of the Canadian proportion. Moreover, Canada was clearly exporting a high proportion of GDP long before entering into the FTA.

This comparison with the United States may not be entirely appropriate: the U.S. economy is diversified and, as a result of protectionist policies pursued at various times in the past, is fairly closed. Comparing Canada with similarly staple-oriented economies may seen more useful, but such comparisons still point to the same conclusion. In 1991, Australia exported 17.8 per cent of GDP, New Zealand 29.3 per cent, South Africa 25.0 per cent, and Argentina 7.7 per cent. Thus in this context, Canada still stands out as having an unusually open economy.[14]

The openness of the Canadian economy is evidenced not only by trade flows but also, as already demonstrated, by capital flows. Capital flows into Canada have taken the form of foreign direct investment (FDI) (witness the flood of foreign take-overs in the late 1980s, and see Semple, this volume, chap. 14) portfolio investment, and investment in government bonds. Kierans (1990) reports that by 1990 total FDI in Canada was $110 billion, and in the United States $329 billion, which (given the ten-fold difference in population) points to a rate of penetration by foreign capital roughly three times higher in Canada.

CANADA—U.S. TRADE

To the extent that roughly three-quarters of Canadian trade is with the United States, it follows that much of the national overview presented above is also a description of the main outlines of Canadian-u.s. trade. The resulting domination, noted at the outset, leaves Canada little room to manoeuvre. It would, for instance, have been very difficult for Canada to resist u.s. overtures for free trade, even if the government had wished to do so. Two aspects of this trade form essential parts of the current review. First, in many respects Canada and the United States are competitive producers. And second, the immediate and long-term impact of the FTA and its successors will be addressed.

Complementary or Competitive Trade?

Canadians are frequently reminded that the dominant orientation of North American geography is north-south; the Rockies, the Great Plains, and the Appalachians all trend along the meridians and cross the 49th parallel. An oft-repeated maxim suggests that the two countries are united by their geography but divided by their history. The geographical corollary to this point seems rarely to have been understood. If the geographies of British Columbia and Washington state, of Saskatchewan and North Dakota, of Michigan and southern Ontario, and of Maine and New Brunswick are broadly similar, then their climate, topography, soils, and vegetation must permit production of similar staples. Indeed, the presence of similar resource endowments goes much further. On both coasts, offshore resources are similar. Also, the hydrocarbon resources that underlie the Great Plains and the Appalachians are present in both countries. The result, in numerous instances, is a highly competitive, if not a charged situation. On the east coast, the Georgian Banks became the subject of a long and acrimonious fishing dispute eventually settled in the World Court. Similar disputes over the fisheries, and threats of trade retaliation, occur quite regularly.

Forest product industries in British Columbia and the United States are likewise regularly in dispute over subsidies and prices, with recent examples including confrontations over the pricing of cedar shingles and shakes and over stumpage fees. Grain farmers are now locked in dispute because Canadian producers have been squeezed out of many overseas markets by subsidised American producers. For beef, hogs, eggs, and even wine, these kinds of disputes have been replicated from time to time. This competition spilled over into end products in 1992, with u.s. car producers demanding an audit (under FTA) of the Canadian content of several Canadian auto assembly plants. Thus Canada's dominant trading partner is a competitor, producing domestically many of the staples on which Canada continues to base its export policy. Yet the u.s. government has protected many of its own staple producers. Given that the two economies are highly competitive, and that Canada is often the higher-cost producer, one may ask about the wisdom of recent Canadian trade policy focusing on closer continental integration.

FTA and NAFTA

An enormous amount was been written on the FTA and NAFTA, much of it speculative and rhetorical. There is little point in repeating the arguments here. What will be discussed is their effect on trade and investment in the period immediately following their implementation.

The FTA created a continental free trade area to which Mexico was attached in 1993 through NAFTA (the maquiladora zones were already incorporated). It was the culmination of a long-continuing trend toward trade liberalization with the United States. For example, 61 per cent of Canada's merchandise trade imports from the United States entered duty free in 1976 (the duty-free percentage for the rest of the world was also 61 per cent). By 1987, the duty-free proportion had risen to 73 per cent for the United States but had dropped to 47 per cent for the rest of the world. In many key sectors, including motor vehicles, aircraft, agricultural machinery, pulp and paper, and books, Canadian-U.S. trade had long been either duty free or subject to duty remissions. The FTA progressively removes the remaining tariff barriers, and so its impact will be highly sectoral.

The 1988 federal election was fought essentially over the FTA. The Conservatives, who won, predicted substantial economic benefits while acknowledging that some painful restructuring would occur in certain protected and inefficient sectors. The agreement began to take effect in 1989. Since it was a year of general economic prosperity, a case can be made that many industrial adjustments made during that year were attributable to the FTA. The early evidence was not encouraging: in its annual review of 30 December 1989, the *Toronto Star* ran the headline "Free Trade: A 'Devastating' First Year." It reported a far quicker impact than expected, with widespread layoffs and plant closures and a substantial decline in Canada's trade surplus with the United States, higher interest rates, and a much higher exchange rate on the Canadian dollar. The 250,000 new jobs predicted by Prime Minister Mulroney did not materialize. The fish-processing, farming, food-processing, dairy, furniture, auto, and appliance industries were hit worst. On the non-merchandise side, Harrington (1989) suggested that the effect on trade in business services (which was also liberalized) would be harder to measure. A year later the Canadian Labour Congress stated that the FTA had led to the loss in its first two years of 226,000 Canadian jobs.[15]

Those favouring the agreement regarded the article in the *Toronto Star* as misguided. They attributed most of the negative consequences of the FTA to its unfortunate timing (at the end of a business cycle) and to adoption of inappropriate monetary policies – an overvalued Canadian dollar and high interest rates. Optimists saw the 1992 downward adjustment in the exchange rate as a necessary short-term change preceding the longer-term gains that they expect to appear during the next economic upswing, as the Canadian economy becomes more efficient and competitive with U.S. producers. Indeed, the report written by Daniel Schwanen for the C.D. Howe Institute and published in October 1992, though it confirmed the continued decline of exports in sunset industries such as furniture, clothing, and household products, reports early evidence of increasing exports to the United States of goods in categories liberalized by the FTA. Specifically, exports of certain high value–added end products, especially of office, telecommunication, and precision equipment and of "equipment and tools" rose after 1988, while exports to the rest of the world declined. Sales of business services to the United States (another liberalized category) were also reported to have risen (though this turned out to be a short-term swing – after 1990 the deficit in business services widened rapidly; see Figure 3 above).

These trends confirm two widely predicted effects of the FTA: it gave impetus to economic restructuring, and it buttressed the domination of Canadian trade by the United States as economic integration gathered momentum. However, the cautious optimism voiced by the C.D. Howe Institute on the reported growth of exports of certain end products and of business services needs tempering with three remarks. First, the improvements were recorded in sectors which, as already demonstrated, still have very serious aggregate trade deficits. Second, the gains made in the United States in these categories of end products have been offset by deterioration in the trade balance with the rest of the world.[16] And third, such bilateral links will count for little in the long run if Canada locks itself into a "fortress North America" within which the United States itself is failing to be a competitive producer in the larger global setting.

A more considered assessment is presented by Bruce Campbell (1990). He stresses that the FTA is not just a free trade agreement but a plan for sweeping economic integration; hence to attribute the negative consequences to other factors is to ignore its scope. The agreement has begun to corrode the social standards of many Canadians by encouraging their employers to insist on cuts in benefits and/or real wages in order to remain competitive with U.S. producers or those located in Mexico. It has triggered a wave of take-overs totalling around $40 billion in 1989, half of them by foreign corporations (three times the 1988 level). It has made the Canadian economy increasingly cyclical and susceptible to any economic impulse coming from the United States (Norcliffe 1994). It led to a wave of branch plant closures (probably over 100 in 1989), with production being relocated either to the United States or to Mexico. The Economic Council of Canada's projection of 250,000 new jobs being created by the FTA and NAFTA in their first 10 years now seems difficult to achieve: the CLC (which kept a detailed inventory of job losses) estimates that job dislocation resulted in the loss of 72,000 jobs by the end of the first year of the FTA and 226,000 by the end of the second year (note that its definition of a job loss was quite comprehensive). Thus the immediate evidence is of rapid and painful adjustment in several sectors and rather few gains. One reason for this dismal start is that in many sectors, Canada and the United States are competitive producers: with an 85-cent dollar, Canada's cost advantage began to evaporate, and only when the Canadian dollar was allowed to drop to around 72 cents U.S. did Canadian manufacturing regain a competitive edge.

CANADA-JAPAN TRADE

The differences between Canadian-U.S. and Canada-Japan trade are startling. As Blain and Norcliffe (1988) explain, Canada's role as a staple exporter is thrown into high relief by Japan. Taking items with a trade worth greater than $100 million in 1984, every one of Canada's twelve leading exports to Japan was a crude or a fabricated material, whereas every one of Japan's eleven leading exports to Canada was an end product. If comparative advantage and complementarity are the name of the game, then Japan is Canada's natural trading partner.

There is also a difference in the style of Japanese FDI. The main goal of Japanese investment in Canadian industries, at least up to the mid-1980s, was to guarantee supply or to gain access to the North American market, rather than to own the industry outright. Between 1965 and 1984, the Japanese initiated 103 major projects, providing only 39.5 per cent of the capital, many of the projects being joint efforts with Canadian partners. However, more recently there has been a large increase in Japanese direct and indirect investment in projects in Canada (Edgington 1993), many of these being wholly Japanese-owned. As Morris (1989; 1991) stresses, recent Japanese investment heralds a radical shift in worldwide industrial patterns. It is aimed at transferring to Canada (and elsewhere) Japanese patterns of production, with an emphasis on teamwork, workforce and production flexibility, and advanced technologies. Japanese firms frequently choose "green-field" sites to build new facilities rather than taking over existing plants, and they invest a great deal in selecting and training a new workforce. Manufactured inputs begin by being sourced largely in Japan, but local content then increases; it seems likely that the local content of Japanese car makers located in Canada will in due course reach the same level as that of U.S. car makers. However, there are two important drawbacks. First, little Japanese research and development (R&D) has been transferred to Canada. And second, except for exports to the United States, these manufacturing plants do not have an export mandate. Edgington (1990) suggests that the short-term effects of the FTA on Japanese investment in the two countries appeared to be neutral.

A FOURTH OPTION

In 1984, Anthony Westell reviewed three trade options facing Canada. The first was the status quo in U.S.-Canadian trade relations. The second, which subsequently became the main economic policy of the Mulroney government, was to move toward free trade with the Americans and eventually to economic integration. The third, which the Science Council of Canada advocated and which Jean Chrétien's Liberal government seems to be pursuing, is to strengthen domestic technology and industry with a view to developing an export base in end products and thereby diversifying Canada's trading partners. The line of argument presented in this chapter leads to the suggestion of a fourth option.

The argument began by showing that Canada's new staple economy is permeated by FDI, with very heavy U.S. investment. The resulting domination of the Canadian economy makes it very cyclical, as trade is led by foreign (U.S.) demand rather than by the innovativeness of domestic producers. It has resulted in a massive outflow of investment income. Current levels of trade deficits are simply unsupportable in the long run. Failure to nurture exports of end products within the context of a broad industrial development strategy has created a technology gap, so that Canada's technology rents are very small. The most recent consequence has been a large inflow of capital – mainly in the form of indirect investment needed to balance the massive trade deficit. Such capital can be attracted only at the price of maintaining interest rates higher than those of the United States.

A fourth option is to link an open development strategy with diversification of trade in the Pacific Rim, and with Japan and China in particular. The latter two economies are highly complementary with Canada's, though in different ways. Both import the products of Canadian resource industries. Being a low-wage economy, China exports to Canada low-technology products that cannot be produced competitively in Canada. Meanwhile, to the extent that the Japanese are prepared to share technology and enter into long-term and stable partnerships in Canada, Canada could move relatively quickly into a range of sophisticated end-product industries in certain high-wage sectors. Turning these into export industries via economic restructuring will require some hard bargaining, but the Japanese will recognize that their own export success has grown out of domestic industrial policies. There are clearly many unknowns in such an option, but it does have the virtue of diversifying Canada's trade linkages, and of developing new, technologically–sophisticated industries, while simultaneously tackling social and regional problems. Canadian trade in goods and services under an export-led continentalist option promises chronic deficits, progressive subordination to U.S. interests, and increasing pressure for economic and political integration.

NOTES

1 This sometimes-overlooked part of Canada's history is discussed by Quinn (1961) and Ruddock (1966). Ruddock notes that after 1480 fishermen of the Hanseatic League effectively blocked Bristol's fishermen from Icelandic fishing waters; there is convincing evidence that the latter then switched to the Grand Banks but kept quiet about their activities to keep competitors in the dark.

2 There is a rumour that to break an impasse in the final stages of the FTA negotiations, a secret understanding was reached to establish a floor on Canadian exchange rates (Campbell 1990, 7). Circumstantial evidence, in the form of a higher exchange rate and high real interest rates pursued by the Bank of Canada from the date the FTA came into effect until 1992, lend some credence to this rumour.

3 An open-development strategy would also have desirable domestic economic benefits, including fuller employment and improved income distribution both regionally and among groups.

4 By the end of 1990, the Uruguay Round of negotiations in Brussels had become deadlocked over protection of agriculture. But otherwise progress had been made in trade liberalization. This deadlock was finally broken in 1993.

5 1993 is the most recent year for which Canadian trade data are available at the time of writing. I chose 1978 as a moderately prosperous year, at the peak of the OPEC oil price rise, and preceding the 1979 mini-recession and the serious recession that followed in 1981.

6 The proportion of Canadian trade with the United States has surged since the FTA was implemented. For example, from 1991 to 1993 the U.S. proportion of imports rose from 63.8 per cent to 67.0 per cent, and exports from 75.3 per cent to 80.4 per cent.

7 Regional breakdowns for non-merchandise trade are not published. However, the merchandise trade reported in Table 3 accounted for 73 per cent of Canada's total trade in 1989.

8 It is for this reason that the monthly deficit on merchandise trade of $421 million in October 1989 was extremely disquieting. This was the first monthly deficit on merchandise trade since 1975.

9 A classic example of this sequence is Connaught Laboratories of Toronto, which was not only highly competitive but also a leading innovator in immunology and related fields of bio-technology. This type of firm might be assumed to generate considerable technology rents. Such a successful Canadian company is a prime target for a take-over, which recent federal governments have shown little inclination to overrule. Thus in 1989 Connaught Laboratories was taken over by l'Institut Mérieux, a leading French company, in which the French government holds a substantial stake.

10 In an article entitled "Canadian Firms Losing Labour Cost Edge" in the *Globe and Mail*, 5 June 1990, James Rusk reports that in 1989, for the first time in 20 years, Canadian real labour costs were higher than those in the United States due to increases in the value of the Canadian dollar and a rise in Canadian wage rates. A 15.6 per cent Canadian advantage in manufacturing's unit labour costs in 1985 had become a 3.6 per cent disadvantage in 1989.

11 To give an indication, in *Financing Confederation* the Economic Council of Canada (1982) estimates (Table 4.1) that the total resource rent generated in Canada in 1980 was $28.8 billion, or 10 per cent of GNP. Nearly half of this total was redistributed under a variety of mechanisms, including the Equalization Program.

12 For example, the *Globe and Mail* (24 March 1990) reported that in January 1990 non-residents (especially Japanese investors) invested $1.77 billion in Canadian bonds.

13 These data are drawn from United Nations (1993), *National Accounts Statistics: Main Aggregate Tables and Detailed Tables, 1991* (New York: United Nations).

14 There are significant interprovincial trade barriers within Canada, which undoubtedly contribute to the openness of the economy.

15 *Globe and Mail*, 15 Dec. 1990, p. A1.

16 The short-term trends reported by Schwanen for telecommunications and for office equipment seem accurate. However, the reported trends for business services hardly correspond with the data that follow – see Statistics Canada, Cat. No. 67–001 (1993), Table 20 and 21 – for the balance of trade in business services (in current $billion).

	1988	1989	1990	1991	1992	1993
Balance with the United States	−3.2	−3.5	−3.4	−3.7	−4.1	−4.2
Balance with the rest of the world	−0.1	0	+0.2	−0.1	−0.4	−0.5
Total balance	−3.3	−3.5	−3.2	−3.8	−4.5	−4.7

REFERENCES

Blain, R., and Norcliffe, G. 1988. "Japanese Investment in Canada and Canadian Exports to Japan 1965–1984." *Canadian Geographer*, 32, 141–50.

Bradbury, J.H., and St. Martin, I. 1983. "Winding-down in a Quebec Mining Town: A Case Study of Schefferville." *Canadian Geographer*, 27, 128–44.

Britton, J.N.H., and Gilmour, J.M. 1978. *The Weakest Link*. Ottawa: Science Council of Canada.

Campbell, B. 1990. "In the Image of the eagle: Remaking Canada under Free Trade." *Canadian Dimension*, 24 no. 2, 6–11.

Cohen, M. 1991. "Exports, Unemployment, and Regional Inequality: Economic Policy and Trade Theory." In D. Drache and M. Gertler, eds., *The New Era of Global Competition*, 83–102. Montreal: McGill-Queen's University Press.

Cohen, S.S., and Zysman, J. 1987. *Manufacturing Matters: The Myth of the Post-Industrial Economy*. New York: Basic Books.

Cordell, A.J. 1971. *The Multinational Firm, Foreign Direct Investment and Canadian Science Policy*. Ottawa: Science Council of Canada.

Economic Council of Canada, 1982. *Financing Confederation*. Ottawa: Supply and Services.

Edgington, D.W. 1993. *Japanese Investment in Canada: Recent Trends and Prospects*. Vancouver: B.C. Geographical Series.

– 1990. "Japanese Perceptions of the Canada-U.S. Free Trade Agreement." *Canadian Journal of Regional Science*, 13, 349–66.

Harrington, J.W. 1989. "Implications of the Canada–United States Free Trade Agreement for Regional Provision of Producer Services." *Economic Geography*, 65, 314–28.

Kierans, T. 1990. "Fear of Foreigners." *Report on Business*, Feb., 37–9.

Morris, J. 1989. The Changing Industrial Structure of Canada in the 1980's: The Role of Japanese Foreign Direct Investment. Final Report to Canadian High Commission, London.

– 1991. "A Japanization of Canadian Industry." In D. Drache and M. Gertler, eds., *The New Era of Global Competiton*, 206–28. Montreal: McGill-Queen's University Press.

Norcliffe, G.B. 1994. "Regional Labour Market adjustments in a Period of Structural Transformation: An Assessment of the Canadian Case." *Canadian Geographer*, 38 no. 1, pp. 2–17.

Ohmae, K. 1985. *Triad Power: The Coming Shape of Global Competition*. New York: Basic Books.

Quinn, D.B. 1961. "The Argument for the English Discovery of America between 1480 and 1494." *Geographical Journal*, 127, 277–85.

Raynauld, J. 1987. "Canadian Regional Cycles and the Propagation of U.S. Economic Conditions." *Canadian Journal of Regional Science*, 10, 77–89.

Ruddock, A.A. 1966 "John Day of Bristol and the English Voyages across the Atlantic before 1497." *Geographical Journal*, 132, 225–33.

Smith, M.R. 1989. "A Sociological Appraisal of the Free Trade Agreement." *Canadian Public Policy*, 15, 57–71.

Stone, F. 1988. "GATT: From the Tokyo Round to the Uruguay Round," Discussion Paper, Institute of Public Policy, Ottawa.

Westell, A. 1984. "Economic Integration with the USA." *Economic Perspectives*, Nov.–Dec., 22.

Williams, G. 1983. *Not for export: Towards a Political Economy of Canada's Arrested Industrialization*. Toronto: McClelland and Stewart.

Wolfe, D.A. 1989. "Technology and Trade." Ontario NDP Economic Policy Review.

Young, R.A. 1989. "Political Scientists, Economists, and the Canada-US Free Trade Agreement." *Canadian Public Policy*, 15, 49–56.

CHAPTER THREE

External Shocks: Regional Implications of an Open Staple Economy

TREVOR J. BARNES

Canada is a classic example of a small, open economy. In 1992 exports represented 24.8 per cent of its GDP, while the comparable figures for the United States and Japan were only 10.5 per cent and 10.4 per cent, respectively. More significant perhaps, over four-fifths (85 per cent in 1992) of that trade was carried out with a single, very large, and relatively autarkic nation, the United States. The consequence, and one that is a central theme of this chapter, is that Canada is continually susceptible to external shocks, which are often conveyed through trading relations with its southern neighbour.

Not surprisingly, there is a history of sustained theorization among Canadian social scientists about the relationship between national development and trade. Best represented by Innis's (1930) staples theory, developed in the 1930s, the core idea of such work is that the motor of Canadian economic history is a series of externally traded primary resources (staples). Though this approach is refined to take into account such factors as multinational corporations, finance capital, and class relations (Clement and Williams 1989), the *spatial* entailments of staple production remain relatively underdeveloped. The purpose of this chapter is to rectify such an omission. By using a staples approach broadly defined, the chapter seeks to explore the differential geographical impact within Canada of its open economy. The effects of external trade do not wax and wane at a single, dimensionless point but are geographically variegated in accordance with distinct regional divisions of labour and their associated institutional structures (Norcliffe 1988). In its most extreme form, such a process is manifest in the form of single-commodity resource towns that come and go with the vicissitudes of international commodity markets.

The chapter has four sections. First, I briefly review the lineaments of staples theory and its subsequent amendments. Second, I look at the history of Canadian external trade in staples since 1950, and its differential regional effects. I suggest that to understand those effects it is useful to graft on to staples theory recent theoretical ideas about Fordism, post-Fordism (flexible specialization), and the transition between the two. By doing so, we can connect international trade and exchange with

Table 1
Canadian merchandise exports and imports by value, 1960–92

Commodity type	Exports % share*				Imports % share*			
	1960	*1970*	*1980*	*1992*	*1960*	*1970*	*1980*	*1992*
Food, beverages, tobacco, and live animals	18.8	11.4	11.1	8.5	10.6	8.0	6.9	6.2
Crude materials (including petroleum)	21.2	18.8	19.9	12.5	13.6	8.4	16.4	5.3
Fabricated materials	51.9	35.8	39.5	30.3	24.4	20.7	18.4	18.6
Finished end products	9.1	33.8	29.3	46.6	49.7	61.8	57.2	67.0

Source: Statistics Canada, *Summary of Canadian International Trade* Cat. No. 65-001.
* Because of rounding errors and omission of the "special transactions" category, column totals do not sum to 100.

specific characteristics of production on the ground. Third, to illustrate the vulnerability of Canadian regions to changing trade relationships, I discuss the recent experience of British Columbia. Finally, to bring together the global trends of trade with their local effects, I examine the resource towns that litter the Canadian landscape, focusing on two specific examples – Chemainus and Port Alberni on Vancouver Island.

STAPLES THEORY

In Mel Watkins's (1963, 144) exposition of Innis's staples model, exports of primary resources function as "the leading sector of the economy, and set the pace for economic growth." In the optimistic version of the theory (associated especially with neoclassical economics), this sector then stimulates diversification through its various links, eventually leading to full industrialization (Baldwin 1956). In the pessimistic account, associated with the Canadian political economy school, of which Watkins was later to become a leading member, the country is ensnared in a staples trap (Williams 1983). In this view diversification is blocked because of such factors as an export mentality among producers, domination of the economy by a few, large, and often foreign-owned multinational corporations, and a truncated industrial branch-plant structure that minimizes development of higher-order control and research functions (Britton and Gilmour 1978). Given the prolonged reality of Canada's continued high export of staples products (Table 1) and its relatively low proportion of world export markets in manufactured end products the pessimistic view appears most borne out by the evidence. The result, to use Innis's terminology, is that Canada is something of a hinterland economy, one whose fate is strongly tied to events in foreign metropoles.

Of course, this still does elucidate the precise mechanisms by which staples production caused Canada's "arrested industrialization" (Williams 1983). The argument is that there is something about the very production and export of primary commodities that provokes instability and dependence and in turn places Canada on the periphery. First, the market for staple commodities much more approximates perfect competition than does that for manufactured goods. Canada is a price-taker in a market where price volatility is the norm. In particular, the bulk and crude exports that are the basis of Canadian trade tend to be vulnerable to demand shifts in markets that are both highly competitive and price-elastic. Second, because domestic sales of staples are relatively small, international market volatility has direct and strong effects, thereby producing the characteristic boom-and-bust economy of resource-producing regions. Third, for a variety of reasons (technological advances that reduce resource inputs for production, growth of synthetic substitutes, and low long-run income elasticities of demand; Webb and Zacher 1985), the terms of trade for primary commodities are increasingly less favourable to staples-producing areas. Finally, resource production tends to be undertaken increasingly by big, often foreign-owned multinational corporations. Spry (1981) argues that this is a direct consequence of the large capital expenditures and production indivisibilities associated with staples. Foreign multinational penetration in staples production can create a number of problems, however, including appropriation of economic rents that would otherwise would go to the province because of the deliberate undervaluing of resources (Gunton and Richards 1987); the absence of (value-adding) resource processing prior to export (Webb and Zacher, 1985, 128–33); limited technological development (Britton and Gilmour 1978; Hayter 1988); lack of local control; and weakened ability to direct trade through explicit policy because of the high degree of intra-corporate transfers (Science Council Industrial Policies Committee 1981).

In sum, from the perspective of Canadian political economy, there is a direct relationship between the type of trade in which Canada engages and its historic inability to become a fully industrialized nation. This connection is not one that traditional (neoclassical) economic theory would ever make. It would suggest instead that Canada most benefits from specializing and trading in those commodities in which it has a comparative advantage – namely, primary resources. But in drawing on this theory, as Innis (1956, 3) wrote some sixty years ago, orthodox economists "attempt to fit their analysis of new economic facts into ... the economic theory of old countries. ... The handicaps of this process are obvious, and there is evidence to show that [this is] ... a new form of exploitation with dangerous consequences." Innis designed his staples theory precisely to circumscribe such exploitation – its object was always a set of peculiarly Canadian issues.

CANADA'S REGIONS AND THE OUTSIDE WORLD

Precisely because Innis's work is a general framework for analysing continued staples production, it is less useful in understanding the particularities of the accompanying geography. Those particularities are vital, however. Specifically, the argument that I put forward in this chapter is that the detailed timing, effects, consequences, and responses of each region to changing conditions of external trade will hinge, first, on the region's commodity specialization and hence its particular division of labour, and, second, on the affiliated political, social, and economic institutions. It is these two spatially variable factors that cause external change to affect different places in different ways.

More formally, associated with each of these two spatially variable causes – the division of labour and institutional structure – is a different mechanism that shapes a given region's response to outside change (Norcliffe 1988, 201). The first is called the industry-mix effect and is associated with the division of labour. Here, as Carmichael (1986, 7) writes, "the existence of industrial structures that vary across the country's regions creates the possibility of levels of economic performance that vary widely." For example, the effects of rising oil prices will differ in Alberta and Ontario because of the industrial structures of each region. In this case, a given external perturbation induces different effects in different places because of variations in commodity specialization and thus in the spatial division of labour.

The second mechanism is called the competitive effect. Because capitalists are not all equally competitive across different regions in producing the same product, more competitive regions are likely to do better than less competitive ones for a given "external shock." In this case, local institutional structures can actually shape a region's competitive edge.

Both competitive and industry effects operate simultaneously and at different scales. But for Canada, at least, a number of studies suggest that at the regional level the industry-mix effect is the most important in determining the consequences of external change (Miller 1987; Raynauld 1987; Norcliffe 1988). This is not surprising, given the earlier discussion about staples. Resources are location specific, and this fact implies a high degree of regional specialization, and thereby a strong industry-mix effect.

The result of the dominance of the industry-mix effect, one that is so clear in the Canadian space economy, is a patchwork quilt of unique regions, each responding differently to outside change because of differential lead and lag times associated with their staples product(s). This implies not that each region is autarkic, only that each is different. Inter-sectoral and inter-urban linkages connect regions, and such ties help to transmit change in one part of the system to other parts. Nonetheless, because of the marked economic specialization and distinct divisions of labour that exist among Canada's regions, a given external shock will create more diversity than uniformity.

Though regional responses may differ, all are affected, albeit in a unique way, by general, global international trade patterns. There have been three such patterns over the last forty years: first, the era of Fordism from 1950 to the early 1970s; second, a period of recession and transition until the early 1980s; and, finally, since then, an emergent post-Fordist, or flexible-specialization phase that is still under formation. Using these three distinct periods as a guide, I examine the differential geographical experiences of Canadian staples-producing regions as they respond to the outside world.

Fordism (1950–73)

The first phase is that of Fordism – the ensemble of economic and social relations that held in North America from immediately after the second World War until the recession of the 1970s (for the Canadian experience, see Jenson 1989). I discuss Fordism further below, but this so-called golden age, during the 1950s and 1960s, saw unprecedented growth of trade. The beggar-thy-neighbour trading policies of the pre-war period were curbed by the General Agreement on Trade and Tariffs (GATT, signed in 1947), which in turn prompted unfettered global exchange. More specifically, between 1960 and 1973 Canada's exports grew by 10 per cent per year, with the ratio of merchandise exports to GDP increasing from about 14 per cent to over 21 per cent during the same span. During these years there were also two other significant changes: first, Canada moved away from Great Britain as an export market and toward even greater reliance on the United States (in 1954, the United Kingdon bought 16.9 per cent of Canada's exports, compared to the United States' 59.8 per cent, but by 1984 the figures were 2.2 per cent and 76.3 per cent, respectively); and second, there was a shift in composition of trade toward finished manufactured goods (as a proportion of merchandise exports, they increased from 7.8 per cent to 29.2 per cent over the period 1960–74). This shift, however, arose from a single act, the signing of the Auto-Pact (see Holmes, this volume). In fact, if we exclude the exports of motor vehicles and parts, there was little change 1950–73 in the share that staples represented within total exports. Furthermore, as I argue below, the logic behind the shift to manufactured end products was the same one that drove changes in the resource sector during the same period – that of Fordism.

Holmes (1987) provides a comprehensive account of Fordism's leading characteristics. It is associated with large, vertically integrated, oligopolistic firms that derive economies of scale from mass producing long runs of a standard product. In addition, machinery is often dedicated to a limited range of tasks, as are individual workers, who are organized along Taylorist principles. Fordism needs both a large and expanding market and access to the resources necessary for uninterrupted mass production. Canada provided U.S. Fordism with both requirements. Through development of an increasingly continental system of production in the 1950s, Canada provided U.S. firms with an enlarged market and a resource hinterland to exploit. In particular, the 1952 Paley Report commissioned by U.S. President Harry Truman singled out 13 "key resources" that were to be obtained from Canada, preferably through direct investment by U.S. corporations (Clark-Jones 1987, chap. 1). The logic of Fordism therefore dictated that Canada be part of a system of "continental resource capitalism" (Clark-Jones

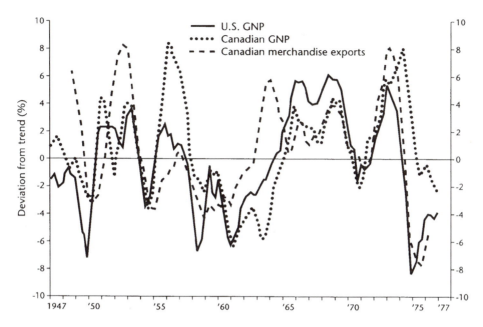

Figure 1

The relationship among U.S. GNP, Canadian GNP, and Canadian merchandise exports, 1947–77. *Source*: Conference Board of Canada.

1987). In this light, the Auto-Pact of the mid-1960s simply represented an extension of that same logic. Central Canada was to be integrated into a continental production system just as the resource periphery had been a decade earlier.

Though the concept of "continental resource capitalism" helps place Canadian staples production within a particular era, in accordance with our earlier arguments, it is less useful in understanding the changing regional economic geographies that were so created. It is to these that we now turn.

Apart from a downturn in activity in the late 1950s and early 1960s, the Canadian macro-economic trend tended ever upward, until 1973 (see Figure 1, which shows the close relationship between U.S. GNP, Canadian GNP, and Canadian exports). In accordance with the earlier arguments, however, there were strong regional contrasts. The Atlantic provinces, especially in the 1950s, suffered employment declines in primary industries, and much new investment was highly capital intensive and increasingly controlled by multinationals (Burrill and McKay 1987). The result was that the region remained a periphery on the periphery. The western provinces fared much better. Both Manitoba and Saskatchewan were aided by good crop yields in both the mid-1950s and mid-1960s, with particularly high export sales and prices in the latter period. Mining exploitation and consequent exports in the 1960s – potash, uranium, and oil and gas in Saskatchewan, and nickel mining in Thompson, Manitoba – also helped. However, British Columbia and Alberta had the greatest growth in staples production and trade. Large-scale investment was directed into Alberta's embryonic oil and natural gas industry, while through a combination of government-funded mega-projects, and often foreign investment, British Columbia's plethora of staples were exploited along Fordist lines, which process culminated in the rapid growth of Vancouver's service economy.

There are, however, considerable local variations in this broad picture. First, as studies by Swan (1972) and Thirsk (1973) showed, there were wide regional differences in the volatility of response to outside change: British Columbia reacted most violently, followed by the Atlantic provinces, Quebec, and Ontario, with the position of the prairies varying over time. Both authors attributed this order to industry-mix effects. Furthermore, though outside shocks were primarily imported from the United States (Figure 1), the speed at which they arrived depended on both the type of product in which the region specialized and the intersectoral linkages (Tomczak 1978). Second, there were significant intra-regional differences, especially among different-sized cities. In particular, Marchand (1986) and Tomczak (1978) – whose work covers 1972–76, slightly outside our time reference – both show that larger cities are more immune to the effects of major business cycles than smaller cities. This is attributed mainly to the greater diversification of activities and demands found in larger urban areas (Marchand 1986, 251).

The upshot was considerable spatial variation within the Canadian space economy during the Fordist era; there were regions within regions, each marked by a unique spatial division of labour and generating a unique set of responses to outside change. This is not to deny links, but the links themselves were of particular kinds, and they could speed up, slow down, or even counterbalance external effects.

Recession and Transition

1973 was a cusp point in recent economic history, marking the beginning of a decade of global recession and massive restructuring. Afterward the golden age was lost. A number of factors contributed: the breakup of Bretton Woods in 1971–73, which led to international currency instability, the OPEC price rises, and the consequent emergence of petro- and Eurodollar markets, satiated markets for Fordist products in North America, falling productivity in Fordist industries accompanied by rising wages, and increased international competition. The growth of world trade declined significantly – in Canada, down to 5 per cent between 1973 and 1983.

Two major trends emerged during the period. The first was the beginnings of a new international division of labour associated with multinational corporations investing in low-wage, newly industrialized countries. The second trend was emergence of the Japanese both as exporters of sophisticated manufacturing products and increasingly as foreign investors in North America (Blain and Norcliffe 1988). Both trends changed Canada's competitive advantage (Webb and Zacher 1985, 95).

A propos of the economic geography of this transitional phase, there were two periods of recession: the first 1974–75, following the first OPEC price rises, and the second, and most severe, from 1981 until about 1983–84.

The first phase until 1980 was mixed. The Atlantic provinces and central Canada
suffered initially because of high energy prices. The west did better, but even that depended on the local area. Thus export prices of mineral and forest products were variable. Maximum prices for different staples rarely coincided in time (Figure 1),
reinforcing the broader point about distinctive regional economic experiences. For example, in British Columbia exports of lumber (the most important product) declined
from $923 million in 1973 to $569 million in 1975, only to rise precipitously to
nearly $2 billion by the end of 1978 (Marchak 1983, 4). In general the prairie prov-
inces, especially Alberta, achieved the most consistent gains. Thus net farm income
during the mid-1970s in Saskatchewan and Manitoba reached record levels because
of high wheat prices and large exports (McLeod 1988), while in Alberta the economy
boomed for at least a decade as a result of high petroleum export prices.

Alberta's experience reveals the potential, or lack thereof, for directing the economy away from traditional staples by using, in this case, greater oil revenues to promote diversification. Between 1973 and 1982 net provincial oil and gas revenues
increased from $349 million to just over $5 billion, which in turn reflected primarily
the price hike in oil from $29 to $155 per cubic metre (Anderson 1988, 169). In 1976
Premier Lougheed set up the Alberta Heritage Savings Trust, to which 35 per cent of
annual oil revenues were to be directed. That trust was to be used primarily to prompt
diversification within the economy. Though it accumulated $11 billion by 1982, the
combination of rapidly falling world oil prices and enactment of the National Energy
Program in 1981 caused a severe jolt to Alberta's economy. Furthermore, as Hayter
(1986) argued, there was no real diversification. Expansion of the manufacturing sector took place only in those industries that were already tied to oil and gas, and so
when the latter faltered so did the former. Brodie (1990, 214) writes: "most available
evidence indicate[s] that after a decade of province-building, Alberta had become
more, not less, dependent on primary extractive industry." As Innis indicated over
forty years earlier, the staples trap offers little hope of escape.

The second period of recession began in 1981, when all regions, including Alberta,
plummeted into deep recession. But in accordance with the earlier argument about the
importance of geographical variation, not all regions felt the effects equally severely,
nor did they all recover at the same rate. Moreover, and also in accordance with the
earlier argument, Norcliffe (1987, 158) argues: "the key factor [in determining a region's response to recession was] ... the demand for, and price levels of, a region's
main exports." Norcliffe breaks his analysis down into two periods: downswing and
upswing. Those regions that suffered most tended to be those specializing and exporting mineral, forest, and energy products and also those focused on producing motor
vehicles. This included British Columbia, Alberta, much of the Maritimes, and the
north shore of Lake Erie. In contrast, the regions that fared best were those that were
diversified (Canada's "Mainstreet") or that had special compensating factors, as with

the prairies, which experienced bumper crops and grain prices. In terms of recovery, central Canada led the way, especially with its revivified automobile industry. In contrast, the hinterlands of western and eastern Canada remained laggards.

In sum, 1973–83 was a difficult decade for Canada. Exports of staples were subject to severe shocks, affecting regions and sub-regions differentially. Such shocks clearly necessitated new strategies to deal with a changed global scene, with increased internationalization of production and emergence of new economic powers in the Pacific, especially Japan. It is to these new strategies that we now turn.

POST-FORDISM AND FLEXIBLE SPECIALIZATION

Definitive definitions of post-Fordism, and the accompanying idea of flexible specialization, remain elusive, so use of those concepts here is as much heuristic as anything. For the most part post-Fordism and flexible specialization represent a move away from mass production of a standardized product toward much shorter runs of specialized, "niche" products. Instead of vertically integrated plants operating on the basis of gains from economies of scale, the norm under flexible specialization is vertical disintegration, in which firms are embedded within a tight network of often-informal subcontracting interlinkages. Further, rather than employing "dedicated" Fordist machines which are restricted in the tasks that they can perform, post-Fordist firms seek flexible techniques of production that incorporate robotics and CAD-CAM systems. Literally by the press of a button such machines not only facilitate a wider range of tasks and product designs compared to their Fordist counterparts but are also able to process a greater variety of input types. Such versatility is further enhanced by post-Fordist labour practices. Gone are the days, at least for some workers, of waiting on an assembly line in order to complete a single, routinized task. The post-Fordist norm is one of multi-tasked workers, each working with others in a team, or "quality circle," completing a range of operations.

Not all is golden in this brave new world of the computer chip, however. Arrival of post-Fordist production methods heralded both significant lay-offs and marked polarization between workers defined as part of the "core" – those who were "functionally flexible" – and those who were deemed only "numerically flexible" and part of the periphery (Atkinson 1985).

Clearly these post-Fordist characteristics do not apply to all firms, but they cover at least enough of them that it is now clear that part of what occurred during at least the 1980s recession was a consequence of the working out of the transition from Fordism to flexible specialization. Specifically, both labour shedding in western Europe and North America and the increasing division between those workers who had full-time jobs and those who became part-time, temporary, or on contract were a result of fundamental changes in the organization of work, production, and technology.

Such changes clearly affected world trade. The growth of merchandise exports continued to be sluggish between 1983 and 1987, rising at only about 3.5 per cent per annum. Canada fared slightly better, at 4.4 per cent. In this slowly growing world export market, which also saw increased foreign competition, the United States in particular was concerned about the degree of subsidies and protectionism afforded to certain staples entering its market (e.g., potash, lumber, and cedar shakes and shingles). This led some U.S. politicians to call for trade barriers. Partly to forestall potential isolation, Canada in January 1989 signed a free trade agreement (FTA) with the United States; the North American Free Trade Agreement (NAFTA), signed four years later, added Mexico. In effect, both documents formalized and extended the process of economic continentalization that was already well entrenched.

Within this wider, more formalized setting of trade relations, flexible specialization methods arrived first in the manufacturing sector, especially the Canadian automobile vehicle and parts manufacturers. They made an immediate difference to exports (OECD 1989, 33). They also affected the economy of central Canada, which was one of the first regions to recover from the 1980s recession and subsequently led the nation in growth. Recovery was more sluggish in the resource peripheries.

Flexible specialization was slower to emerge in the resource sector. There are examples, however, and this chapter contains two case studies of the use of such methods in the forest products industry – changes that were prompted directly by export concerns. To make this discussion more concrete I turn now to detailed examination of British Columbia, a classic example of a staples regional economy, historically buffetted by external shocks.

BRITISH COLUMBIA AS A STAPLE ECONOMY

For British Columbia export of staples remains the stubborn fact around which the province's economic history and geography are organized. Its diverse range of staple exports includes forestry products, minerals, fish, and energy in the form of coal, natural gas, oil, and hydro-electric power. Since the Second World War, however, the "green gold" of forest products has been the principal source of wealth and work.

Following Innis's work, a metropole–hinterland relationship in the province's staple-producing economy is easily recognized. From the start, Vancouver quickly emerged as the principal control centre for resource extraction. That role has become, if anything, even more entrenched with development of a distinct producer-services complex in the downtown area (Ley and Hutton 1987). In contrast, the remaining hinterland outside the Georgia Straits region continues to be a vast repository of staple production carried out by large firms (there are over a hundred single-industry towns in the province, accounting for one-third of the non-metropolitan population; Bradbury 1987, 430). Though the metropole tends to fare better than the hinterland in times of recession because of greater diversity of activities and the presence of higher-order functions (Davis and Hutton 1989), even it is not entirely immune from the instability of staple production. For example, during the early 1980s, when there were a number of changes of ownership within the forest sector (Barnes,

Hayter, and Grass 1990), MacMillan Bloedel shed over 700 employees from its head office in Vancouver. This was a result of the effects of both recession and change in control; the company had in 1981 been taken over by the Toronto-based corporation Noranda, which later reduced severely the staff at the former headquarters. More generally, British Columbia, like much of the country, remains ensnared in a staples trap, where an indigenous manufacturing sector never really developed, thereby thwarting diversification and making the province prone to external perturbations (Marchak 1988, 180). To examine further the openness of the BC economy I offer the examples of fisheries and forest products.

Over 70 per cent of the total BC catch of salmon, herring roe, and halibut is exported, principally to the lucrative markets of Japan, the United States, and countries of the European Union (fish exports represent between 3 and 4 per cent of total BC exports). Much of the actual fishing is carried out by independent operators, though fish processing since the Second World War has been increasingly undertaken by a few, large companies (in 1984 nine firms accounted for over 80 per cent of the fish processed; Standing Senate Committee on Fisheries 1987, 24). Such companies, which during the 1970s became prone to foreign control, practised Fordist production techniques, especially in salmon canning (Proverbs 1978). In response to changes in consumer demand since the late 1970s away from canned products toward more select fresh and frozen fish products, smaller, more specialized processors have recently arisen (Pinkerton 1987, 70).

Unlike other primary resources, the fish industry has a raw product that is literally on the move. For this reason there are tremendous variations in annual output among species (Pearse 1981, chap. 2). Furthermore, there is little relationship between quantity of fish caught and value. The consequence, as Pearse (1981, 18) writes, is that "the economic circumstances of the [Pacific] ... fishery are conspicuously unstable." One example is that of the herring fleet. Overfishing was so egregious that in 1967–68 the fishery was actually closed down. In the early 1970s, with the recovery of stocks, there was a switch to roe herring, which is marketed almost exclusively in Japan. But this export market proved extremely volatile. After rapid increases, the value of the market peaked at over $125 million in 1979, a year when landings had fallen dramatically. But by the following year, a combination of a weak market and a strike by fishers produced a landed value of less than $25 million (Pearse 1981, 19–20). Since then, the market has improved, but never to the dizzy heights of the late 1970s. The herring fishery is indicative of the broader industry. The instability there created by volatile market and supply conditions produces an uncertain environment for both fishers and shoreworkers and the often small coastal communities in which they live and work (Marchak, Guppy, and McMullan 1987).

In the wood products industry, it was not until after 1945 that the very large forest companies emerged in British Columbia (see also Hayter, this volume, chap. 6). Some were American firms, such as Weyerhaeuser, though the largest operating in the province was the initially BC-controlled MacMillan Bloedel (the result of a merger in 1951). The pace of corporate concentration within the forest sector was rapid: in 1940

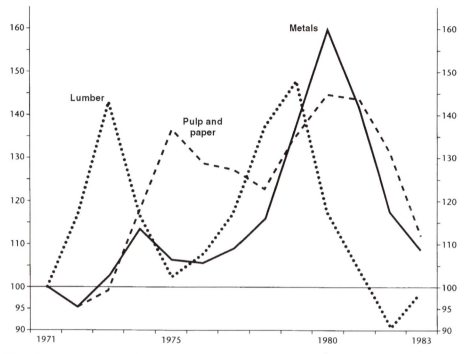

Figure 2

Relative export prices, 1971–83 (1971 = 100). *Source*: OECD *Report: Canada, 1983/84.*

the largest 58 companies controlled 52 per cent of timberland, but by 1974 the top eight controlled 82 per cent. Aiding and abetting the Fordist production methods adopted by such firms was the local state, which provided both necessary infrastructure investment and a tree-licensing scheme favouring larger firms (Marchak 1983, 30).

In the forest-products sector the two main products for export were construction-grade lumber and kraft pulp – standard products that could be mass produced using assembly-line methods, thereby realizing economies of scale (Marchak 1983, 47). Lumber was more valuable, representing about 30 per cent of the value of BC exports. The demand for lumber varied greatly with the state of the U.S. housing market (Figure 2). Similarly, the largest external market for pulp was the United States.

After the energy crisis of 1973, forest products experienced great volatility during the following decade (Table 2), but no one was prepared for the deep recession of the early 1980s. With rapidly falling demand for all forest products, 23,000 workers were laid off between 1979 and 1982, and profits were negative for some years. All areas of provincial life were affected. In response to perceived diminishing revenues from taxes and stumpage fees, the provincial government instituted a period of "restraint"

Table 2
Value ($million) of BC lumber and pulp exports, 1973–84

	1973	1974	1975	1976	1977	1978	1979	1980	1981	1982	1983	1984
Lumber	1,280	1,046	781	1,340	1,923	2,513	3,035	2,461	1,911	1,846	2,632	2,668
Pulp	564	937	870	1,192	1,022	1,093	1,498	1,898	1,709	1,520	1,427	1,762

Sources: British Columbia, *Financial and Economic Review* (various years).

severely cutting back welfare services, education, and medical services. (though Marchak 1988 argues that "restraint" was in effect an attempt to bully workers into accepting the changed labour practices of flexible specialization).

Since the recession of the 1980s, British Columbia has slowly regained its economic position. But it has done so only by becoming even more a staples economy than it was before. From a sample of 269 plants in the manufacturing sector, mainly engaged in processing resources, Hayter and Barnes (1990) found that the export/sales ratio had significantly increased, from 40.1 per cent in 1981 to 56.6 per cent in 1986. In addition, in 1986 over 90 per cent of total provincial merchandise exports of $12.7 billion were traditional staples; the U.S. market predominates (48 per cent), but Japan is also becoming very important (25.7 per cent).

The changed position of Japan (in 1950 it bought less than 5 per cent of BC exports) has led some commentators to argue that British Columbia is entering a new era, one in which it is part of the Empire of the Pacific (Resnick 1985). However, as Resnick (1985, 45) writes: "the rise of Japan ... helps lessen an overriding dependence on one power [the United States]. But the empire of the Pacific is an even frailer reed than the American empire ... on which to build an economic and political strategy. Moreover, integration into the Pacific empire on Japan's terms will simply perpetuate the old Canadian resource exports and manufacturing imports."

Specifically, the recent turn by BC resource firms, especially forest companies, to the "frailer reed" of Pacific markets has gone hand in hand with a move toward flexible specialization. The result is not only perpetuation of the old problems associated with the staples trap but also new ones bound up with the shift to post-Fordism. Both sets of difficulties come together in the single industry community.

RESOURCE TOWNS AND TRADE: CHEMAINUS AND PORT ALBERNI

The backbone of any staples production is the resource town; it is the fulcrum connecting shifting global economic forces with the process of resource extraction on the ground. After briefly discussing resource towns in general, I focus on two – Chemainus and Port Alberni – and try to make concrete the connection among the central themes of the chapter: staple production, trade, and the transition from Fordism to flexible specialization.

As a type, single-industry towns possess some unique features (Marchak 1983). Bradbury (1979, 163), for example, writes: "Canadian resource towns ... exhibit extreme dependence on one single enterprise, the operations of which are effectively exercised from company offices located in such distant metropolises as Toronto or New York. Each of these single industry settlements is highly dependent on the interrelations of an unstable world market, on fluctuating market prices, on foreign investment and control, and on large corporations whose multinational character means that each settlement is a small but dependent sector of an international corporate empire."

The narrowness of their economic base makes most resource towns acutely sensitive to outside change. The ghost towns that pepper the Canadian west are testimony of the susceptibility of single-enterprise communities to the forces of a changing international economy.

In a well-known typology, Lucas (1971) charts the development of single-industry towns through four phases – construction, recruitment, transition, and maturity. Such a scheme, however, is unremittingly progressive in conception; the reality is often one of failure. Bradbury and St Martin (1983) add two further stages to Lucas's model – winding down and abandonment – and apply them to Schefferville, Quebec. Theirs seems a useful modification to Lucas's original typology.

One might also analyse more closely the internal variations within each of the four stages. Below, I demonstrate such internal variability by focusing on the last stage, the mature one. Using Chemainus and Port Alberni as examples, I argue that this phase can include massive restructuring – in this case, prompted by the move from Fordism to flexible specialization in response to external change. The resource town remains intact after such restructuring (it is not abandoned, as in Bradbury and St Martin's scheme), but it is very different.

In my own two case studies we can also compare differences in places' responses to change. My argument is that while Chemainus and Port Alberni are bound together as staples communities and were subject to the general forces of change, their responses have differed because of internal, place-bound institutional factors, in this case revolving around labour. In many ways, this argument goes back to the earlier one about the two causes of spatial variation. While this chapter emphasizes the spatial division of labour, institutional differences, such as variations in labour practices, can also affect the outcome of a place in response to external change. Though such institutional differences can be couched at the regional level (for example, there is now a burgeoning body of literature on regional modes of labour regulation; see, Peck 1992 and Rutherford, this volume, chap. 21), they typically operate at much finer spatial scales. So while Chemainus and Port Alberni might share a common regional division of labour as forest product communities, they are also separated, as we see below, by their localized institutional structures – in this case those around labour – which makes a difference in their response to exogenous shifts.

From the beginning, Chemainus was at the hub of milling in the forest industry, recording its first sale of wood in 1862. The mill, owned by the locally controlled Victoria and Lumber Manufacturing Co., and the town experienced cycles of boom and bust throughout the first 50 years of this century as its lumber market switched from British Columbia to the prairies and then increasingly to the United States. In 1950 the mill was acquired by MacMillan Export Co., which one year later became Mac-Millan Bloedel Corp. From the 1950s on, the Chemainus mill employed typical Fordist techniques of production.

Its principal product was construction-grade timber destined for both the U.S. and Canadian housing markets. It was a standard product produced in large quantities using dedicated machinery. Workers were organized according to Taylorist work principles. The majority were engaged in relatively unskilled, repetitive tasks, such as working on the "green chain" (grading the raw logs), though there were some highly skilled workers, such as sawyers and planars.

In 1981 MacMillan Bloedel was taken over by Noranda. This event precipitated massive changes at the mill, and for Chemainus more generally. Citing declining profitability, MacMillan Bloedel closed down the mill in 1982, laying off 654 workers. In 1983, the firm announced that a new, state-of-the-art facility would be built. Flexible specialization, and its labour practices, had arrived at Chemainus.

When the mill reopened in 1985, working practices had completely changed. Under flexible manufacturing systems, many of the operations were now controlled by a computer. The "green line" had disappeared in favour of the automated "J" bar, planing was now subcontracted out to non-unionized local shops, and there was no need for sawyers "to read" the raw lumber because there were now "laser-sensing devises that ... profile logs [by] measuring their length, width and height, and assessing the best way to cut logs so as to maximize recovery and revenue" (Stanton 1989, 111). In addition, the mill also instituted work teams, and job tasks were recomposed (Stanton 1989). All workers now needed to learn "the manual"; it was not enough to specialize in one single task.

MacMillan Bloedel was attempting to reconfigure production toward new, and more profitable export markets. They wanted in particular to move away from the standard products of Fordism and the concomitant reliance on the U.S. and Canadian housing construction, toward more specialized and valuable timber products saleable in Japan and the EC. This goal was realized in Chemainus through a flexible manufacturing system, which increased the value of the lumber produced from $45 million ($25 million exported) in 1981 to $68.5 million ($65 million exported) in 1986.

The effect of these changes on Chemainus was severe. All workers were laid off for two years, and when labour was rehired only 125 people were offered jobs, which rose to 145 in 1991 (Table 3). Furthermore, because of the length of closure, a clause in the contract with the International Woodworkers of America (IWA) allowed MacMillan Bloedel *not* to rehire on the basis of seniority. As a consequence (Stanton 1989), many unemployed workers were forced to engage in informal work, and then income fell drastically. Furthermore, their spouses needed also to find employment, often as low-paid workers in the town's tourist industry, which centred on a series of giant outside murals celebrating the community's past. Following Harvey (1987), one can argue that

Table 3
Employment levels at MacMillan Bloedel's Port Alberni forest complex and Chermainus mill

| | Number of employees | | |
	1981	1986	1991
Chemainus	682	125	145
Port Alberni			
Woodlands	1,700*	1,090	1,060
Somass sawmill	1,064	588	509
Alpac sawmill	650	533	476
Plywood	450	377	0
Pulp and paper	1,522	1,316	1,340

Source: Fieldwork.
* Estimate.
Employment figures are year-end levels and do not include part-time workers.

in this way Chemainus mirrors yet another aspect of post-Fordism, with "employment growth ... based on the organization of spectacle ... and the recuperation of history and community" (277). But given the poor working conditions and pay, such post-Fordist employment was no match for the heyday of Fordism that it replaced.

Port Alberni, like Chemainus, has been a historical fixture on Vancouver Island. Intermittent sawmill operations began in the 1860s, and the combination of aggressive purchasing of timber land and investing in infrastructure by Bloedel, Stewart & Welch Co. made Port Alberni in 1935 the second-largest BC lumber town second only to Vancouver (Hay 1993, 57). Fordist production methods and labour practices came early and were evident in the Somass sawmill and the associated pulp mill, will that first began integrated operations in 1947. Four years later most of the activities at Port Alberni, which by then also included plywood, shingle, paperboard, and newsprint production, were consolidated under a single owner, MacMillan and Bloedel (Hay 1993, 58).

    For the next 30 years, the combined operations at Port Alberni provided the principal source of jobs within not only the town but also the region. In 1980, just before the downturn, MacMillan Bloedel employed almost 6,000 people and provided the main economic support for a community of almost 20,000 (Table 3).

    When the crash came, employment losses were proportionately smaller but absolutely greater than those at Chemainus. By 1986 more than 2,000 workers had lost jobs, and by 1991 600 more had been laid off (Table 3). In just a decade total employment at the mill had fallen by approximately 43 per cent, throwing the town into a profound crisis (Hay 1993).

    As in Chemainus, much of the subsequent permanent loss of employment resulted from changing staples markets and a shift to flexible specialization. In pulp and paper, there was a move toward specialized production, such as telephone directory paper, and away from the standardized products of newsprint, paperboard, and kraft pulp. Indeed, the (Fordist) paperboard mill was permanently closed in 1982, while the kraft pulp facility was reduced to only one line. The decade saw the shedding of around 200 workers in pulp and paper.

Sawmilling and plywood experienced much more employment loss (Table 4). The plywood operations served principally the Canadian construction industry, but with a declining supply of large "peeler" logs, growing competition from cheaper particle board, and the inability of MacMillan Bloedel to secure a $2-an-hour wage cut from the woodworkers union, the company sold the plant to its workers in 1990 for $1. Perhaps inevitably, by the end of 1991 plywood production there had ceased altogether. The Somass saw-mill became flexibly specialized in the mid-1980s, following a $37 million investment. Many of the techniques pioneered at Chemainus were introduced there. In addition, in accordance with "functional flexibility," the company required that new employees have grade 12 education and current employees who did not were to be retrained. The firm made less new investment at the Alpac mill ($10 million) and less attempt to implement new work practices.

Port Alberni is still reeling from the events of the 1980s. There have been efforts, often involving government funds, to initiate new businesses, but they have conspicuously failed to generate permanent, full-time jobs. Even the second best of tourist employment has not occurred. This situation clearly differs from Chemainus's. Whereas the latter had entrepreneurs who, in combination with the local state, provided the catalyst for alternative forms of economic activity, no such group emerged in Port Alberni. While Chemainus became known as "the little town that did" (Barnes and Hayter 1992), Port Alberni is becoming known as "the little town that died."

In summary, the staples communities of Chemainus and Port Alberni were transformed through the 1980s as they moved from a regime of Fordism to flexible specialization, largely because of changes in external markets. During this period, both towns remained "mature communities" in Lucas's terms, but the nature of that maturity changed spectacularly over a brief period because of broad changes in external trade that led to reconfiguring of production methods.

Despite similarities in the manner of change, there are also differences, a result partly of institutional variations around capital-labour relations. At Chemainus, because of the wholesale change and the length of closure, which altered seniority rights, a complete transformation to flexible specialization, including its labour practices, took place. In contrast, at Port Alberni the change was piecemeal, because of the diversity of operations, because the union local at Port Alberni was more aggressive (witness its refusal to agree to a $2 cut in wages to save the plywood mill), and because seniority rights in lay-offs and rehiring were never lost by the union, and so workers who remained tended to be older and more committed to traditional unionist and work practices.

We can explain the different fates of Chemainus and Port Alberni in terms of their institutional structures.

CONCLUSION

The task of this chapter was to indicate the vulnerability of the Canadian space economy to perturbations in the external trade of staple commodities. Because of the lack of symmetry across primary commodity prices and the complexities of spatial links, external change produced very different effects in different regions. While Alberta boomed in the mid-1970s Ontario suffered, but by the mid-1980s Ontario was surging and Alberta was the laggard. As Massey (1984) puts it: geography matters. Moreover, it seems that geography matters particularly in an economy such as Canada's, where regional specialization is so marked and susceptibility to outside change so acute. These two features have perhaps been the only constants in the whirligig of change that has beset Canadian economy and society over the last forty years.

The only way that Canada's vulnerability might be reduced is if policy makers attempt both to diversify the economy away from staples production and exert stricter controls on foreign investment. Given the nature of the FTA and now also NAFTA, neither appears likely, and so Canada can expect more of the same for the foreseeable future.

ACKNOWLEDGMENTS

I would like to acknowledge the very helpful comments of John Britton, Roger Hayter, and Dan Hiebert. Their suggestions greatly improved the chapter. Any errors remaining are my responsibility alone. In addition, some of the results reported here come from a joint research project with Roger Hayter that was funded by an SSHRC grant.

REFERENCES

Anderson, F.J. 1988. *Regional Economic Analysis: A Canadian Perspective*. Toronto: Harcourt Brace Jovanovich.

Atkinson, J. 1985. "The Changed Corporation." In D. Clutterbuck, ed., *New Patterns of Work*, 13–34. Aldershot, Hants.: Gower.

Baldwin, R.E. 1956. "Patterns of Development in Newly Settled Regions." *Manchester School of Economics and Social Studies*, 24, 161–79.

Barnes, T.J., and Hayter, R. 1992. "The Little Town that Did: Flexible Accumulation and Community Response in Chemainus, BC." *Regional Studies*, 26, 647–63.

Barnes, T.J., Hayter, R., and Grass, E. 1990. "MacMillan-Bloedel: Corporate Restructuring and Employment Change." In M. de Smidt and E. Wever, eds., *The Corporate Firm in a Changing World Economy*, 145–65. London: Croom Helm.

Blain, R., and Norcliffe, G. 1988. "Japanese investments in Canada and Canadian exports to Japan, 1965–84." *Canadian Geographer*, 32, 141–50.

Bradbury, J.H. 1979. "Towards an Alternative Theory of Resource-Based Town Development in Canada." *Economic Geography*, 55, 147–66.

- 1987. "British Columbia: Metropolis and Hinterland in Microcosm." In L.D. McCann, ed., *Heartland and Hinterland: A Geography of Canada*, 2nd edn., 401–40. Englewood Cliffs, NJ: Prentice Hall.

Bradbury, J.H., and St-Martin, I. 1983. "Winding Down in a Quebec Mining Town: A Case Study of Schefferville." *Canadian Geographer*, 27, 128–44.

Britton, J.N.H., and Gilmour, J.M. 1978. *The Weakest Link: A Technological Perspective on Canadian Industrial Underdevelopment*. Background Study No. 43. Ottawa: Science Council of Canada.

Brodie, J. 1990. *The Political Economy of Canadian Regionalism*. Toronto: Harcourt Brace Jovanovich.

Burrill, G., and McKay, I., eds. 1987. *People, Resources and Power: Critical Perspectives on Underdevelopment and Primary Industries in the Atlantic Region*. Fredericton, NB: Gorsebrook Research Institute.

Carmichael, E.A. 1986. *New Stresses on Confederation: Diverging Regional Economies*. Observation No. 28. Toronto: C.D. Howe Institute.

Clark-Jones, M. 1987. *A Staple State: Canadian Industrial Resources in Cold War*. Toronto: University of Toronto Press.

Clement, W., and Williams, G. 1989. *The New Canadian Political Economy*. Montreal and Kingston: McGill-Queen's University Press.

Davis, H.C., and Hutton, T.A. 1989. "The Two Economies of British Columbia." *BC Studies*, 82, 3–15.

Gunton, T., and Richards. J., eds. 1987. *Resource Rents and Public Policy in Western Canada*. Halifax, NS: Institute for Research on Public Policy.

Harvey, D. 1987. "Flexible Accumulation through urbanization: Reflections on 'Post-modernism' in the American City." *Antipode*, 19 no. 3, 261–86.

Hay, E. 1993. "Recession and Restructuring in Port Alberni: Corporate, Household and Community Coping Strategies." MA thesis, Department of Geography, Simon Fraser University.

Hayter, R. 1986. "Export Performance and Export Potentials: Western Canadian Exports of Manufactured Products." *Canadian Geographer*, 30, 26–39.

- 1988. *Technology and the Canadian Forest Product Industries: A Policy Perspective*. Science Council of Canada Background Study No. 54. Ottawa: Supply and Services.

Hayter, R., and Barnes, T.J. 1990. "Innis' Staple Theory, Exports and Recession: British Columbia 1981–86." *Economic Geography*, 66, 156–73.

Holmes, J. 1987. "Technical Change and the Restructuring of the North American Automobile Industry. In K. Chapman and G. Humphrys, eds., *Technical Change and Industrial Policy*, 121–56. Oxford: Basil Blackwell.

Innis, H.A. 1930. *The Fur Trade in Canada*. Toronto: University of Toronto Press.

- 1956. "The Teaching of Economic History in Canada." In M.Q. Innis, ed., *Essays in Canadian Economic History*, 3–16. Toronto: University of Toronto Press.

Jenson, J. 1989. " 'Different' but not 'Exceptional': Canada's Permeable Fordism." *Canadian Review of Sociology and Anthropology*, 26, 69–94.

Ley, D., and Hutton, T. 1987. "Vancouver's Corporate Complex and Producer Services Sector: Linkages and Divergence within a Provincial Staple Economy. *Regional Studies*, 21, 413–24.

Lucas, R.A. 1971. *Minetown, Milltown, Railtown: Life in Canadian Communities of Single Industry*. Toronto: Toronto University Press.

McLeod, A.D. 1988. "You Have to Be Tough to Farm in Saskatchewan." In G.S. Basran and D.A. Hay, eds., *The Political Economy of Agriculture in Western Canada*, 27–34. Toronto: Garamond Press.

Marchak, P. 1983. *Green Gold: The Forest Industry in British Columbia*. Vancouver: University of British Columbia Press.

– 1988. "Public Policy, Capital and Labour in the Forest Industry." In R. Warburton and D. Coburn, eds., *Workers, Capital and the State in British Columbia*. Vancouver: University of British Columbia Press.

Marchak, P., Guppy, N. and McMullan, J. 1987. *Uncommon Property: The Fishing and Fish-Processing Industries in British Columbia*. Toronto: Methuen.

Marchand, C. 1986. "The Transmission of Fluctuations in a Central Place System." *Canadian Geographer*, 30, 259–54.

Massey, D.B. 1984. *Spatial Divisions of Labour: Social Structures and the Geography of Production*. London: MacMillan.

Miller, F.C. 1987. "The Natural Rate of Unemployment: Regional Estimates and Policy Implications." *Canadian Journal of Regional Science*, 10, 63–76.

Norcliffe, G. 1987. "Regional Unemployment in Canada in the 1981–4 Recession." *Canadian Geographer*, 31, 150–9.

– 1988. "Industrial Structure and Labour Market Adjustments in Canada during the 1981–84 Recession." *Canadian Journal of Regional Science*, 11, 201–25.

OECD 1989. *Organization of Economic and Cooperative Development Report: Canada 1988/ 1989*. Paris: OECD.

Pearse, P.H. 1981. *Conflict and Opportunity: Towards a New Policy for Canada's Pacific Fisheries*. Vancouver: Commission on Pacific Fisheries Policy.

Peck, J.A. 1992. "Labour and Agglomeration: Control and Flexibility in Local Labour Markets." *Economic Geography*, 68, 325–47.

Pinkerton, E. 1987. "Competition among BC Fish Processing Firms." In P. Marchak, N. Guppy, and J. McMullan, eds., *Uncommon Property: The Fishing and Fish-Processing Industries in British Columbia*, 66–91. Toronto: Methuen.

Proverbs, T.B. 1978. *Foreign Investment in the British Columbia Fish Processing Industry*. Vancouver: Economics and Statistical Services, Fisheries Management Pacific Region.

Raynauld, J. 1987. "Canadian Regional Cycles and the Propagation of US Economic Conditions." *Canadian Journal of Regional Science*, 10, 77–89.

Resnick, P. 1985. "B.C. Capitalism and the Empire of the Pacific." *BC Studies*, 67, 29–46.

Science Council Industrial Policies Committee. 1981. *Hard Times, Hard Choices: Technology and the Balance of Payments*. Ottawa: Science Council of Canada.

Spry, I.M. 1981. "Overhead Costs, Rigidities of Productive Capacity and the Price System." In W.H. Melody, L. Salter and P. Heyer, eds., *Culture, Communication and Dependency: The Tradition of H.A. Innis*, 155–66. Norwood: Ablex Publishing Corp.

Standing Senate Committee on Fisheries. 1987. *The Marketing of Fish in Canada: An Interim Report on the West Coast fisheries*. Ottawa.

Stanton, M. 1989. "Social and Economic Restructuring in the Forest Products Sector: A Case Study of Chemainus, British Columbia." MA thesis, Department of Geography, University of British Columbia.

Swan, N.M. 1972. "Differences in the Responses of the Demand for Labour to Variations in Output among Canadian Regions." *Canadian Journal of Economics*, 5, 372–85.

Thirsk, W.R. 1973. *Regional Dimensions of Inflation and Unemployment*. Ottawa: Prices and Incomes Commission.

Tomczak, M. 1978. "The Impact of a Business Cycle on the Canadian Urban System: Changes in Manufacturing Employment 1972–76." Department of Geography Discussion Paper No. 23, University of Toronto.

Watkins, M.H. 1963. "A Staple Theory of Economic Growth." *Canadian Journal of Economics and Political Science*, 29, 141–48.

Webb, M.C., and Zacher, M.W. 1985. "Canadian Export Trade in a Changing International Environment." In D. Stairs and G.R. Winham, eds., *Canada and the International Political/ Economic Environment*, 85–150. Toronto: University of Toronto Press.

Williams, G. 1983. *Not for Export: Towards a Political Economy of Canada's Arrested Industrialization*. Toronto: McClelland and Stewart.

CHAPTER FOUR

Shifts in Canadian Direct Investment Abroad and Foreign Direct Investment in Canada

ALAN D. MACPHERSON

Over the last 15 years, outflows of Canadian foreign direct investment (FDI) have typically exceeded inflows, suggesting a growing role for Canada as an international investor. Canadian firms currently invest as much abroad as foreign companies invest in Canada, and few analysts expect this position to change appreciably over the long run (Investment Canada 1992; Rugman and Verbeke 1989). Despite a marked slowdown in Canadian outward FDI over the recent recession (late 1990–early 1992), major Canadian manufacturers such as Northern Telecom continue to employ more production workers in the United States than in Canada, domestically owned resource companies such as Alcan and Noranda have become familiar names in southeast Asia, and such retailers as Safeway and People's Jewelers now earn significantly more from their foreign operations than from indigenous sources.

Much of the impetus behind Canada's recently acquired investment strength can be traced to several decades of capital accumulation based on domestic resource exploitation. Faced with limited growth potential at home (and rising trade barriers abroad), many of Canada's major resource companies have come to regard FDI as the single most profitable route toward corporate expansion. Today, however, Canadian FDI is no longer restricted to resource-oriented ventures, nor is inward investment confined to the miniature replicas of an earlier era. The contemporary pattern is more complex, not least because of the growing internationalization of the world economy. In this regard, several authors have argued that the small scale of the Canadian home market has become a progressively less potent constraint on indigenous corporate growth (Lipsey 1990, 1991; Morici 1990; Porter 1991). In addition, Canada's increasing openness toward inward FDI portends a major restructuring of the domestic corporate environment, notably through international take-overs, mergers, and joint ventures (McNaughton 1992).

A further element in the recent growth of Canadian FDI stems from the global proliferation of non-tariff trade barriers, which virtually all the major industrial nations have put in place. For many Canadian firms, moreover, physical proximity to foreign customers has become a central requirement for continued sales growth (McConnell 1980; 1983; Porter 1991; Root 1990), especially for manufacturers (Menz and

Stevens 1991). New and expanding investment opportunities have emerged in the fast-growing economies of the Pacific Rim, including Taiwan, South Korea, Singapore, and Japan. At the same time, further integration of the European economies has stimulated increased Canadian participation inside the European Community (now European Union), notably since the mid-1980s. Closer to home, Canada's post-1975 FDI performance has been heavily influenced by market developments in the United States. Despite the existence since 1989 of the Canada–United States Free Trade Agreement (FTA), which provides easier export access to U.S. customers, the United States remains the top destination for Canadian capital. While Canada is still an important target for inward investment, especially from the United States, outflows of FDI are now an equally prominent element in Canada's external capital relations.

CANADIAN INWARD AND OUTWARD FDI: AN OVERVIEW

Canada's external balance in year-end flows of FDI moved from deficit to surplus in 1975, indicating a significant change in Canada's foreign investment position. Net flows have generally been outward ever since, averaging more than $3 billion per annum over 1975–91 (Investment Canada 1992). While part of this turnaround can be attributed to periodic repatriations of foreign capital by U.S. subsidiaries, notably since 1975, a more important explanation lies in the investment behaviour of Canadian firms (Figure 1). Since the early 1970s, Canadian direct investment abroad (CDIA) has grown faster than foreign direct investment in Canada (FDIC). While FDIC has followed a more or less positive tack since 1975, the superior record of CDIA deserves special mention. From 1987 to 1988, for example, gross outflows of CDIA grew from $9.3 billion to $12.9 billion (an increase of almost 40 per cent), while gross inflows fell from $10.2 billion to $8.1 billion (a decline of over 20 per cent). Rapid changes of this magnitude can be explained almost entirely by the ongoing international trend toward large-scale corporate mergers.

While Canadian acquisitions abroad have recently exceeded foreign acquisitions in Canada (often by a wide margin), the prevalence of large-scale acquisitions in both directions is worth noting. In 1988, for instance, at least half of Canada's gross outflows were generated by the much-publicized purchase of Federated Department Stores of the United States (valued at U.S. $6.6 billion in 1988) by Canada's Campeau Corp. Two years earlier, Toronto's People's Jewelers became the world's largest jeweller when Irving Gernstein acquired Zale Corp. of the United States in a multi-billion-dollar hostile take-over. On the inflow side, moreover, a good deal of the post-1985 FDIC can be traced to a relatively small number of major acquisitions, many involving well-known Canadian companies such as Mitel (telecommunications and electronics), Connaught Laboratories (pharmaceuticals), and de Havilland (aerospace).

In addition to industrial acquisitions, of course, inward and outward FDI has also involved finance, insurance, and real estate. The Royal Bank of Canada's encroachment on the commercial banking markets of the U.S. northeast is one example. Cana-

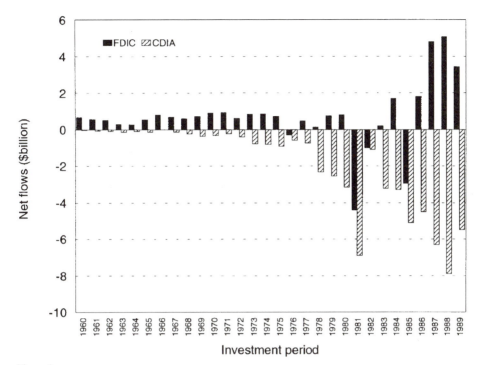

Figure 1
Net flows of inward and outward FDI ($billion), 1960–89. *Source*: Statistics Canada, *Quarterly Estimates of the Balance of Payments*, Cat. No. 67-001.

dians are aware, moreover, of the difficulties encountered by several Toronto real-estate companies that have been active in the world property market. Case histories of Campeau and of Olympia and York, for example, would doubtless dispel any doubts concerning the extent to which Canada has been involved in major real-estate transactions abroad.

FDI, however, continues to account for only a small proportion of Canada's overall balance of payments (Table 1). Though FDI has recently been growing faster than merchandise imports and exports, we should not overstate the extent to which it has become a substitute for external trade. In 1987, for example, gross outflows of CDIA amounted to only 9.6 per cent of Canada's merchandise exports, and gross inflows amounted to only 11.3 per cent of merchandise imports. While FDIC is frequently accompanied by expanded imports (external sourcing), CDIA may not play a commensurate role in domestic export growth. As an example, Honda Canada imports intermediate and finished products from Honda Japan and Honda USA, whereas Noranda Canada exports very little to Noranda USA or Noranda East Asia. Given the resource-based character of Canadian outward investment, the balance of trade creation that stems from FDIC/CDIA may well be import weighted.

Table 1

Canadian foreign investment and merchandise trade, gross outflows and gross inflows, 1980–90

	1980	1982	1984	1986	1988	1990
Exports ($billion)	67.7	71.2	90.2	89.7	116.1	129.0
CDIA ($billion)	4.9	4.4	5.3	8.6	11.4[†]	8.3[†]
Imports ($billion)	61.0	56.7	76.1	83.5	106.0	119.1
FDIC ($billion)	4.2	4.1	3.7	8.4	9.1[†]	8.9[†]
CDIA (as % of exports)	7.2	6.1	5.8	9.5	9.8	6.4
FDIC (as % of imports)	6.8	7.2	4.8	10.0	8.6	7.4
FDI (as % of trade)	7.0	6.6	5.4	9.8	8.2	5.6

Sources: *Investment Canada Newsletter* (various years).

[†] Estimates from the IMF (*International Financial Statistics*, July 1992).

Table 2

Inward and outward FDI ($million), including retained earnings, 1970–82

Year	Inward investment			Outward investment			
	Flow	Retained	Total	Flow	Retained	Total	Balance
1970	905	830	1,735	315	342	657	1,078
1971	925	1,335	2,260	230	333	563	1,697
1972	620	1,545	2,165	400	−65	335	1,830
1973	830	2,165	2,995	770	571	1,341	1,654
1974	845	2,730	3,575	810	493	1,303	2,272
1975	725	2,553	3,278	915	487	1,402	1,876
1976	−300	2,744	2,444	590	199	789	1,655
1977	475	2,971	3,446	740	995	1,735	1,711
1978	135	3,720	3,855	2,325	811	3,136	719
1979	750	4,783	5,533	2,550	1,570	4,120	1,413
1980	800	5,442	6,242	3,150	808	3,958	2,284
1981	−4,400	3,671	−729	6,900	−350	6,550	−7,279
1982	−1,000	1,247	247	1,075	−2,779	−1,704	1,951
Total	1,310	35,736	37,046	20,770	3,415	24,185	12,861
Proportion	4%	96%	100%	86%	14%	100%	–

Source: Rugman (1987).

Unfortunately, published data on annual flows of FDI exclude retained earnings (undistributed dividends) that are reinvested in new physical capital. Since retained earnings are not counted in the balance of payments, annual flow estimates for FDI potentially understate the real scale of foreign investment (in either direction) for any given year. Table 2 shows the effect of adding retained earnings to the annual flow data presented in Figure 1. While the revised data are for a shorter and much older time series (owing to compatability problems between different sets of data), two interesting patterns emerge. First, retained earnings constitute the main source of FDI in

Table 3
Stock value of CDIA by investment region, 1975–88

Region	1975		1980		1984		1988	
	$million	%	$million	%	$million	%	$million	%
North America	6,493	61.7	18,262	70.6	32,107	77.0	49,642	72.6
Europe	1,865	17.7	4,386	17.0	4,669	11.2	8,993	13.1
Asia	317	3.0	1,140	4.4	2,218	5.3	4,163	6.1
Latin America	1,190	11.4	1,037	4.0	1,390	3.3	3,200	4.6
Australia	485	4.6	751	2.9	1,039	2.5	2,061	3.1
Africa	167	1.6	277	1.1	302	0.7	369	0.5
Total	10,526	100.0	25,853	100.0	41,725	100.0	68,428	100.0

Source: Statistics Canada, *Canada's International Investment Position*, Cat. No. 67-202 (1989).

Canada, whereas direct outflows constitute the main source of Canadian investment abroad. Clearly, Figure 1 significantly understates the extent to which foreign corporations invest in Canada. More recent estimates for 1983–91 suggest that fully 57 per cent of all U.S. investment in Canada came from the reinvestment of profits earned by U.S. branch plants (Investment Canada 1992). Second, if we include retained earnings in the FDI account, Canada looks like a net capital importer rather than a net exporter. Here, of course, use of import/export terminology in the context of retained earnings does not accord with accepted balance of payments accounting. Data on balance of payments convey only a partial impression of the real scale of international investment activity (compare Figures 1 and 2).

With or without retained earnings, recent trends in CDIA stock values indicate a consistent role for the United States as a target for Canadian investment (Table 3). While several Pacific Rim countries have received major inflows from Canada since 1975, the accumulated dominance of the U.S. market is unequivocal. Though the United States ranks second to Asia in terms of recent CDIA growth rates (annual flows 1975–88), its absolute importance is increasing. Approximately 50 per cent of CDIA in the United States remains concentrated in manufacturing, reflecting a tradition that goes back to the early 1960s. In 1985 (the latest year for which disaggregated statistics are available), CDIA in U.S. manufacturing was dominated by only 4 sectors – non-ferrous metals (28.6 per cent), chemical and allied products (23.4 per cent), wood and paper products (20.5 per cent), and metal goods (7.3 per cent). Taken together, these four sectors account for over 90 per cent of CDIA in U.S. manufacturing, and a broadly similar pattern prevails for CDIA to the rest of the world.

The sectoral profile of Canada's investment position (Table 4) shows that the structures of CDIA and FDIC are remarkably similar, though accumulated FDIC stock values have usually exceeded CDIA stock values in all major sectors but utilities. While good time-series data on the composition and destination of FDIC and CDIA are not available, partial evidence from Investment Canada suggests that CDIA in manufacturing is dominated by industries that feed directly from the resource sector (notably,

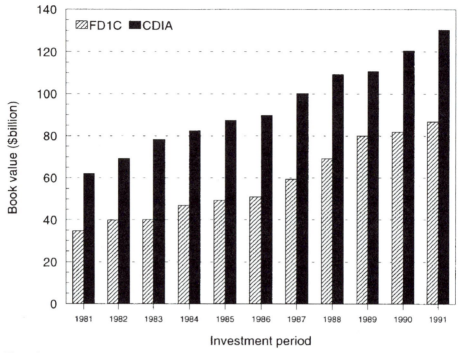

Figure 2
Book value of inward and outward FDI, 1981–91.

industrial chemicals, forest products, beverages, and non-ferrous metals). The data for FDIC suggest, in contrast, a more prominent role for high value-added manufacturing.

Beneath the aggregate flows, however, a series of micro-shifts suggests potentially new directions for Canadian FDI. To begin with, the fastest growth rates for CDIA are concentrated outside North America; per-capita income is growing fastest outside the United States. Second, several of Canada's largest and most innovative firms (Northern Telecom, Bombardier, Fleet Aerospace, CAE Industries) are currently looking at new regional targets in western Europe and the Pacific Rim. Third, many of Canada's small and medium-sized manufacturers are considering FDI as a possible avenue for escaping high export-marketing costs. Taken together, these three factors imply a potentially new complexion for future CDIA, especially in the sectoral structure of outward flows.

Significant changes have also been taking place on the inflow side. While the regional focus of inward investment continues to mirror the distribution of GNP by province (indicating a dominant role for Ontario and Quebec), recent capital inflows to all of Canada's regions have tilted toward technology-intensive activities (Investment Canada 1989; 1992). At the metropolitan scale, Toronto has been the preferred target for external investors of virtually all nationalities, suggesting a key position for southern Ontario and a relatively peripheral role for most other regions (McNaughton

Table 4
1985 stock values of canadian direct investment abroad (CDIA) and foreign direct investment in Canada
(FDIC)

Sector	CDIACDIA		FDIC		Net balance (CDIA–FDIC)
	$	%	$	%	
Manufacturing	24,823	49.8	34,684	41.5	−9,862
Merchandising	2,566	5.1	6,342	7.6	−3,776
Mining/smelting	2,850	5.7	5,228	6.3	−2,378
Petroleum/gas	7,469	14.9	18,577	22.3	−11,108
Utilities	1,731	3.5	659	0.8	1,072
Financial Services	8,392	16.8	14,558	17.4	−6,166
Other	2,077	4.2	3,426	4.1	−1,349
Total	49,909	100.0	83,474	100.0	−33,565

Source: Statistics Canada, *Canada's International Investment Position, 1985*, Cat. No. 67-202.
Note: "Stock" represents the year-end book value of long-term capital (long-term debt and equity, including retained earnings) owned by a foreign investor in a Canadian business (FDIC). The same definition applies to CDIA.

1992). While metropolitan areas such as Vancouver have recently received considerable inflows from East Asia (notably Hong Kong), the aggregate data continue to highlight southern Ontario in general, and metropolitan Toronto in particular. In short, inward investment may help to sustain and/or accentuate existing regional disparities within Canada, casting significant doubt on the extent to which peripheral locations, other than Vancouver, can attract new FDI on the basis of lower factor costs, public subsidies, or other price advantages.

NEW DIRECTIONS IN CANADIAN OUTWARD AND INWARD INVESTMENT

CDIA has long been dominated by large multinational firms, especially those with domestic experience in primary production. A glance at Canada's 20 largest industrial corporations reveals a distinct bias toward resource-based activity, and this bias shows up in the sectoral content of direct outflows. Companies such as Alcan and Noranda stand out as familiar examples, while such technology-driven firms as Northern Telecom represent uncommon exceptions. Though few of Canada's largest corporations spend more than 2 per cent of their earnings on industrial research and development (R&D), mature enterprises with relatively small research budgets account for the bulk of Canada's outward investment (Rugman 1987). In light of Canada's resource tradition, of course, this particular investment pattern comes as no big surprise. At the same time, however, the structure of CDIA raises strategic policy questions that parallel those in merchandise trade. Specifically, Canada's comparative advantage in FDI is concentrated in those sectors for which long-run income elasticities of demand are relatively weak. Inflows of foreign capital (and take-overs), in contrast, have recently been shifting toward technology-based ventures, notably in transportation equipment, biotechnology, and microelectronics (Investment Canada 1988; 1989).

The pattern of investment exchange described above would appear eminently appropriate for Canada. Faced with limited opportunities for new resource projects at home, domestic corporations with expertise in the primary sector are well advised to invest abroad, as are technology-intensive firms that face high export-marketing costs (Porter 1991). Moreover, inflows of foreign capital that translate into import substitution are always welcome, especially when technology transfer is part of the investment package. In practical terms, however, the FDI relationship is not quite as simple as that.

On the inflow side, FDIC has recently been characterized by mergers and acquisitions rather than by new plant construction (McNaughton 1992). Over the period 1983–90, for example, the share of acquisitions in the inward investment pie shifted from 8 per cent ($0.3 billion) to 36 per cent ($3.2 billion), peaking in 1988 at 48 per cent ($4.2 billion). Changes of this magnitude draw attention to the age-old policy question of whether an establishment's post–take-over performance matches, surpasses, or falls below that of its pre-FDI record. Entry by acquisition usually involves a less favourable benefit stream than entry by new plant construction. Investors from Japan and other parts of the Pacific Rim appear to prefer the "greenfield" route, whereas European and U.S. corporations are more prone to invest by acquisition. In empirical terms, however, the extent to which the two entry modes differ in their regional economic impact is unclear. To complicate matters, a good deal of Canada's recent inward investment has taken place as a result of the modernization and/or expansion of subsidiary plants already operating inside the country – with some, all, or none of the necessary finance capital coming from retained earnings. This situation poses a number of measurement problems in assessing FDI's effects at the regional scale. It is difficult, for example, to predict the long-run regional impact of Hyundai's 1989 decision to build a $320-million automobile plant at Bromont versus that of General Motor's $420-million upgrading (1987–89) of its existing Boisbriand plant.

While new plant construction is politically more attractive than most other investment paths, greenfield development can itself introduce significant problems. For example, a new factory can be tailored to absorb key imports from foreign sources, leaving domestic suppliers with a smaller market than might otherwise be the case. Truncated assembly plants with limited manufacturing capability have been identified as notable culprits in this regard (Britton and Gilmour 1978; see also Britton, this volume, chap. 14). Serious questions have also been raised concerning job quality, R&D spending, and labour relations (Drache and Gertler 1991; O'Grady 1990).

For Canadians, however, investment by acquisition has become a more serious concern. Technologically innovative companies – Connaught Laboratories, Leigh Instruments, Lumonics, and Mitel, among others – were snapped up at bargain prices, raising questions about the resilience of Canada's domestic corporate structure. As international capital markets become increasingly deregulated, moreover, Canada's weak corporate environment may leave a growing number of innovative (and possibly undervalued) firms on the auction floor. In itself, of course, foreign ownership of formerly indigenous companies is not a significant policy issue. A more urgent concern is that Canada may be transferring indigenous technology to foreign competitors that have no strategic interest in maintaining a significant Canadian presence. While there is no hard evidence to substantiate this proposition, the possibility of technological raiding ought not to be dismissed.

The question of FDI's impact ultimately boils down to the more fundamental question of whether unrestricted investment is good or bad for the Canadian recipient. In this regard, a popular argument against unrestricted FDIC is that foreign plants are typically among the first to disinvest and/or close during hard times, leaving affected communities with structural unemployment, reduced income levels, and a whole spectrum of social problems. While closure of foreign plants is an important economic and social issue, especially at the community level, Figure 1, above, shows that net foreign disinvestment in Canada has not occurred very often (see also Mac-Lachlan, this volume, chap. 11). Indeed, the data indicate that foreign firms (as a group) generally acquire more assets than they divest, leaving Canada with a net gain, which ultimately implies a surplus of jobs created over jobs lost. This interpretation is consonant with the Economic Council of Canada's (ECC) tracking studies of foreign firms (ECC 1988), suggesting a positive role for FDIC. Of course foreign firms periodically divest more assets than they acquire, and FDIC may in some instances replace skilled jobs with unskilled ones (Britton and Gilmour 1978). But asset acquisition typically outweighs divestment by a wide margin.

The picture rendered above implies that large companies are at the heart of the FDI relationship. While this certainly holds true for Canada (Rugman 1987), it would be wrong to assume that small and medium-sized enterprises (SMEs) play a negligible role. A continuing problem with data on CDIA and FDIC is that sectoral and regional investment distributions are typically recorded as year-end stock values or annual flows, but rarely both. The former data show historically accumulated book values; the latter, the absolute magnitude of investment over a specific period (usually a year). Unless both measures are available, it is difficult to pinpoint the real locus of investment growth in specific regions, industries, or sectors. Moreover, for specific size-classes of firms outside the corporate domain, it is difficult to say anything at all.

One consequence of this measurement problem is an average CDIA profile that emphasizes the role of large firms in relatively mature sectors, leaving SMEs in more advanced industries with a seemingly minor slice of the investment pie. Since the latter have yet to accumulate a critical mass of stock values that can be compared with Canada's traditional investment strength, SMEs are perhaps given less attention than they deserve, though they are beginning to be active in CDIA.

One way of illustrating this point is to consider the recent composition of CDIA for a specific set of regions. Ongoing tracking studies at the Canada–United States Trade Centre (CUSTAC) at the University at Buffalo suggest an increasingly diversified pattern of Canadian investment in the United States, especially since the early 1980s (MacPherson and McConnell 1992). In western New York (WNY), for example, recent inflows from Canada have been traced to a wide spectrum of small firms in secondary manufacturing, distribution, retailing, and business services. Since WNY is a geographically convenient entry point for Canadian investors, the fact that recent in-flows do not conform to the average CDIA profile is suggestive. Several new investors have been found in scientific instruments, metal fabricating, electrical products, engi-neering consulting, and real estate (McConnell and MacPherson 1990), suggesting a broader mix of CDIA than one might normally expect. Moreover, very few of WNY's recent Canadian investors are large multinational firms, and this implies a bigger role for the small business sector than many would have us believe.

What about the role of growth markets at the global scale? While the U.S. economy continues to act as a magnet for CDIA, many Canadian firms are turning toward west-ern Europe and the Pacific Rim for new ventures. These two regions are widely viewed as the principal target markets of the 1990s and beyond. Despite obvious and significant differences between the two regions, the scale of investment opportunity for new CDIA is enormous. The threat of a Fortress Europe in the 1990s, for example, has prompted increased Canadian participation inside the European Union (EU). Since this already-large market is expected to achieve above-average growth rates over the next decade (Root 1990), the potential opportunity costs of not investing in Europe are substantial.

Many of Canada's largest high-technology firms, including Northern Telecom, Bombardier, CAE Industries, and Fleet Aerospace, have established a direct presence in Europe since 1985, and a majority anticipate profitable expansion of their EU oper-ations over the long run (Investment Canada 1989). Tariff-dodging is only a fragment of the rationale here. The real problem for non-European firms is one of stable market entry under potentially rigid non-tariff obstacles such as EU technical standards, non-automatic quotas, variable import levies, and discriminatory procurement (Fieleke 1989). Clearly, Canadian companies that depend on exports run the risk of market ex-clusion in an international environment that is becoming riddled with non-tariff re-strictions.

Of course, regional market integration is not an exclusively European phenome-non. While the FDI implications behind the FTA are not yet fully apparent, the ac-cord eliminates at least part of the rationale behind continued operation of U.S. subsidiaries in Canada, and a similar scenario presumably applies to Canadian firms with U.S. branch plants. In essence, however, the robustness of this interpretation depends on the extent to which tariff schedules before the FTA shaped bilateral in-vestment decisions. As Lipsey (1990; 1991) and several others have noted, how-ever, the distribution of FDI by sector before the FTA does not correlate strongly with the historical structure of import duties on either side of the border. While In-vestment Canada and the U.S. Department of Commerce argue that bilateral invest-ment will expand in both directions as a result of the FTA, smoother cross-border relations will increase the attractiveness of the United States as a production base for Canadian firms. Preliminary evidence from McConnell, Chandra, and Steinitz (1989) suggests that this fear is indeed legitimate. Nevertheless, a potentially more significant question concerns the FTA's impact on FDI from outside North America. In particular, free trade may ultimately give some U.S. regions a major competitive edge in attracting investment from Europe, Japan, and the newly industrializing countries of Southeast Asia.

Canadian factor costs are not quite as attractive as those that exist in U.S. border regions (Porter 1991). In the U.S. south, moreover, cheap and pliable labour markets render a Canadian location potentially undesirable for large multinational investors. In addition, given NAFTA, the growing attractiveness of Mexico as an entry point to the North American market may eventually turn Canada into a residual location for international FDI. Significantly, Mexico's maquiladoras are not exclusively populated by U.S. and Canadian firms; European and Japanese companies have been moving there as well (Morici 1990).

While these uncertainties add up to an unpredictable future for FDIC, relative factor prices (as well as fluctuations in the Canadian–U.S. exchange rate) represent only one part of a potentially more complex investment relationship. Unfavourable U.S. Customs rulings regarding the North American content of Canadian automotive shipments to the United States inevitably suggest to Japanese and other East Asian investors that they seek non-Canadian sites for future production. That the U.S. Customs Service can retroactively apply import duties to Canadian shipments also affects the locational calculations of automotive investors when considering their North American locational strategy. Broadly similar concerns may also be pertinent to U.S. automotive producers with Canadian branch plants.

In this connection, plant-specific U.S. divestment in Canada over the 1990s will no doubt be seen by many analysts as evidence of the thrust toward industrial rationalization under freer trade, irrespective of the real mix of corporate motives behind plant-specific decisions (Lipsey 1991). Domestic firms that shift some or all of their production from Canada to the United States or Mexico will no doubt be seen as opportunistic exploiters of NAFTA's liberal investment provisions. For Canadians, of course, such scenarios cast the trade accord in a distinctly unpleasant light.

By and large, however, the geography of North American industrial production in the 1990s will surely reflect a set of basic locational trends that were well under way before negotiations began on the FTA or NAFTA. The subcontinental shift of manufacturing employment from the Rustbelt to the Sunbelt commenced as early as the 1950s (Bluestone and Harrison 1982), while smaller shifts from central Canada to the U.S. northeast have been documented since the early 1980s (McConnell 1980; 1983). Because a growing number of basic consumer and industrial products have become standardized in design, performance, and durability, locational shifts that deprive original host regions of manufacturing employment are not surprising. In this sense, then, it would be difficult to predict NAFTA's impact on Canadian investment in the late 1990s. Even so, the regulatory environment for FDI is likely to affect future investment patterns, and this environment has certainly been affected by NAFTA.

Canada's increasingly open position on foreign investment raises precisely the same set of policy questions that were posed in the years immediately preceding the advent in 1974 of the Foreign Investment Review Agency (FIRA). How are the costs and benefits of FDI distributed? How might Canadian economic and regional policy goals be meshed with current government enthusiasm for more liberal FDIC regulations? These questions hark back to some of the heated FDIC debates of the 1970s, few of which were ever fully resolved. For some academics, the phenomenon of U.S. miniature replicas springing up inside the protected Canadian market prompted legitimate fears concerning possible industrial truncation, technological dependence, loss of economic sovereignty, and a whole range of international demand leakages that ultimately implied a poor employment deal for Canada (Britton and Gilmour 1978). While such issues have not disappeared (see Britton, this volume, chap. 14), the contemporary FDI relationship has a somewhat new twist, in that recent investment activity has been driven by major acquisitions and mergers, many of them controversial from both a political and a business standpoint. For Canada's small and open economy, ongoing shifts in the structure of industrial ownership continue to raise questions concerning the nature and distribution of business control, employment stability, and decision-making power.

Here, of course, issues of economic sovereignty are not restricted to the industrial sector. Large-scale mergers have also taken place in retail, distributive, and financial services. Whether or not these mergers lead to an efficient internationalization of activity, in which all regions benefit, is a moot point. A more certain development from a regulatory standpoint is that U.S. investors can now acquire Canadian firms with greater ease, if only because the threshold for acquisition review by Investment Canada was raised from $5 million to $150 million in 1992. Like it or not, the domestic policy environment has evolved in a direction that renders foreign participation much easier than at any point in Canada's recent history.

Finally, firm-specific advantages are commonly held to be at the heart of FDI decision making (Dunning 1981; Porter 1991), and Rugman (1987) provides a good appraisal of large Canadian corporations from this perspective. Investment motives and market expectations are unlikely, however, to fall into neat behavioural categories across Canada's entire investor population. Several Toronto manufacturers have recently shifted portions of their production activity to western New York, yet many of these firms continue to serve the Canadian market first, and the U.S. market second (McConnell and MacPherson 1990). For some of these firms, the strategic consideration has been the high cost of manufacturing in southern Ontario. Others have sought secure, direct access to U.S. markets. By way of contrast, Northern Telecom aims to capture 10 per cent of the global telecommunications equipment market by the end of the 1990s, primarily through expansion-oriented FDI in the United States. For Northern Telecom, geographical variations in the cost of doing business are less important than appropriate labour supply (and technical inputs), convenient access to U.S. buyers, and direct access to customer goodwill in the large U.S. marketplace. Clearly, investors' motives are likely to vary significantly among sectors, industries, and size classes of firms, and presumably the same can be hypothesized for FDIC. All these international ventures, however, eventually leave a fingerprint on the sub-accounts of Canada's balance of payments, and spatially and sectorally ordered dimensions to international payments accounts have yet to be assigned.

CONCLUSIONS

At the outset, I observed that Canada has become a net exporter of FDI – as measured by the magnitude of annual flows – and that the United States continues to be central in both CDIA and FDIC. In real terms, however, foreign-owned companies invest more in Canada than Canadian companies invest abroad, and this has been the case for a long time. By including retained earnings in the FDI account, Canada becomes a net capital importer rather than a net exporter. By considering book values for FDI, moreover, we see that foreigners own more of Canada than Canada owns abroad. Significantly, almost all the recent growth of FDIC and CDIA can be traced to a merger thrust, based on major acquisitions. The popularity of investment by acquisition reflects an accelerating international trend toward industrial concentration, corporate integration, and oligopolistic competiton. This movement has also been accompanied by a progressive blurring of national political boundaries.

For Canadians, the costs and benefits of these trends are not immediately clear. While recent CDIA acquisitions have made several Canadian enterprises world-scale firms, investments in the other direction have consolidated the already high level of foreign ownership in Canada. At this stage, the impact of intensified international investment activity on the Canadian space economy is difficult to predict. All we can say with confidence is that central Canada will continue as the nucleus of investment exchange, modified only by the potential rise of British Columbia in light of that province's strategic accessibility to East Asia. Finally, the emergence of lucrative investment opportunities outside North America will no doubt hasten the pace of industrial concentration even further. As Europe moves toward greater market integration, and as the industrializing nations of the Pacific Rim expand their economies, innovative Canadian firms will surely become both hunters and prey in a more oligopolistic international environment.

REFERENCES

Bluestone, B., and Harrison, B. (1982). *The Deindustrialisation of America*. New York: Basic Books.

Britton, J.N.H., and Gilmour, J.M. 1978. *The Weakest Link: A Technological Perspective on Canadian Industrial Development*. Science Council of Canada, Background Study No. 43, Supply and Services Canada. Ottawa.

Drache, D. 1990. "The Future of Collective Bargaining in the Post–Free Trade Environment." Paper presented at the Annual Meeting, Association of American Geographers, Toronto, 19–22 April.

Drache, D., and Gertler, M.S., eds. 1991. *The New Era of Global Competition: State Policy and Market Power*. Montreal and Kingston: McGill-Queen's University Press.

Dunning, J. 1981. *International Production and the Multinational Enterprise*. London: George Allen Unwin.

Economic Council of Canada (ECC). 1988. *Managing Adjustment: Policies for Trade-Sensitive Industries*. Ottawa.

Fieleke, N.S. 1989. "Europe in 1992." *New England Economic Review* (May–June), 13–26.

Investment Canada. 1988. *Investing in Canada, Newsletter on Canada's Investment Climate*, 2 no. 2 (Fall 1988).

– 1989. *Investing in Canada, Newsletter on Canada's Investment Climate*, 3 no. 3 (Winter 1989).

– 1990. *Investing in Canada, Newsletter on Canada's Investment Climate*, 3 no. 4 (Spring 1990).

– 1992. *Investing in Canada, Newsletter on Canada's Investment Climate*, 6 no. 1 (Summer 1992).

Lipsey, R. 1990. "The Outlook for World Trade as Viewed from North America. "In *An Economic Profile of North America 1990*, Canada–United States Trade Centre, University at Buffalo, 11–19.

– 1991. "Thoughts on the Canada–US Free Trade Agreement." In F.C. Menz and S.A. Stevens, eds., *Economic Opportunities in Freer US Trade with Canada*, 1–12. Albany: State University of New York Press.

McConnell, J.E. 1980. "Foreign Direct Investment in the United States." *Annals of the Association of American Geographers*, 70, 259–70.

– 1983. "The International Location of Manufacturing Investments: Recent Behaviour of Foreign-Owned Corporations in the United States." In F.E.I. Hamilton and G.J.R. Linge, eds., *Spatial Analysis, Industry and the Industrial Environment*. Arichester: John Wiley and Sons Ltd.

McConnell, J.E., Chandra, B., and Steinitz, J.L. 1989. "Potential Impacts of the Canada–United States Free Trade Agreement upon the Economies of Erie and Niagara Counties." Occasional Paper No. 1, Canada–United States Trade Center, Department of Geography, University at Buffalo, Buffalo, NY, 14260.

McConnell, J.E., and MacPherson, A. 1990. *Canadian Direct Investment in Western New York: An Examination of Corporate Strategy, Business Performance, and Regional Economic Impact*. Canada–United States Trade Center, Department of Geography, University at Buffalo, Buffalo, NY, 14260.

McNaughton, R.B. 1992. "Patterns of Foreign Direct Investment in Canada 1985–1989: Canadian-Based Foreign-Controlled Firms versus U.S. and Overseas Investors." *Canadian Geographer*, 36, 50–6.

MacPherson, A., and McConnell, J.E. 1992. "Canadian Direct Investment in the United States: An Empirical Perspective from Western New York." *Environment and Planning (A)*, 24, 121–36.

Menz, F.C., and Stevens, S.A., eds. 1991. *Economic Opportunities in Freer US Trade with Canada*. Albany: State University of New York Press.

Morici, P. 1990. "Some Ruminations about the Global Environment and US-Canadian Commercial Relations in the 1990s." In *An Economic Profile of North America 1990*, 15–22. Buffalo: Canada–United States Trade Centre, University at Buffalo.

O'Grady, J. 1990. "Labour Market and Industrial Strategy after the Free Trade Agreement." Paper presented at the Annual Meeting of the Industrial Relations Research Association, Buffalo, 2–4 May.

Porter, M. 1991. *Canada at the Crossroads*. Ottawa: Business Council on National Issues.

Root, F.R. 1990. *International Trade and Investment*. 6th edn. Cincinnati, Ohio: South Western Publishing Co.

Rugman, A.M. 1987. *Outward Bound: Canadian Direct Investment in the United States*. Washington, DC, and Toronto: Canadian American Committee.

Rugman, A.M., and Verbeke, A. 1989. *The Impact of Free Trade on Small Business in Canada*. Working Paper No. 12, Ontario Centre for International Business, Faculty of Management, University of Toronto.

– 1990. "The Canada–US Free Trade Agreement and Its Impact on Canadian Business." *Economic Profile of North America 1990*, 23–31. Buffalo: Canada–United States Trade Centre, Department of Geography, University at Buffalo.

CHAPTER FIVE

The Canadian Financial System in International Perspective

GEOFFREY P. DOBILAS

In the last two decades we have seen a radical transformation in the activities and operation of the international financial system, characterized by a persisting and deepening international debt crisis, deregulation of domestic financial markets across the globe, development of more and variegated financial products, technological innovation, and the increased speed and complexity of capital circulation and integration of financial markets on a world scale. These changes in the international financial system have been necessitated by a set of broader economic and political events that began in the early 1970s. The dismantling (1971–73) of the Bretton Woods system of fixed exchange rates, followed by the oil crisis of 1973, shook the very foundations of the international financial order – international debt soared, and financial institutions everywhere scrambled to reduce their exposure to "bad debt" (Strange 1986). The resulting changes in the global financial system have produced a highly integrated system of international finance, with London, New York, and Tokyo emerging as the dominant centres.

The manner in which the London–New York–Tokyo "axis of control" functions shapes the activities of lower-order financial centres. What place does the Canadian financial system occupy in this international context, given that the traditional autonomy of national financial systems has been considerably eroded by deregulation of financial markets, technological innovation, and development of new financial products? Each of these three forces is operating at a global scale and has brought about rapid international integration of financial activity.

The Canadian financial system has not been immune to their effects. International tendencies explain a great deal about recent events that are shaping the Canadian financial environment. This chapter examines some of the recent changes in international finance and attempts to show how they have affected operation of the Canadian financial system.

RECENT CHANGES

Financial Market Integration

The global integration of financial markets began in the 1960s, when development of Euro-currency and Euro-bond markets brought about formation of international financial markets. However, during the 1960s various levels of domestic regulation and exchange controls left links between individual domestic markets quite loose. Global integration has come about only since the 1970s, and particularly over the last ten years, and is largely the result of deregulation of financial markets and of financial and technological innovation.

In banking, the increased integration of markets was spurred by the dismantling of the Bretton Woods system of exchange controls and the explosion of petrodollars following the 1973 oil-price increases. However, other deregulatory measures also hastened integration – for example, relaxation of the (U.S.) Interest Equalisation Act allowed for freer outflows of U.S. dollars (Martin 1994). Similar liberalization was carried out in the United Kingdom in the late 1970s and early 1980s, and several countries eased constraints on capital inflow. These changes facilitated greater movement of capital in the world economy. Canada, however did little to adapt to this increasingly international system.

The growth of foreign banks' operations in domestic markets has also contributed to the closer linking of national and international markets. As well, several national markets have provided a more liberal environment for foreign banks by removing the constraints on their operations in domestic markets, with the most notable example being the "Big Bang" in the United Kingdom (Galletly and Ritchie 1986). In these environments, foreign firms have spun off financial innovations to the host markets as with asset sales originally developed in the United States and now used elsewhere. Foreign banks have also brought domestic activity to the international marketplace – for example, by underwriting securities that their head offices are unable to carry out. This has in turn led to development of new financial instruments such as note issuance facilities (NIFs), which themselves reflect and reinforce integration of domestic and international markets. Few, if any, of these financial innovations originated in Canada, but adoption rates have been quite high here among financial-market participants.

In securities markets, integration has come about with easing of domestic regulations on market participation by non-residents. In Japan, for example, though foreign membership on the Tokyo Stock Exchange was restricted until mid-1988, the access by non-resident borrowers to the Japanese issue and Euro-yen markets were lifted several years earlier. In the Federal Republic of Germany, foreign-owned banking institutions have been allowed to lead-manage Deutschmark-denominated bond issues for nearly fifteen years.

In 1984, the U.S. removal of the withholding tax on interest payments to non-residents, followed by France, the United Kingdom, and West Germany, began the dismantling of the institutional barriers to integration of domestic and international markets. And finally, the reference rates in the floating rate notes (FRN) market expanded to include not only the London Interbank Offer Rate (LIBOR) but also, for example, the U.S. prime or Treasury Bill rates in the mid-1980s. These developments have led to convergence of domestic and Euro-rates, which have in turn contributed to convergence of issuing cost and relocalization of capital-raising operations from offshore markets to major financial centres, especially London, Tokyo, and New York.

Canada did not begin deregulating financial services to allow broader foreign participation by Canadian firms until the late 1980s. This lag may have hurt the international competitiveness of Canadian financial service firms and markets.

Financial Innovation

Creation of new financial instruments has also speeded integration of financial markets by opening new markets to borrowers and lenders, especially in what is called the "swap market." Swaps involve an agreement between two parties to exchange the payment or receipt of income streams. One form involves a straight currency swap, in which parties may differ in their perception of exchange-rate movements; for example, debt raised in U.S. dollars might be swapped to sterling, based on anticipation of a drop in the value of the dollar in relation to the pound. Another form is the interest-rate swap, in which fixed-rate debt is swapped for floating-rate debt (or vice versa), based on expectations about interest-rate movements. Generally, however, companies will raise debt in the market in which they have comparative advantage (where they are known and the risks of their offerings are understood) and then swap it to a currency of demand or to some rate of interest being offered in a different market.

In asset swaps, an income stream on a bond held by an investor is passed to another investor in return for a different form of payment – say, through repackaging of cheap, high-yield issues as assets with different payment structures. While these assets look like bonds, it is their repackaging that distinguishes them. Use of swaps, particularly currency and interest-rates swaps, encourages international market links by increasing the number of markets to which international borrowers have access. Such links were solidified in 1985, with formation of the International Swaps Dealers Association (ISDA), set up to create uniform international documentation for swaps deals.

Note-issuance facilities (NIFs), arranged as back-ups to U.S. commercial paper (promissony notes), have also facilitated global integration. Traditionally, NIFs were backed solely by U.S. banks. More recently, however, groups of international banks have underwritten them. Between 1983 and 1985, use of NIFs grew rapidly, attaining 14.1 per cent of capital-market borrowing among OECD countries. Since then, NIFs have given way to Euro-commercial paper programs, another form of integrative international financial instrument.

In addition, telecommunications and information technology have encouraged development of new forms of options and futures contracts that require twenty-four–hour-a-day hedging. In fact, physical links between the London Stock Exchange and the Philadelphia Stock Exchange and between the Chicago Mercantile Exchange and the Singapore International Money Exchange allowed users of options and futures markets on one exchange to offset their positions on another exchange (Bank for International Settlements 1986).

Technological Innovation

The ability to conduct business in financial markets internationally has been substantially enhanced in recent years by advances in information technology, especially computer network technology. In the financial services sector, the networks in use range from those facilitating interbank transactions (such as SWIFT, CHAPS, and CHIPS), through electronic funds-transfer systems, automatic-quotation systems for equities (for example, SEAQ at the London Stock Exchange and CATS at the Toronto Stock Exchange), and screen-based dealing systems (such as REUTERS), to fully electronic markets, such as the National Association of Securities Dealers Automated Quotations system (NASDAQ), and a wide variety of public and private financial market databases. As Hepworth (1988) has pointed out, these developments have accelerated worldwide integration of financial markets. In fact, this process led Estabrooks (1988) to posit an "integrated service financial supermarket," in which financial activity is managed on a truly global scale.

Unfortunately, the literature tends to focus less on how these technologies link markets than on the idiosyncracies of financial information and trading systems (Hamelink 1983; Veith 1981). However, several scholars have examined how technological solutions to the settlement problem associated with international equities dealing, direct market links, global currency trading, and program trading have led to market integration and simultaneous financial market behaviour, using the stock market crash of October 1987 as a case study (Dobilas 1988). The links that have been formed between stock markets and futures exchanges, for example, demonstrate how use of information technology encourages market integration. While links between futures and equities markets appear to enhance liquidity by increasing the fungibility of positions held (i.e., through hedging), they have also led, it is argued, to arbitrage and profit taking through program trading.

Other new systems allow for risk management twenty-four hours a day. BLEND and STREAM, developed by I.P. Sharp and now proprietory technologies of REUTERS, provide for worldwide currency trading and risk management. According to Maranoff, Tate, and Whitehorse (1987), it has been possible for some time to pass the financial portfolio or "book" between London, New York, and Tokyo. However, REUTERS's systems create a global book, giving currency traders of a particular firm simultaneous access, to a pool of funds residing "on the wires." In other words, the regional allocation of funds available for currency transactions within a given firm becomes insignificant as currency traders anywhere in the world have, at any given time, access to the firm's total stock of funds available for trading (Dobilas 1988). Information on currency movements is shared simultaneously, and traders everywhere have equal access to the global fund pool. Many such systems are in use today – evidence of the increasing intertwining of financial markets.

THE CANADIAN FINANCIAL SYSTEM

Against this backdrop, the Canadian financial system has been reorganized. Still dominated by Toronto, Montreal, and, to a lesser extent, Vancouver, Canada's financial markets are small by international standards; even Toronto is generally considered a "tier-two" market. Table 1 provides a summary of equity trading in the world's top 20 markets.

As a borrower, Canada has turned increasingly to the international marketplace over the last 25 years. The Economic Council of Canada (ECC) (1989) has pointed out that, given the size of their economy, Canadians use international financial markets more than any other industrial country. Nevertheless, the market for personal and small business loans has remained largely domestic. Any increase in borrowing abroad has resulted from debt raising by large firms, crown corporations, and governments, which are generally looking for large sums and at cheap rates. While Canadian borrowers are turning more to international markets, they are also getting non-Canadian institutions to manage their debt-raising activities. Between 1980 and 1988, for example, Canadian financial institutions managed only 32 per cent of all Canadian dollar–denominated Eurobond issues and only 44 per cent of those originating among Canadian borrowers (ECC 1989). Furthermore, their market share in managing most types of international issues appears to be declining.

Increased competition resulting from globalization generally has somewhat marginalized Canadian institutions, some of which have withdrawn from markets where profits were declining or losses were being experienced. While financial institutions in other countries also experienced a profit squeeze, their relatively larger size and capital bases made them better able to deal with increased competition.

The smaller scale of Canadian institutions is a result of government limits on their range of activities. Until recently, the system had operated under a regulatory system defined by the "four pillars." Chartered banks, trust companies, insurance companies, and the securities industry each operated under rules that defined their activities narrowly and allowed for no overlapping. Chartered banks, for example, could accept short-term deposits and provide business loans – and not do much else. The securities industry transacted purchases and sales of secondary equities and underwrote new stock issues. Trust companies managed estates and trust funds and had some ability to accept short-term deposits and to provide mortgage financing, and insurance companies sold insurance.

The regulatory framework, set up after the Depression to revive confidence in the Canadian financial system became problematic with widespread internationalization and integration of financial markets. As financial activity elsewhere was internationalizing and taking on a more "securitized" character (with greater use of bonds and other, newer financial instruments), Canadian chartered banks were excluded. Also,

Table 1
Equity trading ($U.S. million) in the world's top 20 markets, 1990–91

Market	1990	1991
Tokyo	1,403,887.0	2,338,196.1
New York	1,325,332.4	1,542,845.0
Taiwan	787,845.7	980,999.8
London	587,808.1	453,266.5
Germany	554,208.1	365,788.2
Zurich	400,253.2	413,779.5
Osaka	266,463.2	290,550.7
Paris	127,019.3	115,328.6
Korea	74,616.0	119,481.4
Midwest	71,304.4	99,288.2
Vienna	59,313.3	27,619.6
Toronto	55,179.8	72,258.9
Basel	55,005.8	64,810.9
Italy	44,859.5	42,033.3
Amsterdam	44,010.8	49,329.2
Pacific	41,418.4	48,240.9
Australia	39,765.8	44,895.8
Madrid	38,248.9	34,789.0
American	37,715.0	44,683.0
Hong Kong	34,683.7	34,604.0

Source: Official Trading Statistics, Toronto Stock Exchange, Publications and Information Services, 1991.

securities firms were unable to generate capital bases big enough to meet the challenges of international competition.

Some Canadian institutions attempted to circumvent these restrictions by setting up foreign affiliates to engage in activities not permitted of them in Canada. Orion Royal was set up in London as an affiliate of the Royal Bank of Canada to engage in Euromarket activity. Poor capitalization, heavy competition, and the significant drop in activity after the crash of 1987 led to its withdrawal in the late 1980s.

Nevertheless, the new financial instruments blurred the demarcation lines among the four pillars. It became much more difficult to decide within whose jurisdiction a particular activity fell. Each pillar began to engage in activities at home and abroad that might be construed as belonging to others. Eventually governments across Canada began deregulatory action.

In Canada, deregulation has led to restructuring. Since mid-1987, individual financial institutions have been allowed to provide the full range of financial services. It is now commonplace for banks to offer full-service merchant banking, in addition to their personal and commercial banking services. Similarly, investment houses can now establish personal demand-deposit accounts.

The manner of implementing change in Canada reflects the demands of increasing international competition and the problems associated with the previous regulatory

Table 2
Selected investment dealer partnerships in Canada, 1986–91

Original firm	New partner
Geoffrion Leclerc	Bank Indosuez
Housser and Co.	Geoffrion Leclerc
Burgess Graham	Geoffrion Leclerc
McLean McCarthy	Deustche Bank
Gordon Capital	CIBC
Midland Doherty	Gordon Capital
Walwyn Stodgell	Financial Trustco
Brown Baldwin	James Capell
Geoffrion Leclerc	Laurentian Group
Nesbitt Thompson	Bank of Montreal
Macleod Young Weir	Bank of Nova Scotia
Burns Fry	Security Pacific
Marathon Brown	Central Capital Corp.
Capital Group Securities	Bank Indosuez
Gardiner Group	Toronto Dominion Bank
Dominion Securities	Royal Bank of Canada
Bell Gouinlock	Pemberton Securities
Wood Gundy	CIBC
Burns Fry	Prudential Bache Securities Canada
Wood Gundy	Merrill Lynch Canada
Midland Walwyn	Dean Witter

Sources: *Globe and Mail* and *Toronto Star.*

environment. Successful dealing in the markets for the new international financial instruments requires capital bases that are beyond those Canadian merchant bankers had available to them. In an effort to cope, Canadian financial institutions have merged and made acquisitions among themselves, especially through merger of investment banking firms and chartered banks' acquisition of investment banking firms. Investment houses not interested in providing broader personal and commercial financial services have merged with other investment bankers in an effort to expand their capital bases and increase their ability to deal in the international marketplace. Often both parties have had some international experience, but a new, larger entity has a better chance of winning lead-management roles on specific issues. Chartered banks, which traditionally had an interest in investment banking but were limited by regulations and lack of the relevant in-house knowledge, have acquired existing investment banking firms.

In fact, by the end of 1988, all five major chartered banks in Canada had made significant acquisitions in the securities industry (Table 2). Most of this activity has been focused on firms headquartered in Toronto and Montreal. However, this pattern of events has wrought structural change in the Canadian financial services industry and organizational change within Canadian institutions – a particularly important area for further research.

Historically, Canada's financial markets have been open to foreign competition, but the 1967 revisions to the Bank Act restricted foreign banks in Canada to being foreign affiliates, with little scope for service provision. Then, in 1980, the Bank Act was revised to allow foreign banks to set up subsidiaries (Schedule B) in Canada. Though limits were retained on asset growth and on the right to establish retail networks, they could now lend to business. Increased competition led to growth among foreign banks in Canada, largely at the expense of domestic institutions. In 1988, Ontario allowed foreign companies to operate fully in the securities industry, permitting higher levels of foreign penetration. Finally, NAFTA has virtually eliminated all restrictions on the operation of U.S. financial institutions in Canada. These events have led to much more activity by foreign institutions in Canada, greater competition, and the possibility that Canadian institutions may lose their share of domestic markets. Indeed, acquisitions of Canadian investment houses by U.S. ones have been significant over the last three years, while some U.S. institutions – most notably, American Express – have established complete-service operations in Canada.

Finance and the Canadian Space Economy

In terms of the spatial organization of Canadian finance and banking, current developments abroad and Canada's reactions to them tend to reinforce earlier patterns. Comparative advantage, associated in domestic financial markets with timely and accurate information exchange, knowledge of and experience in dealing in new financial products, likelihood of direct contact with major market players, and access to technological innovations, usually favours the already-predominant centres of financial activity in Canada – namely, Montreal and Toronto. In addition, more market depth and broader connections to international markets also favour those cities, despite the federal government's designation of Vancouver, along with Montreal, as an international financial centre. Toronto's larger financial markets and more international orientation are likely to bring greater consolidation of financial activity there. Table 3 shows the value and volume of shares traded in the major Canadian equity markets between 1981 and 1991. By value of trades, Toronto and Montreal increased their share of equity trading in Canada from 86.9 per cent in 1981 to 95.5 per cent in 1991. By volume, the corresponding figures were 49.4 per cent in 1981 and 62.2 per cent in 1991. The high volumes given for Vancouver indicate its emphasis on the trading of highly speculative "penny stocks."

The recent reinforcement of the geography of finance in Canada is probably most evident in the evolution of Canada's futures and options markets. The market for financial futures began in Canada in 1980 with the initiation of two futures contracts (one in Canadian Treasury Bills (T-Bills) and one in Canada Bonds). These were traded primarily from Toronto and Montreal. While the Winnipeg Commodity Exchange began trading in its own T-Bill and Bond contracts in 1981, subsequent additions were focused on the Toronto Futures Exchange – for example, the TSE

Table 3
Equity trading in major Canadian markets (%), 1981–91

Year	Toronto		Montreal		Vancouver		Alberta		Winnipeg	
	Value	Volume	Value	Volume	Value	Volume	Value	Volume	Value	Volume
1981	76.7	43.5	10.2	5.9	11.8	45.4	1.3	5.2	n.a.	n.a.
1982	80.0	47.0	12.5	6.2	7.0	43.0	0.5	3.7	n.a.	0.1
1983	76.5	40.2	12.9	5.2	10.0	51.3	0.6	3.3	n.a.	n.a.
1984	73.8	42.6	19.4	8.0	6.2	45.3	0.6	4.1	n.a.	n.a.
1985	76.5	47.6	18.3	9.3	4.7	39.8	0.5	3.3	n.a.	n.a.
1986	75.2	49.7	18.9	11.2	5.3	35.4	0.6	3.7	n.a.	n.a.
1987	77.3	49.5	16.9	13.5	5.1	32.1	0.7	4.9	n.a.	n.a.
1988	78.4	50.1	17.3	11.7	3.8	31.8	0.5	6.4	n.a.	n.a.
1989	77.0	50.6	18.8	12.7	3.5	30.9	0.7	5.8	n.a.	n.a.
1990	76.1	47.9	18.3	11.6	4.9	34.9	0.7	5.6	n.a.	n.a.
1991	75.2	50.5	20.3	11.7	3.9	31.9	0.6	5.9	n.a.	n.a.

Sources: Official Trading Statistics, Toronto Stock Exchange, Publications and Information Services, 1991.

300 index futures and the TSE 35 index future. The latter has in fact replaced the former as one of the few futures contracts available in Canada. Again, activity in these contracts is concentrated in Toronto. Activity in the Canadian options markets – in particular, bonds and T-Bills – focuses on Montreal, though the TSE 35 index option has shown recent signs of possible success (Kalymon 1989). In addition, activity in the swaps markets is dominated by the larger actors, in Toronto and Montreal. While these innovative markets have evolved in those two centres, the Vancouver Stock Exchange has taken a somewhat different path. Despite its traditional emphasis on trading of resource stocks, it has more recently seen promotion of highly speculative stocks, including everything from cures for AIDS to diamond mines in northern Canada – a development on which provincial regulators have heaped much criticism (Lush 1993).

In a global context, Canada's financial system and financial centres may be come increasingly marginalized. The same comparative advantage operates at the international level as at the domestic level, and internationally it favours London, Tokyo, and New York. Greater market depth, larger players, a historically international orientation, and a propensity for innovation have meant growing concentration there. These factors, coupled with strategic locations vis-à-vis global time zones, have led Winder (1984) to conclude that with rapid internationalization of financial markets and increasing use of high-technology and financial-product innovations, "it is [markets] like Toronto and Johannesburg that will suffer."

With respect to regional development in Canada, factors altering the spatial organization of finance in Canada are a source of growing concern (ECC 1989). While people outside central Canada have historically decried the relatively poor availability of financial services, growing concentration in Central Canada is likely. Not only will Canadian firms be looking increasingly to international markets for investment opportunities, but so too will the new foreign entrants to the Canadian market, as they come to manage the trading of Canadian issues abroad. While consolidating financial market activity in London, New York, and Tokyo, this process would distance small investors and small business outside central Canada even further from the centres of decision making.

Consequences for the Canadian Financial System

The most obvious consequence for Canada of the internationalization of finance is a loss of domestic autonomy in choosing the direction for the finance and banking sector, as reflected in the form that deregulation has taken and in the adoption by Canadian institutions of technology and new financial instruments. The loss of autonomy is best reflected, however, in the stock market crash of October 1987. As Dobilas (1988) has suggested, not only did equities markets around the world, including the TSE, react simultaneously to the shocks emanating from New York on 19 October 1987, but the scope of defence available to individual markets was limited to wholesale suspension of trading. Nevertheless, even those markets that exercised that option experienced significant declines once their markets reopened. Hong Kong was particularly notable in this respect.

Another important consequence of internationalization is its effect on patterns of employment in financial services in Canada. In the evolving global system, the activity generated within any one market is going to mirror the level of activity within the system as a whole, taking the scale of market into consideration. For a second-tier market, activity is to a large extent predetermined by that in the first-tier markets, and patterns of employment will in turn reflect this reality. For example, if the volume of trades being conducted on the Dow Jones or London Stock Exchange experiences a secular decline, as it did after October 1987, then a similar decline might be expected in second-tier markets. Over the medium term such declines are likely to affect employment levels, which in Canada are being determined outside the domestic system. Additionally, because foreign institutions can participate fully in Canadian markets, the level of employment that they maintain in Canada will be shaped by their involvement in Canadian markets. Again, employment levels in Canada are being determined from outside. Not only will aggregate levels of employment be affected, but so will the types of jobs and hence local incomes. Indeed, if the branch-plant syndrome (see also in this volume, Britton, chap. 14; MacPherson, chap. 4; and Norcliffe, chap. 2) eventually comes to characterize financial services, will certain types of jobs – say, high-paying ones in research – become repatriated, as they have in other economic sectors? This is an important research question. Finally, to the extent that Canadian financial services companies involve themselves in foreign financial markets, given the internationalization of financial markets, employment within a Canadian financial institution may grow faster within a subsidiary operating abroad. In 1988, for example, Canadian financial institutions already employed 3,400 people in London (City of Toronto 1990).

CONCLUSIONS

The globalization and integration of financial markets brought about by deregulation of financial markets and financial product and technological innovation have transformed Canadian finance and banking. By participating in the evolving international financial marketplace, Canadian institutions have become increasingly exposed to competition. They have met this competition with varying levels of success. Nevertheless, the Canadian banking community realizes that competing aggressively in international markets is essential. The Canadian Imperial Bank of Commerce announced recently that it, through its affiliate CIBC Wood Gundy, intends to establish a major presence in international derivatives markets over the next two or three years by setting up offices in nine cities around the world, including New York, London, Tokyo, Hong Kong, and Singapore (Partridge 1994). Additionally, the Bank of Montreal has announced its intention to have half of its earnings generated in the United States by 2002 (Haliechuk 1994).

From a policy perspective, there are a number of considerations that arise as a result. First, it appears that the ability of Canadian institutions to capture an increasing share of international activity may be contingent on expansion of their capital bases. Already, this reality has resulted in significant consolidation of financial market players, highly concentrated in Montreal and Toronto (see also in this volume Semple, chap. 19; and Simmon, chap. 17). And it is these two centres that will transmit growth impluses through the urban system. While ongoing consolidation may allow Canadian institutions to compete more effectively abroad, it may reduce competition at home. Indeed, current thinking has it that the large Canadian banks, having already acquired brokerage houses, will soon turn toward the insurance industry. Alternatively, it may be possible to develop policy that encourages development of "niche players," which specialize in specific instruments or market segments. A number of such companies have emerged in the brokerage industry within the last two years as a result of job losses caused by consolidation among larger firms (Hemeon 1991).

Second, opening up of Canadian financial markets and apparent concentration of financial activity in central Canada may reduce availability of financial services in the rest of the country. While this problem has been chronic, it may be exacerbated by recent events. Though use of telecommunications and information technology offers possibilities for spatial substitution in provision of financial services, as seen in the proliferation of automatic teller machines (ATMS), so far there is little indication of such a trend in higher-level financial services.

REFERENCES

Bank for International Settlements. 1986. *International Banking and Financial Market Developments*. Basle: BIS.

City of Toronto. 1991. "Financial Services Discussion Paper." City Plan Background Paper No. 17, City of Toronto Planning and Development Department.

Dobilas, G. 1988. "Information Technology and Simultaneous Financial Markets: The Crash of October 1987." Geography Discussion Papers, New Series, No. 23, London School of Economics.

Economic Council of Canada (ECC) 1986. *Competition and Solvency: A Framework for Financial Regulation*. Ottawa: Ministry of Supply and Services.

– 1989. *A New Frontier: Globalisation and Canada's Financial Markets*. Ottawa: Ministry of Supply and Services.

Estabrooks, M. 1988. *Programmed Capitalism: A Computer Mediated Society*, Armouk, NY: M.E. Sharpe Inc.

Friedman, J., and Wolff, G. 1982. "World City Formation: An Agenda for Research and Action." *International Journal of Urban and Regional Research*, 6 no. 3, 319–44.

Galletly, G. and Ritchie, N. 1986. *The Big Bang*, London: Northcote House.

Haliechuk, R. 1994. "Bank of Montreal Embraces U.S.." *Toronto Star*, 28 Oct.

Hamelink, C. 1983. *Finance and Information*. Norwood, NJ: Ablex.

Hemeon, J. 1991. "Small Brokerage Firms Make Their Mark on Bay Street." *Toronto Star*, 21 July.

Hepworth, M. 1988. "Information Technology and the Global Restructuring of Capital Markets." In T. Leinback and S. Brun, eds., *Collapsing Space and Time: Geographical Perspectives on Communication and Information*, 132–48. New York: Allen and Unwin.

Kalymon, B. 1989. *Global Innovation: The Impact on Canada's Financial Markets*. Toronto: John Wiley and Sons.

Lush, P. 1993. "Regulator Criticizes VSE for Catering to Fads." *Globe and Mail*, 14 July.

Maranoff, J., Tate, P., and Whitehorse, B. 1987. "Around the World in 24 Hours." *Datamation*, 15 Jan., 1987, 75–7.

Martin, R. 1994. "Stateless Monies, Global Financial Integration and National Economic Autonomy." In S. Corbridge, R. Martin, and N. Thrift, eds., *Money Power and Space*, 253–78. Oxford: Blackwell.

Partridge, J. 1994. "CIBC Draws up Game Plan to Be Derivatives Player." *Globe and Mail*, 28 Oct.

Strange, S. 1986. *Casino Capitalism* London: Basil Blackwell.

Veith, R. 1981. *Multinational Computer Nets: The Case of International Banking*. Lexington, Mass.: D.C. Heath and Co.

Winder, R. 1984. "The Final Days of the Trading Floor." *Euromoney*, Oct., 80–91.

PART TWO

The New Staple Economy

Canada has benefited from a sequence of export commodities reflecting highly productive combinations of natural wealth and industrial utility. Thus the Canadian economy, though now one of great diversity, continues its long-term dependence on "staple" export commodities (Innis 1946; Watkins 1977); commodities whose supply or demand has waned have been replaced through exploitation of new resources. Even at late twentieth century, Canada's export trade relies especially on lumber, wood pulp and paper, crude petroleum, metal ores and concentrates, and wheat. Canada has developed industrial depth, acquired sophisticated human resources, and built a strongly metropolitan form of settlement, but it is still the leading exporter of mineral and forest products. Its place in world markets for these and other commodities, however, is threatened by resource exhaustion, technological lags, and new competition. In the past, Canada's international position has been exceptionally powerful in some commodity markets – nickel and uranium, for example – but now, across the resource industries, including fishing and forest products, agriculture, and minerals, production has proved vulnerable to shifts in the exchange rate because of the narrow margins on which Canadian firms operate as price takers in world markets.

The authors of this section suggest that Canada's resource industries are at a decision point. Unless they take active steps to confront a variety of stressful factors they will be much more limited sources of long-term prosperity. New suppliers such as Third World producers of copper (see Wallace, this volume, chap. 7) have heightened competition. In other cases, too, superior resources are the threat, whether through exploitation of new deposits, faster rates of resource renewal, or cheaper production. Shifts in demand result from these cost factors, environmental concerns are constraining opening of new sites of production (despite successful use of modern exploration technology), and new technology in consuming and processing industries is reducing demand.

While shifts by Canadian firms to higher-value products appear, an obvious response, Wallace notes that international tariffs have constrained this tendency in the minerals industry. Nevertheless, resource producers must increase the proportion of output sold that incorporates successful product research. To be competitive in export markets, Canadian resource producers must also cope with Canada's prosperous standard of living, especially its high wage rates and, from time to time, the higher than expected value of the Canadian dollar.

Some resource firms have successfully pursued new process developments in order to remain cost competitive. This trend is associated with reductions in employment, as discussed by Found (this volume, chap. 9) for agriculture, for the forest products industry by Hayter (this volume, chap. 6) and the minerals industries by Wallace. Nevertheless, some firms have sought the benefits that are possible from the faster adoption of technology, including the flexible production and marketing methods more commonly deployed in secondary manufacturing and the services. Movement to higher-value output is occurring in some parts of the forest products industry, but the chapters by Barnes and by Hayter provide clear evidence that the Canadian firms that lead in adoption of new technology are just not sufficiently representative of the majority of producers for us to hold a sanguine view of any increasing future competitiveness of the resource sector.

Rather than growing, Canadian research and development (R&D) have declined in the forest industries, and use of imported equipment has increased, according to Hayter; the consequence seems to be, general lack of success in direct stimulation of related secondary manufacturing – either upstream or downstream industries – by this staple. A similar pattern apparently prevails in minerals, though Wallace notes that the considerable variability in the response of mining firms to business pressures has meant that the more aggressive of them have invested in new technology, continued to maintain large R&D labs, and have acquired strategic alliances with foreign partners to assist in this process.

In years gone by, Canadian agriculture, especially wheat, stimulated domestic design and supply of production equipment. In recent years, however, Canadian manufacture and export of agricultural equipment have fallen: Massey-Ferguson, the industry's leader, has withdrawn from its historical connections, has drastically cut production, and has become the Varity Corp., headquartered in the United States. Furthermore, the state of the industry illustrates a more general Canadian pattern of underdeveloped domestic supply and export of resources-related equipment, paralleled by limited exports of high value–added products based on resources. This modest domestic development of secondary manufacturing directly related to exploitation of resources defines the "staples trap" as identified by Watkins and discussed further by Hayter: it is of fundamental significance for the structure of the Canadian economy. In all regions, too, the locational effects of resource exploitation are sharply defined by the limited lasting development that has occurred based on local resources such as coal in the Atlantic region and minerals in northern Ontario and Manitoba. Public assistance has brought little success, and the failed one-industry towns across the periphery of the ecumene and the developmental and social toll of boom-and-bust cycles of income and work encountered in virtually all resource regions are testimony to that (Bradbury and St-Martin 1983).

The small proportion of the Canadian population directly involved in agriculture (3.7 per cent) belies its economic importance in terms of exports, and the socio-political significance of this sector is considerable, especially because it is the organizational basis for settlement over vast distances in the west. But farming (as explained by Found, this volume, chap. 9) is environmentally precarious (see also Science Council of Canada 1991) and economically marginal. The large proportion of farmers who have become part-time workers in other occupations and the need for major government subsidies of farm receipts indicate the serious state reached by Canadian agriculture under an old trading order. Now, environmental priorities are entering international markets, and NAFTA and the Uruguay Round of GATT indicate the direction that new trading conditions will take. The subsidies to producers made by the public sector (*Economist*, 11 June 1994, 105) in Canada at 32 per cent of value of production (1993) are below the European Union average of 45 per cent but larger than the 23 per cent of the United States. They are considerably greater than those made to farmers in Australia (11 per cent) and New Zealand (4 per cent). Liberalized international trade will almost certainly move Canadian farming toward reduced production support, including existing supply-management systems, despite the resistance of farming interests. Geographic changes in Canadian production will probably follow, and Found indicates that some regions may be economically marginalized further.

Energy production is a complex part of the resource economy: unlike any other staple industry it is both an input to all other economic activities and an exporter. There is a poor regional fit between Canadian energy consumption and production, and an extensive oil and gas pipeline network now serves most of Canada and connects with U.S. markets. By contrast with significant cross-Canada flows of oil and gas, interprovincial sales of electricity are limited, though sales to the United States have long been sizeable. This trade has probably been made easier by removal of barriers with the United States, but, as Maclaren (this volume, chap. 8) explains, appreciation of Canadian benefits from NAFTA seems to vary by province, depending on whether it is a production-exporting location or not.

Canada has major energy resources, and consequently it is the most energy-intensive economy among the industrial group. While energy intensity has dropped since the first oil crisis, Maclaren finds scope for much more. The 30 per cent decline from the early 1970s to the late 1980s occurred mainly as a result of a structural shift in the economy in favour of services and improved efficiency in the transportation, commercial, and residential sectors. In manufacturing the energy-intensive industries have declined in output, while their adoption of new production technology has increased their energy efficiency. The drop in the price of oil by the mid-1980s reduced the domestic cost pressures for greater industrial energy efficiency, but now environmental issues are spurring changes in production methods and consumption. Environmental criteria are modifying acceptable energy investments and hastening pollution reduction by both power plants and other, direct or indirect users of fossil fuels.

The rise of environmental issues is international, and energy trade, particularly within North America, is influenced by the same sort of political-environmental issues that are affecting exports of paper and pulp and trade in agricultural products. These environmental concerns are becoming non-tariff barriers to trade in a variety of commodities and constrain both private and public investment. Environmental concerns therefore are having increasing influence on the technological choices of those enterprises seeking to be competitive in international markets. As forecast by Porter (1991) the process is one of innovation by capital-equipment suppliers, and fast adoption of clean technology is being induced by policies introduced by national, provincial, and local governments. In this connection Maclaren clearly points out the strong environmental position being taken by some Canadian cities.

REFERENCES

Bradbury, J.H., and St-Martin, I. 1983. "Winding down in a Quebec Mining Town: A Case Study of Schefferville." *Canadian Geographer*, 27, 128–44.

Innis, H.A. 1946. *Political Economy in the Modern State*. Toronto: Ryerson.

Porter, Michael. 1991. *Canada at the Crossroads*. Ottawa: Business Council on National Issues and Minister of Supply and Services.

Science Council of Canada. 1991. *Reaching for Tomorrow – Science and Technology Policy in Canada, 1991*. Ottawa: Minister of Supply and Services.

Watkins, W.H. 1977. "The Staple Theory Revisited." *Journal of Canadian Studies*, 12, 83–95.

CHAPTER SIX

Technological Imperatives in Resource Sectors: Forest Products

ROGER HAYTER

Over the past two decades, the Canadian forest product industries have experienced profound changes in competitive conditions. In terms of fibre supply, Canada's traditional advantage of access to plentiful high-quality, long-fibre, first-growth softwood has been reduced by its exploitation, by technologies that have economically extended use of hardwoods, by establishment of fast-growing plantations in other countries, and by increased self-sufficiency, especially for pulp and paper, in Canada's major markets. In addition, environmental interests, broadly representing various 'non-wood' benefits of the forest related to conservation, recreation, fishing, agriculture, and Aboriginal land title claims, have challenged Canadian industry's traditional dominance over forest-use decisions. In terms of demand, as forest-product markets have become more quality conscious and differentiated, market growth in bulk lumber, pulp, and newsprint – Canada's traditional specialities – has levelled off or even declined. This market weakness was exacerbated by recessions in the early 1980s and the early 1990s, which led to substantial declines in employment and profitability. The problems facing Canadian lumber and shingle and shake producers were made even more difficult by the protectionist measures taken by the u.s. government in the mid-1980s. The events of the past decade also reveal the vulnerability of low-value commodity exports to fluctuations in exchange rates.

These changes present a fundamental challenge to the Canadian forest industries: to change themselves from producers of a few bulk commodities whose profitability depends on cost minimization into diversified and flexible manufacturers that serve a wide range of geographical markets and whose profits stem from value maximization (Figure 1; Hayter 1988). The technological nature of this challenge involves recognition and implementation by forest product firms of technology – in the form of research, development, and innovation – as a competitive weapon that can reduce costs and create new market opportunities. This chapter assesses the implications of such a technological imperative for the evolving economic geography of the Canadian forest industries.

The discussion is informed by an Innisian approach, whose point of departure for understanding Canadian resource development begins with the observation that Can-

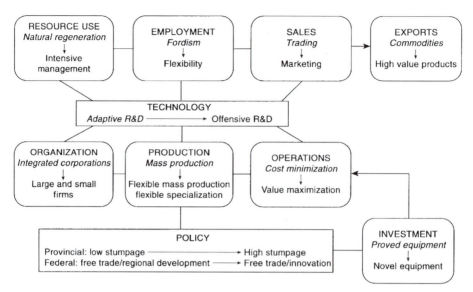

Figure 1
Canadian forest industries: production-system emphases and trends, 1950–90.

ada was developed as a source of resources and primary manufactures ("staple products") for metropolitan countries. Given the importance of staple exports in defining Canada's global role, Innis sought to interpret the Canadian economy's distinctive character in a framework that interwove the themes of geography, institutions, and technology (see Clement, 1989; Hayter and Barnes 1990; Melody, Salter, and Heyer 1981; Rotstein 1977; Watkins 1963). These themes provide the organizing framework for this chapter.

For Innis, geography meant, first, natural resources or more generally the physical geography of a region which provides "the grooves which determine the course of and to a large extent the character of economic life" (Innis 1946, 28). Second, geography meant spatial relationships, particularly with respect to processes creating discrepancies between cores (centres) and peripheries (margins) both between Canada and the outside world and within Canada (Innis 1946; 1967). As for institutional considerations, Innis wished to emphasize the asymmetrical nature of bargaining power among organizations, the role of the state both in providing infrastructure and in setting the conditions of exploitation, and the distinction between market (arm's-length) and non-market (administered) transactions, the latter including transactions within a corporation (Rotstein 1977). As Innis recognized, staple exploitation frequently involved large organizations, many of which were foreign controlled, and to a significant degree Canada's resource trade has been administered within corporations. Innis also explored the close relationships between the characteristics of staples and technological choice (Watkins 1963, 141). While at the local level Innis revealed how the inherent diversity of staple characteristics encouraged a range of technological adaptations, in macro terms he recognised that in many resource industries investments in infrastructure and process equipment were dominated by the approach of exporting large volumes of commodities at minimum cost.

Geography, institutions, and technology, however, are subject to change. Indeed, a central theme of this chapter is that the geographical, institutional, and technological conditions underlying the Canadian forest industries have experienced a profound transformation since the mid-1970s, and especially since the early 1980s. To understand this shift, we should look at the various dimensions of these industries, such as production, organization, forest use, marketing, and research and development (R&D), as an interrelated whole (Figure 1). Policy prescriptions need also to recognize these interconnections.

The rationale for a special consideration of the forest industries can be summarily stated. If employment has declined from 1979 levels (Jacques and Fraser 1989), in 1989 the forest industries (logging, wood processing, and paper and allied) still directly employed over 285,000 Canadians, and indirectly many more; they generate shipments whose value exceeds that of mining, fishing, and agriculture combined; lumber, newsprint, and pulp continue to be among Canada's top export earners; and Canada is the world's leading exporter of forest products. Regionally, they are the most significant industry in the Atlantic provinces, Quebec, and British Columbia, and they are important to the Ontario and prairie economies. Locally, they provide the principal economic base for about 300 small, specialized communities across Canada. Moreover, there are good reasons to believe that these should be an important sectoral priority in a technology policy for Canada (Hayter 1988).

THE GEOGRAPHY OF THE
CANADIAN FOREST INDUSTRIES

Canada's forest resource is vast. Approximately 27 per cent of Canada's total land area of 9.9 sq km, or about 453 million ha, is classified as forest land, which represents almost 12 per cent of the world's productive forest land area. On a world scale, Canada lies within the relatively homogeneous boreal coniferous forest zone, but with notable variations in the species mix of Canada's forest formations. Thus British Columbia's coastal forests are dominated by Douglas fir, hemlock, and western red cedar, and these species are gradually replaced in the interior by lodgepole pine, alpine fir, and other species. Further eastward, across the northern prairies and to the southern Laurentian Shield of Ontario and Quebec, pines gradually give way to white spruce and balsam fir on the better-drained soils and black spruce, tamarack, and jack pine on the less-well-drained soils. In Atlantic Canada spruce, firs, and pines continue to dominate the coniferous forest. Deciduous hardwoods comprise about 24 per cent of Canada's forest volumes. The most important are aspen and poplar, which are particularly extensive in the prairies, and birch and maple, found mainly in central and

Table 1
Log production (000 cu m) for industrial use in Canada, by region, 1974, 1989, and 1992

	1974			*1989*			*1992*
	Softwood	*Hardwood*	*Total*	*Softwood*	*Hardwood*	*Total*	*Total**
Canada	123,488	9,841	133,329	174,640	16,804	191,444	170,188
Atlantic Canada	13,496	2,109	15,605	13,793	2,715	16,508	16,782
Quebec	27,184	1,968	29,152	34,884	4,994	39,878	31,059
Ontario	14,753	3,157	17,910	22,864	6,778	29,642	24,208
Prairies	2,717	411	3,128	4,868	675	5,543	4,679
Alberta	4,718	71	4,789	10,641	1,652	12,293	14,592
BC coast	27,605	250	27,855	29,940[†]		29,940	24,874
BC interior	31,809	311	32,120	57,474[†]		57,474	53,705

Source: Statistics Canada, *Logging Industry*, Cat. No. 25-201.
* Totals for 1992 combine softwoods and hardwoods.
[†] Softwood and hardwood combined.

eastern Canada, and there are numerous other species that are noteworthy locally, including cottonwood in the lower Fraser Valley of British Columbia.

The Forest Harvest

Canada's inventoried amount of merchantable timber in 1986 was in excess of 23 billion cu m, though not all of this resource is economically accessible and one-third is considered immature. With respect to industrial log harvesting, notwithstanding occasional downturns, until 1989 there was a clear trend of increasing softwood and hardwood log production (Table 1). Thus, softwood production increased from 123 million cu m in 1974 to 174,000 million cu m in 1989 – an all-time record. Indeed, after 1989, at least until 1992, harvests declined noticeably with the onset of recession. Except for Atlantic Canada, harvests expanded throughout Canada between 1974 and 1989, and in both years British Columbia accounted for about half the softwood harvest, with Quebec and Ontario providing much of the remainder. Hardwood production is largely concentrated in Ontario, Quebec, and New Brunswick, though use of aspen species has increased notably in Alberta (and northeastern British Columbia), where several pulp mills have been built in recent years.

Since the late 1970s, assessments of Canada's forest resource have raised serious questions about the sustainability of the timber harvest (F.L.C. Reed and Associates 1978), and, outside Alberta, the area of timber harvest (see Table 2) and harvested volumes are unlikely to expand further under prevailing or foreseeable conditions. In most parts of Canada actual harvest levels for softwoods have reached the allowable annual cut (AAC) set by provincial forest ministries (CFIC 1986). Indeed, in British Columbia, especially in the coastal region, it is recognized that soon there will be a "fall-down" effect – the decline in harvest levels that occurs as the large volumes of old-growth, mature and over-mature forests are (gradually) replaced by the smaller

Table 2
Inventoried productive forest land, 1986, and estimates of area harvested and planted, 1975, 1980, 1989, and 1991, by province

	Productive forest area (million ha)	Area harvested (1,000 ha)				Area planted (1,000 ha)			
	1986	1975	1980	1989	1991	1975	1980	1989	1991
Newfoundland	11	16	5	19	21	0	.4	3	3
Prince Edward Island	3	2	3	–	2	.1	.4	1	1
Nova Scotia	4	27	36	36	38	1	4	10	8
New Brunswick	6	94	86	90	92	7	22	20	19
Quebec	55	135	245	252	236	16	14	103	99
Ontario	38	197	243	200	200	30	32	82	83
Manitoba	15	16	25	12	9	1	1	7	8
Saskatchewan	16	18	17	22	17	4	6	6	6
Alberta	25	20	24	47	48	6	9	24	33
BritishColumbia	51	157	188	199	193	63	64	172	199
Other	22	–	–	–		–	–	–	
Canada	244	683	881	875	854	128	152	428	461

Sources: Forestry Canada, *Selected Forestry Statistics, 1990*; *Compendium of Canadian Forestry Statistics 1993*; National Forestry Data Base.

volumes of second-growth forests – even if its precise extent remains unclear (Peel Commission 1991).

Admittedly, the planted forest area across Canada has increased rapidly of late – from 152,000 to 428,000 hectares between 1980 and 1989, the lion's share of it in British Columbia, Quebec, and Ontario (Table 2). Indeed, the area of tree cover regenerated by planting alone increased in British Columbia from 34 per cent to 92 per cent, and in Quebec from just 6 per cent to 41 per cent. British Columbia also continued to expand its planted area in the recession years of the early 1990s. In support of this planting effort, there have emerged a growing silvicultural industry – about 800 million seedlings were raised in Canadian nurseries, with a value of $260 million, in 1989 (Hall, Carlson, and Dube 1990, 141) – and a greater commitment to forest management research. There is, however, a considerable area of degraded forest across Canada, so that recent planting is replacing forests from previous as well as current harvests. Moreover, the Canadian forest industry is facing increasing competition from non-wood values for use of the forest, new and old, so that in many parts of the country wood supply will remain problematic.

The Forest Industries

During the 1950s and 1960s, the principal commodities making up the Canadian forest industries – newsprint, pulp, lumber, and plywood – experienced noteworthy and steady rates of growth and a significant shift westward, especially to British Colum-

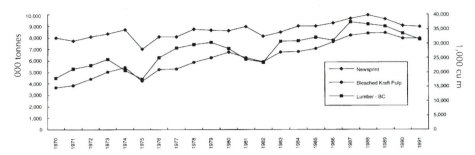

Figure 2
Production trends in selected Canadian forest industries, 1970–91.

bia (Hayter 1976; 1978; 1979). Since the mid-1970s, as wood supply conditions have become more restrictive in Canada and as fast-growing plantations in other countries have increased their share of global industrial wood supply, the rates of growth of Canada's forest industries have declined and fluctuations have been more severe (Figure 2). Thus, following output reductions during the recession of 1982 and 1983, newsprint, pulp, and lumber reached all-time record levels of production during the late 1980s, to be followed again by another deep recession in 1990 and 1991, the results of which are still evolving. This "trend" is particularly well revealed by the BC softwood lumber industry: production was 23 per cent lower in 1982 than in 1979; in 1987 it had recovered to exceed the 1979 and 1982 levels by 24 and 60 per cent respectively; and by 1991, though still higher than 1979, production declined to less than 84 per cent of the 1987 level (Figure 2).

The volatility of the 1980s has necessarily affected employment and corporate profitability. Throughout the Canadian forest industries, the substantial job losses that occurred in the early 1980s were partially retrieved by 1989 (Table 3). Though aggregate figures are not available, in sawmilling and pulp and paper during 1990 and 1991 job loss was again substantial, and major corporations, just as they did in the early 1980s, have faced difficulty achieving profitability. The effects of the 1990s recession are also regionally more widespread than was the case in the early 1980s. In the latter case, recessionary effects were worst in British Columbia, especially in the coastal region (Barnes, Hayter, and Grass 1990; Grass and Hayter 1989; Hayter 1987). During the early 1990s, however, significant profit and job losses in British Columbia have parallels in central and eastern Canada, especially in the newsprint industry.

The relative volatility of the BC forest industries over the past decade relates to their reliance on a narrow range of commodities for export, especially softwood lumber, whereas in Ontario and Quebec is industries have traditionally added more value to wood harvested and have relied more than British Columbia on the paper and allied industries and the domestic market. Indeed, lumber and pulp have significantly greater export-sales ratios in British Columbia than elsewhere in Canada (Table 4). While BC lumber and pulp exports are less reliant on the United States than those elsewhere in Canada, in Ontario and Quebec access to a larger domestic market has

Table 3
Employment in the Canadian forest industries, 1979–92

	1979	1982	1983	1989	1992
Logging	56,614	40,156	45,943	49,185	41,156
Wood industries					
Sawmills/planing mills	68,328	53,698	58,135	58,884	52,031
Shingle/shake	1,986	1,421	1,539	1,801	1,555
Veneer/plywood	13,618	10,344	9,724	9,236	7,728
Sash, door, millwork	25,972	21,294	22,925	40,076	n.a.
Other wood*	12,144	10,293	9,642	13,612	n.a.
Particle and waferboard	2,407	2,280	2,422	3,916	n.a.
Paper and allied					
Pulp/paper mills	87,055	83,104	79,164	80,595	69,689
Other paper	41,863	34,565	35,144	39,511	35,319
Total	307,580	254,875	262,216	292,900	249,750

Source: Statistics Canada, *Selected Forestry Statistics 1990 and 1992*, Cat. No. 25-202.

* The wooden box and pallet industry, coffin and casket industry, and other wood industries. In 1992 total employment amounted to 103,589.

encouraged concentration of more stable value-added activities such as remanufactured wood products and converted paper products. Even so, by 1990, many pulp and paper mills, especially export-oriented newsprint mills, in central Canada had old machines that needed modernization and adaptation to a more favourable product mix.

Forest Communities in Crisis

Adjustment to changing global realities has had particularly profound effects on those many specialized and often remote communities where most of Canada's principal forest product exports are produced. The number and distribution of these communities are impressive. Thus Forestry Canada (1990) estimated that in 1981 there were 348 "forest industry" communities scattered throughout Canada, including 103 in British Columbia. In addition, there are specialized communities, such as Kitimat, where the forest industry provides a major share of the economic base.

The effects of the recessionary crises of the early 1980s and 1990s, and the associated restructuring of production, have varied considerably among communities. In general, the most serious problems of adjustment have occurred in places such as Chemainus and Port Alberni, long dominated by mills with old technology and exporting softwood lumber and newsprint (Barnes, this volume, chap. 3; Barnes, Hayter, and Grass 1990; Barnes and Hayter 1992). In Port Alberni, for example, during the 1960s and 1970s, its ranking among Canadian cities and towns in terms of per-capita income was consistently within the top ten, but by 1987 it had dropped to 89th (Hay 1993). In addition, during the early 1990s, its reliance on the forest industry has

Table 4
Distribution of shipments of newsprint, wood pulp, and lumber, by region, 1990

	Quebec	Ontario	British Columbia	Other Provinces	Canada
Newsprint					
Shipment (000 tonnes)	4,105	1,628	1,754	1,588	9,075
% exported	86.0	89.3	88.0	92.8	88.1
% to United States	75.0	87.9	64.3	49.4	70.8
Wood pulp					
Shipment (000 tonnes)	1,419	1,424	3,942	2,520	9,305
% exported	57.1	74.1	93.7	84.6	82.6
% to United States	26.8	69.8	22.8	53.4	38.9
*Lumber**					
Shipment (000 cu m)	9,378	5,357	32,959	6,664	54,357
% exported	46.9	39.4	72.0	31.7	59.3
% to United States	34.0	n/a	52.5	n/a	43.3

Source: Forestry Canada, *Selected Forestry Statistics Canada 1991* (Ottawa, 1992).

* The small share of shipments this source failed to classify as either for export or domestic markets was divided equally between the two.

been eroded by further rationalization of its pulp and paper activities. Indeed, as with other forest communities experiencing similar problems, Port Alberni is looking to diversify through tourism, education, and health-related activities, secondary manufacturing, and the attraction of retired people. While the success of such activities varies considerably, forest communities across Canada are in a state of transition (Barnes and Hayter 1994).

INSTITUTIONAL STRUCTURE

Traditionally, decisions about use of Canadian forests, and the viability or otherwise of forest communities, have been primarily the concern of what Wilson, in a BC context, labels the "wood exploitation axis" of government, business, and labour (Wilson 1987/88, 7). In practice, government and business have exercised the most influence over the timing, location, and nature of use. While forest resources remain largely under provincial ownership the private sector controls industrial investment and production. In essence, the historic "deal" meant the leasing of large tracts of timber for long periods by provincial governments to corporations in return for relatively small royalties and/or stumpage payments but large-scale investment in industrial facilities. Typically, long-term timber leases have implied "in perpetuity," and only rarely have provincial governments sought to influence the sale (or acquisition) of timber rights among private-sector interests. Nevertheless, in recent years the traditional relationships between government and business (and labour) have been brought into question by a rethinking of forest policy, corporate restructuring, and changing labour relations.

Rethinking Forest Policy

Wilson's observation, of the late 1980s, that BC forest administration has been compliant to the needs of industry and conducted in a rather secretive way, probably applies throughout Canada (Swift 1983). The BC forestry ministry's unannounced introduction of "sympathetic administration" during the early 1980s, allowing companies to ignore regulations to reduce costs or increase revenues, provides a recent illustration of this argument. Such sympathy, once discovered, however, proved controversial and contributed to growing public scepticism about existing forest policy.

In fact, much encouraged by public concern, rethinking of forest policy, led by initiatives in British Columbia and Quebec, is evident. This rethinking is related to the acceptance by industry and government – the evidence has been available for some time (Smith and Lessard 1972) – that forests in Canada cannot be adequately renewed by sole reliance on natural regeneration. Indeed, this awakening led most provinces to identify forestry as a development priority, and subsequent joint agreements with the federal government underlay much of the recent increase in tree planting (Table 2, above). More fundamentally, society now demands that use of forests reflect a much broader range of public values than before. In British Columbia especially, conflicts between environmental interests and the forest industry, in and out of court rooms, have provided the background for two commissions of inquiry in 15 years. Both recommended significant change. Thus the Pearse Commission (1976) led to shifts in forest tenure policy and management responsibilities and to policies that gave small firms greater access to the forest resource. The more recent Peel Commission (1991) formally recommended "enhanced stewardship," designed to secure environmental as well as economic values from provincial forests.

As Baskerville (1990) observes, "sustainability" is the predominant issue in forest management throughout Canada and is no longer exclusively, or mainly, about maintaining timber volumes for industry. Whether or not public expectations are in accord with reality, sustainability is a multidimensional concept that will be difficult to implement. In the meantime, wood costs for industry are likely to increase as logging continues to shift to more remote areas; planting costs are incorporated by individual firms as part of lease arrangements; stumpage is raised – a trend reinforced by the arguments raised in the American countervail actions taken against the softwood lumber industry (Hayter 1992); more stringent harvesting regulations are introduced and enforced; parts of timber leases are removed for conservation purposes; and governments introduce competition for wood or reallocate wood to new users.

There are of course reasons to believe that customary forest policies will prevail. Witness Alberta's allocation of extensive and exclusive areas of timber to firms planning large-scale industrial development.

Corporate Restructuring

In a manner reflecting the thrust of provincial forest policies, the industrial structure of the Canadian forest industries has been dominated by large-scale, commodity-exporting plants controlled by large firms pursuing horizontal and vertical integration (Hayter 1976; Schwindt 1977). In addition, federal and provincial governments have welcomed foreign investment, and the resulting high level of foreign ownership, already significant by the 1930s (Marshall, Southard, and Taylor 1936), has reinforced this industrial structure. Thus foreign firms investing in Canadian forest resources have typically been large, integrated forest product-based companies that have either acquired or invested in large commodity-producing plants, frequently for export to affiliated operations (Hayter 1981; 1985).

Such an institutional structure – featuring corporate concentration, foreign ownership, and a top-heavy size distribution of plants, and underpinned by "free enterprise" forest policies exchanging timber for capital investment – was conducive to impressive rates of growth during the long boom beginning in the late 1940s. Unfortunately, this same structure has proven vulnerable to the vicissitudes of the 1980s and 1990s. Thus the industry's narrow export specialization on standardized commodities in association with high levels of foreign ownership discouraged in-house research and development (R&D) in Canada, especially product R&D. Similarly, since sales of such commodities are either handled at arm's length or administered within units of the same corporation, the industry had no need for sophisticated marketing expertise and networks. Yet over the past decade, as Canada's long-standing specialities of newsprint, pulp, and softwood lumber have become more vulnerable to international competition and falling demand, private-sector deficiencies in R&D and marketing have constrained diversification into the faster-growing, more product-differentiated segments of the forest industries.

Certainly, the recession of the early 1980s constituted a profound shock to forest product corporations in Canada and stimulated restructuring, which is still evolving. Here the capital-intensive nature of the Canadian forest industries is a vital consideration. Thus the recession provided a powerful signal to modernize and develop new product lines while compromising the industry's financial ability to do so. This dilemma was particularly serious in British Columbia, where the industry incurred a net income loss of $500 million in 1982. Elsewhere in Canada, profits were much reduced. Thus corporations had to reduce debt while investing in new faciltes.

Several characteristics of this corporate restructuring are worthy of note. First, in the absence of comprehensive planning by governments, industry continued to restructure through initiatives by individual firms. Second, to reduce debt and raise cash for investment, many large forest product corporations sold operations, including foreign-based ones. The Canadian-owned multinational MacMillan Bloedel, for example, sold not only its head office in Vancouver, a substantial share of its paper-converting operations in Canada, and a newsprint mill in New Brunswick but virtually its entire European and Asian holdings. Similarly, long-established, foreign-owned Canadian subsidiaries, such as Canadian International Paper (CIP) and Crown Zellerbach Canada, were sold by their parent companies. Third, some "Canadianization" of the forest sector occurred, though not because of public policy. To the contrary, foreign investment continued to be solicited, as the acquisition of Crown

Table 5
Capital and repair expenditures ($million) in the Canadian forest industries, by region, 1989 (1992)

	Atlantic Provinces	Quebec		Ontario		Prairies	British Columbia		Canada	
Logging	40	85	(31)	120	(69)	28	369	(183)	642	(313)
Wood	38	231	(245)	217	(138)	103	702	(597)	1,291	(1,175)
Paper and allied	979	1,949	(971)	1,450	(897)	105	2,069 (1,291)		7,497	(4,380)
Total	1,056	2,265 (1,247)		1,787 (1,104)		1,181	3,140 (2,071)		9,430	(5,868)

Source: Statistics Canada, *Selected Forestry Statistics 1990 and 1992*, Cat. No. 25-202.

Zellerbach by Fletcher Challenge of New Zealand and investments by Louisiana Pacific (US) and several Japanese firms in British Columbia and Alberta confirm. Fourth, central Canadian-based conglomerates, notably Canadian Pacific and Brascade Ltd., increased their participation in the forest industries by acquiring such companies as CIP and Noranda and its forest-industry subsidiaries, including (for a time) MacMillan Bloedel. Fifth, the 1980s saw the emergence, largely through acquisition, of several new major corporate entities, including Fletcher Challenge in British Columbia and the Quebec-based Cascade Industries.

Notwithstanding myriad changes in ownership and nationality, big firms and big export-oriented plants, as well as foreign ownership, remain dominant features. The industry remains capital intensive, and in 1989 alone, for example, capital expenditures, exceeded $9 billion (Table 5). Even in the severe recessionary year of 1992, capital expenditures were almost $6 billion. While in the pulp and paper industry, which accounts for most of the expenditures, important investments have been made to combat environmental problems and to build a few new mills in western Canada, most investments have emphasized modernization of existing mills.

The purposes of this huge level of aggregate investment have varied considerably among individual operations. A trend toward replacement of "Fordist" methods of production, which feature mass production of an extremely limited range of commodities, with computerized methods of "flexible mass production" (Figure 1), is emerging. Computer technology, increasingly differentiated market demands, and increasing wood-fibre costs have encouraged more mills to mass produce a wide range of higher-value, specialty products. This trend is evident both with respect to wood processing, especially in the BC coastal forests (Barnes and Hayter 1992), and throughout Canada in the paper industry, where many mills are replacing newsprint production with a growing range of specialty papers. Complementing this trend, albeit on a smaller scale, are strategies of flexible specialization – geographical concentrations of small plants, often owned by small firms and closely interdependent in subcontracting relations. Examples include the development of furniture manufacture in Quebec and wood remanufacturing in the Vancouver metropolitan area (Rees 1993).

Admittedly, the severity of the recession of the early 1990s suggests that the forest industry remains too narrowly specialized on its traditional commodity base. Yet these same recessionary conditions will doubtless give further impetus to value-added trends. A similar point can be made about labour relations: the changes that began in the 1980s will gather momentum from the present crisis.

Changing Labour Relations

The recession of the early 1980s constituted a major shock to labour in the Canadian forest industries, especially in British Columbia, where the crisis was most severe and labour power the strongest. Despite earlier attempts to form unions, union power became consolidated, effective, and committed to collective bargaining only during the boom years following 1945 and was particularly strong in the main commodity industries of lumber, pulp, and paper. Until 1980, employment and union membership, in such unions as the International Woodworkers of America (IWA) and the Canadian Paperworkers Union (CPU), grew steadily. During the 1950–80 period, most lay-offs during recessions were temporary, limited to "blue-collar" workers, and organized on the basis of seniority. Along the coast especially, the early 1980s recession brought an abrupt end to this pattern. Job losses in the big union mills were permanent and affected all segments of the workforce, including management (Grass and Hayter 1989; Barnes, Hayter, and Grass 1990).

It is reasonable to argue that the recession shifted bargaining power in favour of management. Certainly, unions, such as the IWA, once the largest union in British Columbia and the Canadian forest industries, are less important than they once were. Indeed, the typically smaller mills of the wood-remanufacturing industry, which has experienced some growth, as well as some of the new pulp and paper mills in Alberta, are non-unionized. Moreover, even among union mills, work practices are changing. The dominant trend, paralleling the shift toward flexible mass production, is the move from "Fordist" labour relations, characterised by principles of seniority and strong job demarcation, to "flexible operating cultures" (Barnes and Hayter 1992; Hayter and Barnes 1992; Hayter, Grass, and Barnes 1994). This complex trend includes efforts both to produce a multi-skilled "core" workforce and to reduce labour costs, including by relying more on "peripheral" workforces.

TECHNOLOGY

The relationship between technical change and industry evolution is more complex than once thought (Chapman and Humphrys 1987). In the forest industries there is widespread opinion that the pace of technological change has quickened over the last 100 years. Indeed, since 1950 there have been major innovations in tree genetics, logging, pulping, paper making, paper finishing, bleaching, and wood products, as well as extensive application of micro-electronics to all aspects of forest product operations (Hayter 1988; Hopgood 1986; Ofori-Amoah 1990; Schuler 1985; Silversides 1984; Tillman 1985). In addition, incremental technical change has long been an important source of productivity increase (Cohen 1984). Technological innovation over the past four decades has enhanced wood utilization, increased labour efficiency, differentiated and diversified the mix of paper and wood products, reduced environmental damage, and extended forest regeneration capability. Moreover, the potential for further major, even radical technological change has been identified (see especially Jurasek and Paice 1984; Tillman 1985).

The Tradition of Technological Conservatism

In contrast to other advanced forest product nations, in Canada technological attitudes and choices have been conservative. In particular, the forest industries have focused on process technologies consistent with specialization on a limited range of bulk commodites (Hayter 1988; Hopgood 1986; Sector Task Force 1978; Woodbridge, Reed and Associates 1984). In practice, investment has almost invariably emphasized "proven" technology. Before choosing equipment Canadian buyers want to see it in operation. Technological change does occur, but predominantly in response to, rather than in anticipation of, market demands. While the Canadian lumber industry did achieve a high level of efficiency during the 1980s, its commodity orientation kept it vulnerable to American protectionsim, exchange-rate fluctuations, and demand changes (Percy and Yoder 1987). In pulp and paper, by the 1980s many of the older mills were less efficient than modern mills in the U.S. south and other "warm" regions. Moreover, investment decisions by Canadian pulp and paper mills have typically stressed short-run financial considerations and periodical replacement of individual machines, so that modernization rarely means creation of state-of-the-art mills.

However, the size of the Canadian forest industries, the entrepreneurial nature of logging and wood processing, and the highly variable conditions in which forest product activies are located create a huge demand for technology. In a crude way the size of this demand is indicated by investment expenditures (Table 5). This demand is met from foreign and domestic sources and by R&D activities undertaken by forest product firms themselves, then associations' and cooperative laboratories, equipment suppliers, universities, and government ministries (Hayter 1988). In Canada, however, the R&D system reflects the industry's technological conservatism: by international standards it is relatively underfunded, public-sector funding has played an unusually important role, and risky, long-term product R&D is rare.

International comparisions, whether based on origin of patents, R&D employment, or proportion of sales devoted to R&D, invariably confirm Canada's low level of R&D funding and low standing as a supplier of forest product technology. In 1975, for example, Sweden accounted for 5.46 per cent of world pulp and paper production and 8.3 per cent of OECD-based forest product R&D; Finland, 3.88 per cent and 4.14 per cent, respectively; Japan 9.40 per cent and 14.00 per cent; the United States 34.50 per cent and 55.30 per cent; and Canada, 10.5 per cent and 5.91 per cent (Hanel 1985, 57; see also Hayter 1982; Hopgood 1986). Recent data indicate that this situation has not changed or is worse. Thus, in 1986 the industry spent 0.3 per cent of sales on R&D,

Table 6
Selected characteristics of "important" in-house technological developments since 1970 by
six forest-product firms

Cost ($million)	Period of development (years)*	Innovation	R&D costs	
			As % of project cost	$000
1	3 (2)*	Building materials	40	210
1	6 (5)	Paper quality	90	650
1–5	17 (13)	Pulping	90	2,700
1–5	13 (3)	Bleaching	10	300
5–10	8 (4)	Pulping	20	2,000
>10	11 (8)	Pulping	9	2,500
>10	11 (9)	Pulping	19	2,500
>10	5 (2)	Pulping	30	7,000
>10	15 (10)	Pulping	20	3,000
>10	7 (2)	Pulping	10	3,000

Source: Hayter (1988), 51.

* Numbers in parentheses represent years until prototype developed.

compared to 1.25 per cent in the United States, 1.35 per cent in Sweden, and 0.75 per
cent in Finland (Science Council of Canada 1992, 14). In wood processing, Canada
similarly lags behind the R&D leaders.

Remarkably, given the quickening pace of technological change, since the 1960s
levels of R&D in the Canadian forest industries have remained relatively low and may
have actually declined. Thus, in 1968 Smith and Lessard (1972) estimated R&D expenditures by the public and private sectors for the forest industries (excluding equipment supply) to be $54 million. Solandt (1979) estimated that in 1979 R&D
expenditures would have to be $151 million to maintain 1968 levels of activity; the
actual figure was $115 million. In real terms, this effort was maintained through the
first half of the 1980s, though since then several R&D programs in the private sector
have been terminated. If firms such as MacMillan Bloedel and Canfor have maintained and even expanded R&D efforts through the past decade, others, such as Canadian Pacific (formerly Canadian International Paper), have closed down once-large
R&D programs. Similar closures have been implemented by forest equipment firms.
The net effect is that in-house R&D was no bigger, and possibly smaller, in real terms,
in 1992 than in 1981 or 1986, when it was already small by international standards.

The principal goal of R&D in the Canadian forest industries is to create, or adapt,
processing technology to meet specific local needs (see Hayter 1988). During the
1970s and 1980s, in-house R&D groups emphasized development of new pulping processes (Table 6), generally to increase yield and reduce environmental effects within
the resource and production contexts of specific mills. While in two cases the know-
how gained was used to resolve similar problems in affiliated mills (thereby reducing
lengthy development times and costs), the technologies created were seen by firms as
site specific.

Table 7
Selected characteristics of recent in-house innovations for the forest industry by equipment suppliers

Project cost ($000)	Period of development (months)*	Area of innovation	R&D costs as % of total costs
Top six R&D performers			
100–500	13 (8)	Wood processing	99
100–500	36 (12)	Pulping	90
500–1,000	29 (24)	Wood harvesting	40
500–1,000	30 (18)	Wood harvesting	50
100–500	14 (12)	Wood harvesting	70
500–1,000	18 (12)	Electronics	20
Next four R&D performers			
500–1,000	24 (?)	Wood processing	80
100–500	40 (24)	Wood processing	80
50–100†	48 (24)	Paper making	100
50–100	13 (8)	Pumps	57
Others			
100–500	8 (7)	Wood processing	50
50–100	6 (1)	Wood processing	30
5–20	6 (1)	Wood processing	10
50–100	18 (6)	Wood processing	15
100–500	36 (12)	Wood drying	55
20–50	9 (7)	Electronics	80

Source: Hayter (1988), 54.

*Numbers in parentheses represent months until prototype developed.

†Firm terminated its R&D program in 1984.

Among equipment suppliers, most in-house R&D programs focused on wood-processing or wood-harvesting equipment (Table 7). These priorities are also rooted in the need to develop technology to meet distinct regional needs within Canada. In logging, Silversides (1984) has characterized Canadian efforts in terms of the "mousetrap syndrome" – myriad local solutions developed for broadly similar problems. More generally, the Canadian forest-equipment supply industry is made up of small and medium-sized firms oriented to domestic and, to some extent, U.S. markets; large international companies, with major R&D programs, direct foreign investments, and export activities, are based in Sweden, Finland, Germany, and the United States.

Other Canadian organizations conducting forest-sector R&D are also concerned primarily with creating or adapting process technology to meet Canadian conditions. They include a network of federally funded forestry laboratories; Paprican, a pulp and paper research association funded by private-sector members; and Feric (logging) and Forintek (wood processing), cooperative research organizations funded by governments and to a lesser degree by industry. Their research efforts vary. For example, Paprican's long-term research has focused on creating technology, while Feric has concentrated on short-term efforts to modify existing, often foreign technology (Hayter 1988).

In effect, Canada's physical geography – variations in climate, landforms, soil, and vegetation – has shaped indigenous R&D and innovation in the forest industries. Thus, it may be argued that the physical environment has "protected" Canadian technological efforts from foreign competition, most notably in wood harvesting and processing, activities in which distinctive problems have encouraged distinctive expertise. Failed imports further illustrate this point (Hayter 1978; Silversides 1984). Even so, whenever possible, the industry has preferred to buy proven technology, regardless of source, rather than undertake the risks and costs of development. In the case of technologies, such as paper-making machinery and computerized control systems, which require little or no local adaptation, Canadian R&D efforts have typically declined and more equipment has been imported (Ofori-Amoah 1990). Moreover, logging and wood-processing imports have increased – and from countries such as Austria, Italy, and Switzerland, as well as the United States, Sweden, and Finland. As foreign firms have learned more about Canadian conditions, these same conditions provide less effective protection against technological imports (Hayter 1988, 72).

In today's changing competitive conditions, however, reliance on conservative technology strategies, in which foreigners increasingly control the R&D process, raises significant questions.

In-house R&D and Value-added Strategies

In comparision to the United States, Sweden, Finland, and Japan, Canada has placed greater emphasis within forest-product R&D on public-sector laboratories and funding (Table 8). Despite changes in Canada's R&D system over the past decade, investment in in-house R&D continues to be significantly lower here (measured on a per-capita basis) than in those countries. Unfortunately, public-sector, cooperative, and association laboratories cannot substitute for in-house R&D, especially in situations where market success is based on firm-specific product differentiation.

Certainly, those few forest product firms in Canada that do have an in-house R&D program recognize the "firm-specific advantages" that it generates (Hayter 1988, 34–5; see also Cohen and Mowery 1984). These firms claim that there are significant "transaction costs" to subcontracting R&D and that external agencies simply cannot respond to the myriad technical problems facing each firm (and at the time firms want to solve problems). Even more important, public-sector or industry-wide efforts cannot develop products that give individual firms market advantages. In fact, deficiencies in the funding of in-house R&D have reduced its potential impact on the rest of the R&D system, for example, through various kinds of technological exchanges or liaisons. Indeed, Canada's forest-product R&D system is fragmented, and cooperation among organizations limited, compared to Finland, Sweden, and the United States.

Canadian forest product firms often argue that there is no incentive to invest in in-house R&D because the industry is "open" with respect to technology and most new technology can be readily purchased from equipment suppliers, especially foreign ones (Hayter 1982). In this view a local R&D capability is necessary only to the extent

Table 8
Professional R&D employment* in the forest product sector, selected OECD countries, c. 1984

Types	Canada	United States	Japan	Sweden	Finland
Government	423	988	788	n.a.	256
University	161	826	150	365	51
Cooperatives	570	110	100	330	270
Industry	203	3,500	1,500	Strong	Strong
Equipment suppliers	Weak	Strong	?	Strong	Strong

Source: Based on Hayter (1988) and Himli (1986).

* "Professional employment" refers to scientists and engineers.

that new technology has to be modified for Canadian conditions. In fact, it is the industry's commodity orientation and reliance on foreign firms, most of which have centralized R&D outside Canada, that undermines the rationale for in-house R&D in Canada. Moreover, the lack of in-house R&D compromises the industry's ability to reduce its commodity reliance and diversify into faster-growing, value-added market segments. Indeed, the "open to technology transfer" thesis begs the question as to why so many foreign firms invest so much in R&D. The thesis vastly oversimplifies the situation by ignoring the costs, timing, and uncertainties involved in technology transfer and the competitive advantages with respect to lower costs of production and market diversification and penetration that can be achieved by aggressive technology strategies.

The costs and missed opportunities stemming from a low level of private-sector R&D are substantial. First, there is a loss of high-income jobs directly associated with R&D. Second, imported technology in the form of equipment, licences, and R&D services from parent corporations contributes to deficits in both visible and invisible trade. With respect to equipment and machinery manufacture, it is well established that in-house R&D is important for export success and the growth of firms; the meagre levels of R&D by Canadian firms are contributing to their disappointing and declining export performance (Hanel 1985; Hayter 1981). In fact, Canada has a balance-of-payments deficit in the forest-product equipment trade. Third, because of continued use on proven technology to manufacture a narrow range of standardized commodities, Canadian firms have lost market share and position to their competitors. Worldwide, Canada's share of the value of forest-product export markets declined from 25.7 per cent in 1961 to 19.4 per cent in 1989; in contrast, the United States increased its share from 8.7 to 13.1 per cent during the same period. Particularly in paper products, the United States, and indeed Finland and Sweden, have invested more strongly in R&D and have been more innovative in shifting toward new and faster-growing market segments.

In the pulp and paper industry, Finland has been particularly innovative in developing higher-value papers, such as light-weight coated papers and wood-free copy papers (Croon 1988, 271). Elsewhere, the United States has maintained technological leadership in tissue papers and is competing with Europe in coated publication grades. Major Canadian companies have been slower to shift from low- and medium-value commodities, such as kraft pulp and standard newsprint (Woodbridge 1988, 283). Admittedly, value-added shifts are occurring in Canada, typically initiated by firms showing technological leadership. Thus the efforts of MacMillan Bloedel's R&D group, the largest in-house program in Canada, have led to product innovations and efficiencies. The firm, beginning in the 1970s, developed several new specialty papers, most notably light-weight telephone directory paper and papers for advertising flyers, and manufactures them on machines once dedicated to newsprint. In the early 1990s it developed a new form of bulk packaging for liquids such as milk. There are similar trends toward more value-added production in solid wood production.

A shift to value-added production does not, and should not, preclude manufacture of bulk commodities. Nor will such a shift be easy: value-added opportunities typically have to be created by proactive investments in R&D, and in many cases resulting market segments are relatively small and best left to small companies. In addition, Canadian companies will continue to face competition for value-added markets in pulp and paper and in wood products. Indeed, if supply and demand forces are inevitably pushing/pulling the Canadian forest industries along the value-added path, questions may be asked as to whether the speed and extent of this transition are sufficient for the industry to realize the available job and income potentials.

A TECHNOLOGY POLICY

For the past 25 years at least, there have been a litany of pleas for stronger commitment to R&D and innovation by the Canadian forest industries (Hanel 1985; Hayter 1988; Hopgood and Associates 1986; Smith and Lessard 1972; Solandt 1979). When, in December 1987, the federal Government named the forest industries an innovation-policy priority, such pleas seemed to have reached fruition. Indeed, Ottawa did help fund construction of a western laboratory for Paprican in Vancouver, and provincial governments have broadened their thinking about R&D policy to include the forest industries. However, the R&D system remains much the same; technological efforts are largely fragmented and underfunded. Progressive proposals for new forest policies, such as the Peel Commission (1991) in British Columbia, scarcely mention R&D. Whatever government's failure to develop effective technology policy, the private sector has revealed an impressive and consistent reluctance to invest in R&D.

Yet the restructuring of the Canadian economy, which, practically speaking, began with the recession of the early 1980s, clearly implicates the forest industries, and technology is central to these implications (Figure 1, above). The last decade has revealed the forest industries to be at a crossroads. To survive and prosper the industry must shift from cost minimization to value maximization, from specialization to diversification; commodity trading must give way to product marketing; the emphasis on proven equipment should be modified by more innovative machinery; dependent technology strategies need to be replaced by offensive technology strategies; corporate concentration must change to more balanced size distribution of firms; forest management ought to become less extensive and more intensive; and the environmental implications of technology and product choices should become more central.

These trends, of course, are interrelated. Intensive forest management, for example, means increasing the value of the forest, which would reflect and encourage the research, development, marketing, and production of higher-value products. Greater use of the forest resource for non-wood benefits would increase its value to industry, which would further stimulate the shift to value-added production and thinking.

Certainly a strong case can be made for establishing the forest industries as an innovation priority (Hayter 1987, 229–30). Thus innovation is the key to enhancing productivity, adding value, and stimulating change in the supply industries. The forest industries themselves comprise a range of low-, medium-, and high-tech industries and can provide a seedbed and orientation for the retraining of workforces and for Canadian development of newly emerging technologies, such as biotechnology. There is also evidence that rates of return on investment in R&D in forest products can be high, as it can be in many mature industries (Mansfield et al. 1977).

Moreover, if such trends as volatile corporate profits and increasing wood costs create barriers to transition from the old, cost-based production system to the new, value-based production system, the same trends also provide incentives to do so. If the transition is not made, then the job losses of the old system will continue. It is unclear whether governments, especially the provinces, will introduce policies to help with the transition. Provincial governments may wish to consider how to give greater weight to R&D and innovative behaviour as part of contractual obligations in timber leases.

REFERENCES

Barnes, T.J., and Hayter, R. 1992. "The Little Town That Did: Flexible Accumulation and Community Response in Chemainus, British Columbia." *Regional Studies*, 26, 647–63.

– 1994. "Economic Restructuring, Local Development and Resource Towns: Forest Communities in Coastal British Columbia." *Canadian Journal of Regional Science*, 17.

Barnes, T.J., Hayter, R., and Grass, E. 1990. "MacMillan Bloedel: Corporate Restructuring and Employment Change." In M. de Smidt and E. Weaver, eds., *The Corporate Firm in a Changing World Economy*, 145–65. London: Routledge.

Baskerville, G.L. 1990. "Canadian Sustained Yield Management – Expectations and Realities." *Forestry Chronicle*, Feb., 25–8.

Canadian Forest Industries Council (CFIC). 1986. *The Management of Canadian Forests 1986*. Ottawa: Canadian Forest Industries Council.

Chapman, K., and Humphrys, G., eds. 1987. *Technical Change and Industrial Policy*. Oxford: Basil Blackwell.

Clement, W. 1989. "Debates and Directions: A Political Economy of Resources." In W. Clement and G. Williams, eds., *The New Canadian Political Economy*, 36–53. Montreal: McGill-Queen's University Press.

Cohen, A.J. 1984. "Technological Change as Historical Process: The Case of the U.S. Pulp and Paper Industry, 1915–40." *Journal of Economic History*, 44, 775–99.

Cohen, W.M., and Mowery, D.C. 1984. "Firm Heterogeneity and R&D: An Agenda for Research." In B. Bozeman and A. Link, eds., *Strategic Management for R&D*, 197–232. Lexington, Mass.: D.C. Health.

Croon, I. 1988. "The Scandinavian Approach to the Future of Pulp and Paper." In G.F. Schreuder, ed., *Global Issues and Outlook in Pulp and Paper*, 268–75. Seattle: University of Washington Press.

Forestry Canada. 1990. *Forestry Facts*. Ottawa: Ministry of Supply and Services.

Grass, E., and Hayter, R. 1989. "Employment Change during Recession: The Experience of Forest Product Manaufacturing Plants in British Columbia, 1981–1985." *Canadian Geographer*, 33, 240–52.

Hall, J.P., Carlson, L.W., and Dube, D.E. 1990. "A Forestry Canada Approach to Environmental Forestry." *Forestry Chronicle*, April, 138–42.

Hanel, P. 1985. *La technologie at les exportations canadiennes du matériel pour la filière bois-papier*. Montreal: l'Institut de Recherches politiques.

Hay, E. 1993. "Recession and Restructuring in Port Alberni: Corporate, Household and Community Coping Strategies." MA thesis, Department of Geography, Simon Fraser University.

Hayter, R. 1976. "Corporate Strategies and Industrial Change in the Canadian Forest Product Industries," *Geographical Review*, 66, 209–28.

– 1978. "Locational Decision-Making in a Resource-Based Manufacturing Sector: Case Studies from the Pulp and Paper Industry of British Columbia." *Professional Geographer*, 33, 240–49.

– 1979. "Labour Supply and Resource-Based Manufacturing in Isolated Communities: The Experience of Pulp and Paper Mills in North Central British Columbia." *Geoforum*, 10, 163–77.

– 1981. "Patterns of Entry and the Role of Foreign-Controlled Investments in the Forest Sector of British Columbia." *Tijdschrift voor Economische en Social Geografie*, 72, 99–113.

– 1982. "Research and Development in the Canadian Forest Product Sector – Another Weak Link?" *Canadian Geographer*, 26, 256–63.

– 1985. "The Evolution and Structure of the Canadian Forest Product Sector: An Assessment of the Role of Foreign Ownership and Control." *Fennia*, 163, 439–50.

– 1987. "Innovation Policy and Mature Industries: The Forest Product Sector of British Columbia." In K. Chapman and G. Humphrys, eds., *Technical Change and Industrial Policy*, 215–32. Oxford: Basil Blackwell.

– 1988. *Technology and the Canadian Forest-Product Industries: A Policy Perspective*. Background Study No. 54, Science Council of Canada. Ottawa: Minister of Supply and Services.

– 1992. "International Trade Relations and Regional Industrial Adjustment: The Implications of the 1982–86 Canadian-U.S. Softwood Lumber Dispute." *Environment and Planning A*, 24, 153–70.

Hayter, R., and Barnes, T.J. 1990. "Innis' Staple Theory, Exports and Recession: British Columbia 1981–86." *Economic Geography*, 66, 156–73.

– 1992. "Labour Market Segmentation, Flexibility and Recession: A British Columbian Case Study." *Environment and Planning C*, 10, 333–53.

Hayter, R., Grass, E., and Barnes, T.J. 1994. "Labour Flexibility: A Tale of Two Mills." *Tijdschrift voor Economische en Sociale Geografie*, 85, 25–38.

Himli, H.A. 1986. *World Compendium of Forestry and Forest Product Research Institutions*. Rome: FAO.

Hopgood, A. (Enterprises Ltd.) 1986. *The Potential for New Technologies in Canada's Forest Sector*. Ottawa: Ministry of State for Science and Technology.

Innis, H.A. 1946. *Political Economy in the Modern State*. Toronto: Ryerson.

– 1967. "The Importance of Staple Products." In W.T. Easter brook, and M.H. Watkins, eds., *Approaches to Canadian Economic History*, 16–19. Toronto: McClelland and Stewart.

Jacques, R., and Fraser, G.R. 1989. "The Forest Sector's Contribution to the Canadian Economy." *Forestry Chronicle*, April, 93–6.

Jurasek, L., and Paice, M.G. 1984. Biotechnology in the Pulp and Paper Industry. Manuscript report, Science Council of Canada, Ottawa.

Mansfield, E., Rapoport, J., Romeo A., Villani, E., Wagner, S., and Husic, F. 1977. *The Production and Application of New Technology*. New York: Norton.

Marshall, H., Southard, F.A., and Taylor, K.W. 1936. *Canadian-American Industry: A Study in International Investment*. Toronto: Ryerson.

Melody, W.H., Salter, L., and Heyer, P., eds. 1981. *Culture, Communication, and Dependency: The Tradition of H.A. Innis*. Norwood, NJ. Ablex.

Ofori-Amoah, B. 1990. "Technology Choice in a Global Industry: The Case of the Twin-wire in Canada." PhD dissertation, Department of Geography, Simon Fraser University.

Pearse Commission. 1976. *Timber Rights and Forest Policy in British Columbia*. Victoria, BC.

Peel Commission. 1991. *Forest Resources Commission: The Future of Our Forests*. Victoria: Forest Resources Commission.

Percy, M.B., and Yoder, C. 1987. *The Softwood Lumber Dispute and Canada-U.S. Trade in Natural Resources*. Halifax: Institute for Public Research.

Reed, F.L.C., and Associates. 1978. *Forest Management in Canada*. Ottawa: Environment Canada, Canadian Forestry Service.

Rees, K.G. 1993. "Flexible Specialisation and the Case of the Remanufacturing Industry in the Lower Mainland of British Columbia." MA thesis, Department of Geography, Simon Fraser University.

Rotstein, A. 1977. "Innis: The Alchemy of Fur and Wheat." *Journal of Canadian Studies*, 12, 6–31.

Schuler, A. 1985. "Sawmilling in the 80s: A Look at Factors Mitigating Change." *Canadian Forest Industries*, April, 33–6.

Schwindt, R. 1977. *The Existence and Exercise of Corporate Power: A Case Study of MacMillan Bloedel Limited*. Ottawa: Ministry of Supply and Services.

Science Council of Canada. 1992. *Canadian Forest Products*, Technology Sector Strategy Series No. 9. Ottawa: Ministry of Supply and Services.

Sector Task Force. 1978. *The Canadian Forest Products Industry*. Ottawa: Department of Industry, Trade and Commerce.

Silversides, C.R. 1984. "Mechanized Forestry: World War II to the Present." *Forestry Chronicle*, Aug., 231–5.

Smith, J.G., and Lessard, G. 1972. *Forest Resources Research in Canada*. Background Study No. 14, Science Council of Canada. Ottawa: Information Canada.

Solandt, O.M. 1979. *Forest Research in Canada*. Ottawa: Canadian Forestry Advisory Council.

Swift, J. 1983. *Cut and Run*. Toronto: Between the Lines.

Tillman, D.A. 1985. *Forest Products: Advanced Technologies and Economic Analyses*. New York: Academic Press.

Watkins, M.H. 1963. "A Staple Theory of Economic Growth." *Canadian Journal of Economics and Political Science*, 29, 141–58.

Wilson, J. 1987/88. "Forest Conservation in British Columbia, 1935–85: Reflections on a Barren Political Debate." *BC Studies*, Winter, 3–32.

Woodbridge, P. 1988. "Marketing and Production Issues in the Pulp and Paper Industry of the Future." In G.F. Schreuder, ed., *Global Issues and Outlook in Pulp and Paper*. Seattle: University of Washington Press.

Woodbridge, Reed and Associates. 1984. *British Columbia's Forest Products Industry: Constraints to Growth*. Prepared for the Ministry of State for Economic and Regional Development in Vancouver.

CHAPTER SEVEN

Restructuring in the Canadian Mining and Mineral-Processing Industries

IAIN WALLACE

Like the forest sector, the Canadian mineral sector has, in the 1990s, finally recognized that its large resource base is not limitless, and that its traditional ways of doing business are no longer adequate to ensure long-term survival and prosperity. As visibly extractive industries in an era of heightened sensitivity to environmental degradation (clearcutting and creating waste heaps with acidic drainage are both ready sources of emotive imagery in the hands of critics); as industries whose hinterland sphere of operations has situated them in the complex political currents of renegotiation of Native land claims; and as sectors whose firms have only comparatively recently begun to experience significantly intensified global competition, the mineral and forest industries face many similar challenges (see Hayter, this volume, chap. 6). The distinctive characteristics of the mineral sector are equally apparent, however, and form the focus of this chapter. Among them are the continuing importance of Canada as the world's largest and most diversified exporter of mineral commodities; the marked increase in the internationalization of the Canadian mining and mineral-processing sectors (with significant flows of foreign investment both in and out); and the modes of corporate adjustment to the new constraints and opportunities brought about by changes in the global economy.

Canada's extensive territory and diversified geology have provided a rich resource base for the mineral sector; but the industry, like all staples, is essentially the creation of external markets. Though mining began to play a staple role within some regional economies, notably northern Ontario and southern British Columbia, at the close of the nineteenth century, Canada's emergence as a major supplier to world (primarily U.S.) markets dates essentially from the 1950s. The postwar boom and Cold War demand for strategic minerals ended U.S. mineral self-sufficiency and triggered substantial and widespread investment in Canadian mines and related infrastructure. By the 1970s, the growing resource demands of Japanese industry generated a new wave of mineral investments, geographically more localized in western Canada.

But while demand in the Asia-Pacific region held up well as growth of the world economy slowed after the early 1970s, Canadian mineral producers began to experience a major structural shift in global trading conditions. The resource-intensity of industrial production fell, not just with respect to energy inputs, and demand in the traditional u.s. industrial heartland contracted significantly. Partly as a lagged response to the boom of the early 1970s, large new mineral investments came onstream in Australia and many less-developed countries. The urgent revenue needs of the latter made the output of their predominantly public-sector operators relatively price inelastic, depressing the earnings of competing Canadian producers. The recession of the early 1980s in the world's industrialized economies saw base-metal prices fall as low, in real terms, as during the 1930s. As price takers, Canadian mining firms were forced into a radical restructuring of their operations, with some severe regional effects, in order to survive. More recently, the collapse of the Soviet Union created particular uncertainties for Canadian mineral producers, as the potential for the huge mineral resources of this formerly non-market economy to disrupt world commodity markets (notably in nickel, uranium, and aluminum) became evident. Adjustments by the Canadian mineral industry to ongoing changes in global trading conditions and the domestic climate for its operations have involved accelerating internationalization and progressive restructuring of employment.

Most of the conceptualization and interpretation of the radical changes that have characterized the global economy over the past 25 years have focused on particular elements of the manufacturing and service sectors, and hence on certain archetypal urbanized regions, such as southern California, industrial Britain, or "Third Italy." In contrast, resource-based industries, and the hinterland economies in which they are typically found, have been relatively neglected (see Norcliffe 1994). Hence the literature on the geography of economic restructuring provides both themes that are pertinent to analysing the experience of Canadian mineral producers since the 1970s and a counterfoil, helping to focus the distinctive characteristics of a particular sector that continues to shape Canada's industrial space economy. In addition, because the broad structural changes that impinge on the minerals sector do not overdetermine specific outcomes, the following account gives full weight to the differential responses of corporate and institutional actors.

Canada has long been the world's largest mineral exporter and in the early 1990s was a leading producer (ranked in the top three) of ten major mineral commodities. Though mineral exports' share of total exports has halved since its peak of 34 per cent in 1957, and the industry's share of GDP has approximately halved since its peak of 8.4 per cent in 1965, the sector remains one of the most comprehensive in the world, in terms of both its commodity scope and the level of expertise in different branches, and is still a significant element of the domestic economy (IWGMI 1992b). The market power of Canadian producers has, however, undoubtedly declined over the past 30 years, reflecting the reduced resource intensity of global industrial production, associated with the rise of new technologies (especially those based on microelectronics), and continued absolute growth in global demand for non-renewable resources, which has brought many new competing mineral-producing nations onto the world market. As well, it reflects the shifting centre of gravity of the global economy, away from

Canada's dominant market, the United States, toward the Asia-Pacific region. The net result has been a challenge to Canadian mineral firms to increase sales from an exhaustible asset base that constantly needs replenishing, and into markets that are being restructured, both sectorally and geographically.

Despite Canada's prominence among the world's mineral producers, its mineral firms are generally medium-sized. In 1992, Inco and Noranda ranked 9 and 10 among global mine operators in market economies, with 1.3 per cent each of the value of non-fuel production. This made them each half the size of BHP (Australia), a quarter that of RTZ (United Kingdom), and one-sixth that of Anglo American (South Africa), though a revenue measure undoubtedly exaggerates the size of precious metal producers (*Financial Times*, 27 May 1994). Only Alcan (not classified as a mining company) among Canadian-based minerals firms ranks with the global leaders in its sector. The dearth of large firms in Canada is related to, but not fully explained by, the country's geographically widespread resource base and great diversity of mineral commodities, all exploitable, with relatively easy access to U.S. markets. This fostered the emergence of Canadian mining firms that have generally stayed medium-sized, quite regionalized in production sites, and narrowly specialized in outputs. Their corporate geographies and product profiles have been much simpler than those of multinational, multi-commodity firms such as BHP and RTZ. In particular, foreign direct investment has until recently had neither supply nor demand imperatives for Canadian firms commensurate with those that have shaped the sector's growth in other nations.

Industrial restructuring in mining and mineral processing differs from the forms of response seen in most of the manufacturing sector. Strategies such as flexible specialization, vertical disintegration, and niche marketing are not self-evidently appropriate to an industry that remains concerned with technologies of continuous bulk processing, in highly capital-intensive and functionally specialized plant, of variegated raw materials to mass produce a few, relatively homogeneous, intermediate products. Economies of scale in production, and process-related transaction cost savings favouring vertical integration, encourage retention of apparently conventional approaches to industrial organization and maintaining competitiveness (see O hUallachain and Matthews 1994). Indeed, there is "a tremendous cultural boundary … in the metals-producing value chain once it goes beyond commodity production" (Ala-Harkonen 1993, 14).

The business methods and skills required in the downstream markets of differentiated, customized products differ markedly from those that make for success in upstream mining, and only the oligopolistic aluminum transnationals appear to have internalized that divide successfully. The recent history of the Canadian minerals industry has for the most part been one of cost reduction through technical innovation in mining and processing, downsizing and reconfiguring the workforce, and improving the quality of the resource base, rather than one of product innovation or diversification. But both subsectors and individual firms have been sufficiently varied in their specific challenges and responses that a more disaggregated analysis is called for.

BASE-METAL MINERALS CORPORATIONS

On the basis of in-depth case studies of some of the leading global base-metal transnational firms, Ala-Harkonen (1993) suggests a typology of corporate strategies of growth and adjustment. Because market structures differ in other subsectors (such as iron and steel), this classification does not capture all patterns of development in the minerals industry, but the importance of base metals in Canada makes it a useful point of departure. The largest group of firms is made up of "minerals-driven companies." These are particularly competent in mine development and sometimes also in metallurgy, often having been founded on a particularly rich deposit. They grow primarily by searching for new resources: their corporate culture tends to be oriented to production, rather than to finance. The next largest group of companies is "technology-driven." These usually grow and diversify "mainly vertically and downstream, to metals processing and manufacturing ... [but into] other technologically attractive areas as well, driven by the value they place on innovativeness and on new technological learning" (Ala-Harkonen 1993, 6). Despite this forward-looking culture, such firms tend to suffer commercially from overdiversification and lack of financial resources to participate in large-scale mining projects. The smallest group consists of "business-opportunity–driven companies," usually with competence in (and cultural orientation to) finance. Growth comes via acquisition of business opportunities often beyond the minerals sector, with subsequent attention to potential technological or commercial synergies. With the usual caveat about the danger of applying "ideal types," and a note that Ala-Harkonen recognizes that firms may move from one category to another, we may employ this framework to interpret the corporate evolution and geography of some leading Canadian firms.

Cominco

For most of its life, Cominco has closely corresponded to a "minerals-driven" firm. After being a subsidiary of Canadian Pacific for 80 years, in 1986 it became part of a complex group (Figure 1) that is a major force in the world's lead and zinc industry. The geography of the firm's core operations, feeding concentrates to the Trail, BC smelters, has been remarkably simple and stable, reflecting the quality of the deposits that it has developed. Until 1966, ore came almost exclusively from the nearby Sullivan mine, which still has some years' life after a century of operation. Additional supplies were then brought onstream from the Pine Point mine, NWT, which ceased shipments in 1991. Raw material from these sources has increasingly been replaced by output from the Red Dog mine, Alaska, which began commercial production in 1990, with a projected 50-year life. Pine Point and Red Dog highlight another dimension of Cominco's "minerals-driven" profile – expertise in developing high-latitude operations – evident also in the mines that it opened in Greenland (Black Angel, sold in 1986) and Little Cornwallis Island (Polaris), which ship concentrates to Europe.

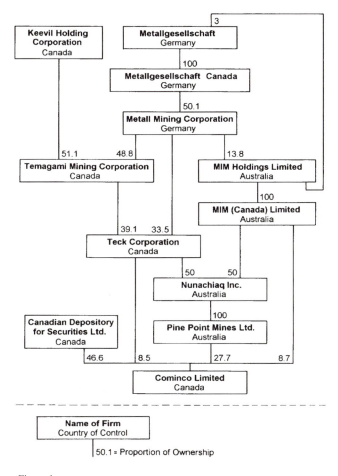

Figure 1
Ownership structure of Cominco Ltd, 1994. *Source*: Statistics Canada (1994).

However, purchase of a controlling interest in Cominco by Vancouver-based Teck Corp. in 1986 demonstrated some of the dangers of a production-driven corporate culture and a narrow market profile. It reflected the comparative financial health of the two firms at the end of the early-1980s recession and the more astute business strategy that had helped Teck grow. Cominco had accumulated a debt of $1.1 billion by 1985, but its future depended on heavy capital expenditures to upgrade the Trail smelters and bring the remote Red Dog mine into production. Teck, a smaller base-metals company in a much stronger financial position, capitalized on its links with Metallgesellschaft (the leading German mining corporation, which had earlier been brought into Teck, holding an equity interest and providing three out of 14 directors) and MIM (a major Australian mining firm, with one representative on the Teck board) to aquire control of Cominco without resort to any new bank loans. An immediate start on developing the Red Dog site was the first benefit to Cominco of the new management.

The union of Cominco and Teck also facilitated significant rationalization of the three major, adjacent copper mines in Highland Valley, south of Kamloops, BC. Developed independently in the early 1980s by Cominco, a Teck-headed group (Highmont), and Lornex (owned 68 per cent by RTZ subsidiary Rio Algom and 22 per cent by Teck), each with its own concentrator, these mines were "stacked" over a vertical distance of 350 m up the valley side. A rationalization agreement between Cominco and Lornex, made just prior to the Cominco-Teck union, allowed closure of Cominco's dated mill and processing of its richer ore body through the more efficient Lornex concentrator. Once Teck had control of Cominco, it could rationalize the trucking of ore to concentrators throughout the vertical range of all three mines, which was accomplished in 1988 by dismantling the Highmont mill, moving it 7.5 km downslope, and re-erecting it adjacent to the ex-Lornex plant. On the basis of its economies of scale and operating efficiencies, the Highland Valley complex now ranks among the world's lowest-cost copper mines.

Beyond limited involvement in refining and fabricating precious metals derived from its lead/zinc ore feed, Cominco's diversification was restricted to creating a market for what otherwise was an environmental liability – its smelter gases. Thanks to Trail's location and topographic setting, these gases became the focus of precedent-setting international transboundary pollution-control legislation in 1929. By capturing them to produce sulphuric acid, as a basis for fertilizer production, Cominco forged a link between non-ferrous smelting and fertilizer manufacture that was subsequently replicated across Canada (see below).

From a phosphate plant constructed at Trail in 1930, Cominco expanded in the 1960s into natural gas–based nitrogenous fertilizer (in Alberta and the United States) and potash mining in Saskatchewan, before selling off its fertilizer division as an independent corporation in 1992. Though its independence from Canadian Pacific and association with Teck have broadened the firm's focus, Cominco retains the profile of a minerals-driven corporation. In renewing its Trail smelter in the late 1980s, it opted for innovative but untested (at a large scale) German technology – a costly mistake. Its growing international involvement in the 1990s (partner in a giant new Chilean copper mine – at 4300 m, one of the highest in the world – and purchase of a Peruvian zinc smelter) deploys its traditional corporate strengths.

Sherritt (Gordon)

Sherritt is the clearest Canadian example of a "technology-driven" minerals firm: in 1990 it had the highest spending on R&D as a percentage of sales in the industry (Science Council 1992b). Indeed, between 1987 and 1994 it ceased to have any direct involvement in mining. The firm originated in 1927 as a copper/zinc mine in northern Manitoba. Its postwar profile took shape with development of a major nickel mine at Lynn Lake in 1953 and the decision to research a more efficient processing technology than conventional methods. Corporate R&D resulted in innovation of a non-smelting, ammonia leach process, which required plentiful supplies of cheap natural gas. A refinery was therefore built at Fort Saskatchewan, Alberta, 1350 km by rail from the mine, together with a fertilizer plant based on the by-product ammonium sulphate. No Canadian capital sources would commit funds to the project, and financing was arranged only through acquisition by Newport Mining Corp. (New York) of a 40 per cent equity interest in Sherritt. Active pursuit of downstream, value-added markets (including a mint) and continued metallurgical research became characteristic of the company.

With the decline of output from Lynn Lake in the 1970s and failure to find a replacement deposit, Sherritt became dependent on limited-term and sometimes farflung (including Australia) ore-supply contracts to maintain the throughput of its nickel refinery. Its remaining mining operations in Manitoba, after generating record profits in 1979, produced continuous significant losses after 1980 and were closed or disposed of by 1987, when Sherritt ceased to be a mining company. Fertilizers and metal refining and fabricating remained as its core businesses, but planned expansion of the refinery was postponed in 1989 because of continuing shortages of feedstock. (Inco failed to renew a 10-year refining contract for Thompson ore, which had been using 60 per cent of the Fort Saskatchewan capacity.) With fertilizer markets weak in the late 1980s, all the more attention focused on the firm's research-intensive operations. Acquisitions included a Richmond, BC, metallurgical testing and consulting firm and a British firm supplying specialized metal products to the European aerospace sector. This emphasis on high technology was given tangible geographical expression in 1989 with the movement of Sherritt's head office from Toronto to Edmonton, in conjunction with a $40-million contribution from Alberta to the cost of expanded research laboratories and establishment of a joint federal-provincial-corporate venture to research advanced industrial materials.

Even a company with this profile, however, illustrates the inertial strength of commitment to core competences and investments in capital-intensive processing technology. The firm's inability to tie down secure sources of feed to keep the Fort Saskatchewan nickel refinery operating at an economic capacity led to a shareholder revolt in 1990 and installation of a more adventurous management. Its strategy, despite the political and commercial risks, has been steadily to increase Sherritt's business in Cuba, home to one-third of known nickel deposits. Increasing purchases of Cuban concentrates (which involved adapting the refinery to treat laterite ores in addition to the sulphides that it was designed to handle), together with flows from new nickel mines coming into production in Canada not under the control of competitors, enabled Sherritt to restore and stabilize refinery revenues. With the demise of managed trade with eastern Europe, Cuba has been keen to find new nickel markets, and in 1994 its government concluded a joint venture with Sherritt, giving the firm a direct stake in the Moa Bay nickel/cobalt mine.

Noranda

Noranda is one of the large transnational firms identified in Ala-Harkonen's (1993, 15) study as "business-opportunity–driven." Management styles in this group are "entrepreneurial ... and also very finance-driven and cash-driven." Noranda Mines started life in 1927 as a copper producer in northwestern Quebec and operator of the Horne smelter. It pursued a policy of local mine acquisitions from the start, and in the 1950s and 1960s it expanded by developing or buying into other copper mines in its home region, Gaspé, British Columbia, and New Brunswick (which added zinc). It continued to diversify its mineral interests (gold in Ontario and Quebec, potash in Saskatchewan, and molybdenum in British Columbia), and from the late 1960s expanded into aluminum smelting and fabrication (in the United States), forest products (in British Columbia), and oil and gas (in Alberta), by which time its corporate structure was undoubtedly the most complex in the Canadian minerals industry (Mars 1976). Following the appointment of Alfred Powis, a former financial analyst, as president, in 1968, further "aggressive tactics, including a chain of takeovers, were key contributors to the company's success" (Hast 1991, 165). Despite diversification, however, the recession of the early 1980s left the firm financially weakened. This provided an opportunity for the Bronfman family's financial empire to acquire control in 1981. As the natural resource arm of this conglomerate, Noranda became the vehicle for purchase of control of the large forest products firm MacMillan Bloedel in 1983, by which time only 12 per cent of corporate sales were derived from copper. In 1988, jointly with the Swedish minerals firm Trelleborg, Noranda bought Falconbridge, a leading nickel producer, whose rich Kidd Creek (Timmins) zinc/copper mine had long been a desired acquisition to secure feed for the Horne smelter. To pay down debt in the early 1990s, Noranda sold its interest in MacMillan Bloedel and some of its interest (while consolidating control) in Falconbridge. Despite its corporate vision of the late 1980s, "to be the world's premier diversified natural resource company" (*Annual Report*, 1988, 2), one cannot claim that Noranda really realized the synergies and counter-cyclical stability that are among the objectives of "business-opportunity–driven" firms.

MARKET FORCES SHAPING OTHER
MINERALS SUBSECTORS

Canadian base-metals producers supply global markets in which there are many active players. Contract sales predominate, but "world" prices are established by spot trading on the London Metals Exchange. Only in nickel did Canada once have the market dominance to allow Inco to set a producer price, but that power eroded in the 1970s. The development of Canada's other mineral resources has reflected the particular market structure of each subsector.

Uranium, of which Canada is the leading producer, has been the most subject to non-commercial influences. Strategic military considerations led to public control of mining and processing from the 1940s and development of mines (in Northwest Territories and Ontario) that by the 1970s had significantly higher costs than a new generation of much richer mines coming onstream in northern Saskatchewan. The latter, developing deposits discovered under the impetus of growing civilian demand for nuclear power in the industrialized nations, and with significant participation by European and Japanese mining interests, also sell into a sector whose economics are strongly shaped by public policy.

Ontario Hydro's fuel procurement illustrates well the tensions that have resulted. In 1977, when spot prices for uranium were at their "post-OPEC" peak, it entered into a 40-year contract guaranteeing the original Elliott Lake mines a base price of "cost plus $5/lb" (McKay 1983, 75). Almost immediately, world prices began to drop and the Saskatchewan mines started production, but it took the utility until 1990 to renegotiate the Elliott Lake contracts to allow for complete closure of the Ontario mines by 1996. The pace of mine development in Saskatchewan, meanwhile, has been slowed by the uncertain future of global nuclear power generation.

The historical geography of the Canadian coal industry reflects the fact that reserves are peripheral to the industrial heartland of southern Ontario and Quebec, where more accessible U.S. Appalachian coal dominates the market, and extensive BC and Alberta deposits found limited markets before the 1960s. (Nova Scotia's coal production has been maintained by a long history of government subsidies.) Mine-mouth generation has become the basis of Alberta's growing electricity supply in recent decades, providing a steady market for a number of single-product coal firms. But it was increasing Japanese demand for metallurgical (and later thermal) coal that allowed Canada to become a major coal exporter after 1970 and stimulated construction of large open-pit mines in the Kootenays and subsequently in northeastern British Columbia.

Even though these mines were developed by companies with considerable financial and technical resources, the structure of the market into which they sell is comprehensively shaped by the buyers. The major Japanese steel producers have fostered development of coal deposits around the Pacific Rim (notably in Australia, Canada, and more recently Indonesia) by a variety of separate mining firms with which they enter into long-term delivery contracts. They collectively negotiate annual prices with these fragmented and competing suppliers. They have thus been able to exert continuous downward pressure on prices and effectively to impose reductions in contracted shipments as necessary to suit their own commercial interests (Anderson 1987). The decade ending in 1993, spanning the opening of the northeastern BC coalfield and the radical restructuring of the Kootenay operations, illustrates these pressures.

Japanese interest in seeing a second major coalfield open in western Canada was significantly shaped by concern over the captivity of Kootenay mines to their transportation infrastructure (CP Rail and Roberts Bank export terminal). The attraction of developing the northeastern BC reserves was that their shipments would underwrite investment in a second major export terminal at Prince Rupert. Packaged as a "developmental" megaproject (despite concern in some quarters that it would create overcapacity to the detriment of existing BC coal producers), and substantially underwritten by the provincial and federal governments, the coalfield came into production in 1983, initially protected by premium prices to recover infrastructure and development costs. However, "world" coal prices were already declining significantly, soon leading to Japanese pressure for concessions, while one of the two mines (Quintette) ran into such severe difficulties that as early as 1986 its half-owner and operator (Denison Mines) wrote off its investment and attempted to refinance the $650-million project. By 1990, Quintette was technically bankrupt, and it was eventually restructured, but within two years the Kootenay industry had almost collapsed. Westar, the largest producer, was bankrupt (as much because of business failures in other sectors as because of coal prices), and the other major producer was into an eight-month strike triggered by demands for wage concessions. Westar's mines resumed production under new corporate ownership in 1993, but with reduced tonnage and employment halved.

Market forces shaping the Canadian iron and steel industry are more continental than global, though investment from outside North America has increased since conclusion of the Canada-United States Free Trade Agreement (FTA). It was primarily U.S. steel industry demand that brought the Quebec/Labrador ore deposits into production in the 1950s, and despite contraction in output since the late 1970s, ownership of the mines still features vertical integration into American steel companies, but with greater involvement of Canadian and off-shore steel producers. Canada's major steel firms have traditionally expanded production capacity conservatively, aiming to maintain profitability in the domestic market through business cycles. But growing market opportunities in the United States from the mid-1970s (reflecting net disinvestment by major U.S. producers) made the industry more sensitive to export sales, which peaked at 30 per cent of Canadian shipments in 1987 (Science Council 1992a).

Crucially, the prospect of a free trade agreement promised relief from the involuntary export restraints that Canadians had started to practise so as to keep within manageable limits the strong protectionist pressure generated in Washington by the U.S. industry. Experience in the early 1990s did not fulfil these expectations, however. A stronger Canadian dollar at the beginning of the decade, a new round of investment in the U.S. steel industry, and, despite the FTA, imposition of anti-dumping duties on certain products compounded the effects of the domestic recession for Canadian firms. Both Stelco and Dofasco were in a financially critical state by 1991, the latter in particular. Having taken over Algoma Steel (Sault Ste Marie) in 1988 to boost its raw steel capacity, Dofasco was forced to write off its problem-plagued investment, leaving the heavily indebted subsidiary to be restructured in a provincially supported buyout by employers.

North American integrated steel producers (including Stelco and Dofasco) have been facing, on the one hand, much stronger competition from mini-mills and, on the other, stronger pressure from customers to invest to meet higher product-quality standards. The former has been most evident in the United States, where innovative (and non-union) firms such as Nucor have captured market share in an increasingly wide range of product lines. The customer pressure has come primarily with arrival of Japanese-owned auto plants, demanding inputs that match the specifications of their home-country supplies. Both capital and technology have been transferred into North American steel firms through a number of joint ventures involving Japanese steel industry partners. So, for instance, in 1989, Dofasco borrowed $350 million from Mitsui to build a cold-roll mill and collaborated with Nippon Kokan and its half-owned subsidiary National Steel (US) to establish a jointly owned galvanizing plant at Windsor. A similar plant in Hamilton was built as a joint venture between Stelco and Mitsubishi Canada. Through such projects, and the demand for higher quality through the auto-components sector and beyond, Canadian steel producers have been forced (and helped) to keep pace with global "best practice."

TECHNOLOGICAL CHANGE AND EMPLOYMENT RESTRUCTURING

Between 1981 and 1991, employment shrank in Canadian mining by 35 per cent and in smelting and refining by 29 per cent. The number of coal miners was the same in both years, but employment in nickel/copper/zinc mines dropped 44 per cent and in iron mines 54 per cent (Canada 1994). These declines, which vastly exceed reductions in output (and in some subsectors mask increased production), indicate the speed and extent of technological change in what is usually regarded as a mature, "sunset" industry, few of whose firms are noted for innovation. In fact, much stronger price competition from a widening array of competitors, and challenges associated with the maturity of the Canadian industry (such as increasing depth of mines), have forced Canadian minerals firms to reassess their use of, by world standards, expensive human resources. Tighter controls on environmental effects, particularly in smelting, have given further impetus to technological change. As hinted above, innovation has focused on extraction and processing technologies more than on product development, and workplace restructuring has not always been accomplished without strife.

Some of the most dramatic changes have taken place within the nickel/copper industry of the Sudbury Basin, home of Inco and Falconbridge and the centre of Canadian hardrock underground mining. Local employment in the subsector was cut from its all-time peak of 25,700 in 1971 to 10,500 in 1991 (Saarinen 1992). This period saw the transition from traditional mining techniques, based on manual working in horizontal drifts and stopes (excavated chambers), to highly automated bulk mining, which leaves less ore stranded as pillars to protect access to the workings (and is considerably safer). Inco's initial trials of this technology recorded "a 300 per cent improvement in productivity (in terms of tonnes per person per shift at a reduced capital cost)" (Science Council 1992b, 21), and it has become standard practice in the Sudbury region. This transition has involved a move away from pneumatic to electrical and microelectronic-based equipment, with applications of sensor technology for machine monitoring, control, and guidance, often requiring considerable product adaptation to the operating environment of deep mines (Science Council 1992b, 22).

Processing productivity increases usually come from incremental improvement of installed technologies, but the threat to Inco of curtailed production capacity if the Ontario government's progressively more stringent sulphur dioxide – abatement regulations could not be met prompted a major R&D effort. This culminated in 1991 in the commissioning of an oxygen flash-smelter, which greatly improved recovery of the smelter gases and saves a lot of energy in treating nickel (Scales 1988; Inco *Annual Report* 1991).

Productivity-enhancing innovation is usually less widely replicable in the minerals sector than in manufacturing, being geology and/or commodity specific. While miners of massive sulphide deposits were moving to bulk methods in the 1980s, the fastest-growing minerals subsector on the Shield was gold mines, most of them of limited size and capitalization and laggards in technological innovation. For large open-pit mines, characteristic of BC coal and copper producers, computer applications leading to more efficient handling of materials have been the principal source of productivity gain (as at Highland Valley, noted above).

Whatever its immediate form, reduction of employment in the minerals sector has been accompanied by shifts in the labour-force profile and by management's search for multi-skilling and more flexible labour deployment (Chaykowski 1992). The amount of labour conflict resulting has varied by subsector and firm. Following a prolonged and high-profile strike in 1978–79 by the United Steelworkers of America, the culmination of years of conflict at Inco (Clement 1981), the firm made a serious commitment to improved labour relations, and it has restructured its workforce without any major disputes. In 1993, the miners at Falconbridge, last remaining local of the fractious Mine, Mill and Smelterworkers Union, voted to join the Canadian Auto Workers Union (CAW). Similar mergers in the early 1990s have given the well-funded CAW, with a reputation for its opposition to wage rollbacks, a presence in the mining industry. Nevertheless, a 16-month dispute over proposed reductions in pay and seniority rights at the Westmin copper mine, Vancouver Island, involving some of the highest-paid miners in Canada, ended in 1994 with the union accepting concessions in compliance with an arbitration ruling (*Globe and Mail*, 12 May 1994, 17 Aug. 1994). Restructuring of the Westar coal mines (see above) was made more contentious by its timing, just ahead of BC labour legislation that would have protected prior bargaining rights (*Globe and Mail* 31 Dec. 1992).

PROSPECTS

"Base metal reserves have been declining since the early 1980s. Many mines are scheduled to close [in the 1990s] ... [m]ineral exploration in Canada has declined ... [M]any Canadian companies [are concentrating] on international mineral development opportunities ... Canada's mineral sector could actually shrink" (IWGMI 1992a). These unsettling trends, identified by a federal/provincial/industry task force, signal a new era for the mineral industry in Canada, but they are not the whole story. Certainly, Canada's relative attraction as a mineral-producing nation has declined since the 1970s. Higher costs of access to new resources, erosion of processing cost advantages (notably electricity rates in Ontario), increasing imposition of provincial government fees (such as BC water charges), increasingly comprehensive regulation (as with sulphur dioxide emissions and acid mine drainage) or development freezes (notably, the Windy Craggy and Kemano Completion projects in British Columbia) on environmental and/or Native land claim grounds, together with a worsening fiscal climate, have made the economics of minerals production in Canada significantly less attractive than they were during the postwar boom. Conversely, mineral-rich nations in Latin America (in particular) and other parts of the world have become much more inviting locations for investment by Canadian firms as a result of greater political stability, better economic prospects (often linked to IMF structural adjustment programs, as in Ghana), and a more business-oriented administrative regime (which in most cases where Canadian firms are active is not merely a euphemism for pollution havens and repressive labour laws).

During this time of a changing global geography of mineral development opportunities, Canadian firms of all sizes and from many elements of the sector's input-output matrix are deploying their capital and expertise internationally on an increasing scale. Examples include the evolution of Yorkton Securites from a penny stock promoter on the Toronto Stock Exchange to an international mineral finance syndicator; the growth of Finning Ltd. from a supplier of heavy equipment to BC resource industries to being one of the biggest such firms globally, and dominant in Chile; and increased exports of mineral-related design and project management services by firms such as SNC-Lavalin and of geophysical and related expertise by many medium-sized Canadian firms. All this is in addition to (and often stimulated by) substantial foreign investments in mineral extraction by almost every Canadian mining firm of any consequence (see also MacPherson, this volume, chap. 4).

This all suggests that Canadian minerals-related firms are responding quite effectively to the locationally differential opportunities of contemporary globalization. Corporations built on minerals extraction in Canada, or nurtured on linkages to this domestic staple, are now diversifying their operations geographically, and often functionally, in arenas that are more profitable. But the earnings and home-country employment flowing from this activity tend to benefit primarily Canadian metropolitan regions.

Mining and mineral processing in Canada remains concentrated in the hinterland, and its vital signs are less healthy. Sudbury has so far weathered the decline of its traditional economy fairly well (Saarinen 1992), and Elliott Lake is marketing its reasonably priced housing and good communal infrastructure as a haven for the growing cohort of active seniors. Restructuring of the mineral industry will not be as manageable in many other parts of the country, however. The 1980s saw abandonment of mining towns such as Schefferville (Bradbury and St-Martin 1983) and Uranium City, and the founding (at Lynn Lake, Manitoba) of the Canadian Association of Threatened Single Industry Towns. As in other spheres, capital and labour have unequal mobility in the face of change. Drucker's (1986, 768) observation, that "the primary-products economy has become 'uncoupled' from the industrial economy," is not an obituary for the Canadian minerals industry, but it is a truth that catches its reduced status.

REFERENCES

Ala-Harkonen, M. 1993. "Corporate Growth and Diversification Paths within the Minerals Industry." *CRS Perspectives* (Kingston, Ont.), 44, 2–17.

Anderson, D.L. 1987. *An Analysis of Japanese Coking Coal Procurement Policies: The Canadian and Australian Experience*. Kingston, Ont.: Centre for Resource Studies.

Bradbury, J.H., and St.-Martin, I. 1983. "Winding down in a Quebec Mining Town: A Case Study of Schefferville." *Canadian Geographer*, 27, 128–44.

Canada. 1994. *Canadian Minerals Yearbook, 1994*. Ottawa: Natural Resources Canada.

Chaykowski, R.P. 1992. "The Challenge to Industrial Relations in the Mining Industry: Developing a Competitive Strategy." *CRS Perspectives* (Kingston, Ont.), 39, 2–14.

Clement, W. 1981. *Hardrock Mining: Industrial Relations and Technological Changes at Inco*. Toronto: McClelland and Stewart.

Drucker, P.F. 1986. "The Changed World Economy." *Foreign Affairs*, 20, 768–91.

Hast, A., ed. 1991. *International Directory of Company Histories*, Vol. IV. S.v. "Noranda." Chicago: St James Press.

Intergovernmental Working Group on the Minerals Industry (IWGMI). 1992a. *The Canadian Mineral Industry in a Competitive World*. Ottawa: IWGMI.

– 1992b. *The Importance of the Minerals and Metals Industry to Canada: Background Study No. 1*. Ottawa: IWGMI.

McKay, P. 1983. *Electric Empire: The Inside Story of Ontario Hydro*. Toronto: Between the Lines.

Mars, P.J. 1976. *Noranda Mines Limited: A Corporate Background Report*. Royal Commission on Corporate Concentration, Study No. 9. Ottawa.

Norcliffe, G. 1994. "Regional Labour Market Adjustments in a Period of Structural Transformation: An Assessment of the Canadian Case." *Canadian Geographer*, 43, 2–17.

O hUallachain, B., and Matthews, R.A. 1994. "Economic Restructuring in Primary Industries: Transaction Costs and Corporate Vertical Integration in the Arizona Copper Industry, 1980–1991." *Annals of the Association of American Geographers*, 84, 399–417.

Saarinen, O. 1992. "Creating a Sustainable Community: The Sudbury Case Study." In M. Bray and A. Thomson, eds., *At the End of the Shift: Mines and Single-Industry Towns in Northern Ontario*, 165–86. Toronto: Dundurn Press.

Scales, M. 1988. "Sudbury Smelting: Major Overhaul Ahead." *Canadian Mining Journal*, June, 49–55.

Science Council of Canada. 1992a. *The Canadian Iron and Steel Sector*. Sectoral Technology Strategy Series, No. 3. Ottawa: Science Council of Canada.

– 1992b. *The Canadian Nonferrous-Metals Sector*. Sectoral Technology Strategy Series, No. 8. Ottawa: Science Council of Canada.

Statistics Canada. 1994. *Inter-corporate Ownership*. Cat. No. 61–517. Ottawa.

CHAPTER EIGHT

Redrawing the Canadian Energy Map

VIRGINIA W. MACLAREN

Canada is the world's most energy-intensive industrialized nation. If we define energy intensity as total primary energy requirements[1] divided by GDP, then Canada in 1970 ranked third among OECD countries, after Luxembourg and Belgium (McLachlan and Itani 1991). By 1989, its energy intensity had decreased because of structural changes and improvements in energy efficiency. Its ranking changed to sixth, though it still ranked ahead of major economic powers, such as the United States (13th), the United Kingdom (16th), Germany (17th), France (18th), Italy (21st), and Japan (23rd), and its energy intensity was about 30 per cent higher than that of the United States and about 175 per cent greater than Japan's.

Comparisons of energy intensity are difficult because of differences in efficiency of energy use, but there are other complicating factors as well (McLachlan and Itani 1991; National Energy Board 1991). For example, Canada fares poorly because of its cold climate, the long distances between its centres of population, an industrial structure that has a high proportion of energy-intensive industries, and historically low energy prices, which have slowed efficiency improvements in all sectors.

Since Canada uses more energy per unit of output than any other major industrialized country, it should be more sensitive to interruptions of energy supply and increases in energy prices. However, it has had a strong energy industry, abundant resources, and interventionist government pricing policies which have enabled it to withstand external energy shocks. Within the last several years, however, these conditions have changed considerably. Declining supplies of conventional domestic energy resources in western Canada and the advent of the Canada–United States Free Trade Agreement (FTA) and the North American Free Trade Agreement (NAFTA) may affect the resilience of the Canadian sector. In addition, the sector is facing new pressures arising from concerns about the environmental implications of energy use.

This chapter examines trends in Canadian energy supply and demand over the last 20 years and relates them to the historical development of energy policy. It also explores the impact of some of the major forces shaping energy policy in Canada.

A RECENT HISTORY

The 1970s and 1980s saw three international events that dramatically affected Canadian energy policy. The first was the Organization of Petroleum Exporting Countries (OPEC) oil embargo of 1973, which initiated a rapid increase in the world price of oil. The second was the Iranian revolution in 1979, which pushed world oil prices up even higher. The third was the collapse of world oil prices and the OPEC cartel in the early 1980s.

Canadian energy policy since the 1973 crisis can be characterized by three policy regimes (Chapman 1989). During the first phase, 1973–78, the domestic price for oil and natural gas more than doubled, gasoline prices rose by 67 per cent and electricity prices increased by 91 per cent. Though the domestic price of energy increased substantially, government intervention moderated the impact on the Canadian economy of a 500 per cent increase in international oil prices. By 1978, domestic oil prices were about $10 lower than the import price of $25.70 per barrel.

In 1973 the government restricted oil exports and imposed an oil export tax that was used to fund an oil import compensation program. This program reduced the cost of crude oil purchases for refineries east of the Ottawa Valley, since those facilities relied on international crude oil as feedstock, and made the cost equivalent to that paid by refineries using domestic oil.[2] In 1975, Petro-Canada was established, and the Petroleum Administration Act gave the federal government control over oil and gas pricing. By 1976, an extension of the Interprovincial Pipeline to Montreal allowed Quebec to reduce its dependence on foreign oil. Ottawa also launched a number of energy conservation initiatives, relying heavily on education and monitoring rather than on subsidies or grants.

Except for gasoline prices, which rose by about 120 per cent, the period 1979–83 saw increases in domestic energy prices similar to those between 1973 and 1978. The federal government's response to a doubling in the international price of oil in 1979 and 1980 was the National Energy Program (NEP) of 1980. The NEP's three objectives were to decrease Canadian dependence on the world oil market, increase domestic ownership in the oil and gas industry, and increase the "fairness" of pricing and revenue sharing for oil (Energy, Mines and Resources 1980). One of the policy instruments used was a plan for a blended oil price, which would allow domestic oil prices to rise to no more than 85 per cent of world prices and set natural gas prices at 67 per cent of domestic oil prices. The NEP also included conservation and renewable-energy programs, off-oil programs that encouraged conversion of space heating from oil to other sources, reductions in the cost of investment for Canadian-owned and -controlled companies, and incentives to increase oil exploration activity in the Canada Lands.[3]

Debate over the pricing policies and exploration incentives of the NEP created considerable acrimony between the oil-producing provinces and Ottawa. Alberta in particular was angered by the prospect of new price controls and claimed that previous controls had resulted in the provision of a $17-billion subsidy by Alberta to Canadian

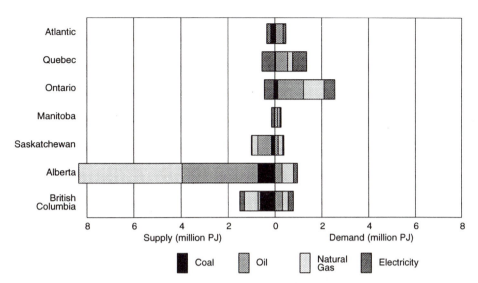

Figure 1

Energy production and demand, Canada, 1993. PJ = petajoules = joules x 10^{15}. *Source*: Statistics Canada (1994).

energy consumers since 1973 (Doern and Toner 1985). Memories of this debate resurfaced during the Canadian-U.S. free trade negotiations several years later and influenced both the FTA's energy provisions and the support given to those provisions by the oil-producing provinces.

The second oil crisis triggered an international economic recession that reached its most severe levels in 1982. The combined effects of the recession, conservation efforts, and high energy prices led to stabilization and eventual decline in the international price of oil (Chapman 1989). With the decline in oil prices, the pricing provisions of the NEP became obsolete, and the Conservative government deregulated oil and natural gas prices in 1985. Since that time, international oil prices have experienced only minor fluctuations.

MATCHING SUPPLY AND DEMAND

Canada's major energy-producing and energy-consuming regions are quite distant from each other (Figure 1). The country has therefore had to develop an extensive energy distribution network for its most important energy resources – oil and natural gas. The existing network serves all of Canada except the Atlantic provinces and parts of eastern Quebec. It also extends southward into the United States.

Distribution networks for Canada's other main energy resources are much more limited. The only significant interprovincial trade in hydroelectricity is that between Newfoundland and Quebec. Since coal is expensive to transport by land, western

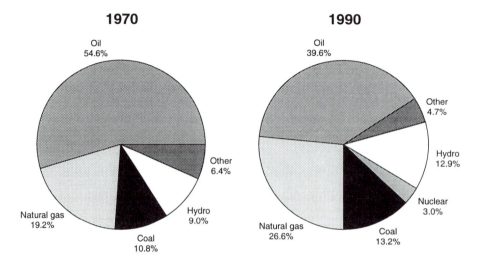

Figure 2
Domestic demand for primary energy, Canada, 1970 and 1990. *Source*: Statistics Canada (1992).

Canada exports most of its coal by sea to Japan and South Korea. It supplies only 25 per cent of the demand for coal in Ontario (National Energy Board 1991); the bulk of the remainder comes from coal mines in Appalachia. Similarly, Quebec imports 85 per cent of its coal from the United States and receives the rest from Nova Scotia.

Changing Patterns of Demand

Though end-use demand[4] for energy has increased in Canada by about one-third over the last 20 years, the second oil crisis and its repercussions have helped slow that growth. The average rate of increase in the demand for energy declined from 3.8 per cent between 1970 and 1980 to 1.1 per cent between 1980 and 1990 and was actually negative in 1980, 1981, and 1982. The structure of energy demand, by fuel type, has changed dramatically. Demand for primary energy, by fuel type, between 1970 and 1990 shifted considerably (Figure 2). Oil's share of the energy pie has declined, while all other energy sources, most notably natural gas, have increased their market shares. Two-thirds of the oil-share decline occurred after 1980, reflecting the impact of the off-oil provisions of NEP. Between 1980 and 1990, for example, the proportion of households using oil to heat their homes declined from 37.3 per cent to 17.6 per cent (Statistics Canada 1991). Two other notable shifts in the past 20 years have been the rise of nuclear power, with 86 per cent of nuclear production based in Ontario (Energy, Mines and Resources 1991), and the switch to a limited amount of renewable energy.

In contrast, there has been little change in the relative shares of energy end-use demand held by the major sectors. Industrial energy use has maintained the biggest proportion, at 34–39 per cent, followed by transportation at 20–25 per cent, residential at 27–30 per cent, and commerce at 12–14 per cent.

Changes in Energy Intensity

Energy intensity can alter with changes in the efficiency with which energy is used or in the structure of economic activity. Monitoring and understanding these changes are useful because of the competitive advantage that can accrue to countries with low energy intensities. Flavin and Durning (1988) posit an opportunity cost associated with energy consumption in energy-intensive economies. Since nations with low energy intensities tend to have a lower proportion of GDP allocated to energy consumption, investment that would otherwise be necessary for energy purchases can be used in other areas and result in lower costs of production. In addition, energy-intensive economies are liable to produce more energy-related pollution and incur higher environmental costs.

As noted above, Canada's energy intensity is relatively high. Though the decrease since 1973 has not matched that of the other leading industrialized nations, the kinds of changes that have occurred may help us to find opportunities for additional improvements.

An analysis by Marbek Resource Consultants (1989) of changes over the period 1973–87 found a decline in energy intensity of over 30 per cent. The greatest portion of this decline was the result of improvements in energy efficiency, but structural changes, largely attributable to the growth of energy-extensive service industries, also helped.

In the residential, commercial, and transportation sectors increased energy efficiency was the key factor. In the residential and commercial sectors, it resulted from fuel switching, improved thermal efficiency in new housing, and behavioural changes. In transportation, fuel economy increased for both private and commercial vehicles. The majority of the decline in the industrial sector can be attributed to structural changes, which are discussed in the next section.

Industry

Energy consumption in the industrial sector in Canada has long been dominated by a few energy-intensive[5] industries whose share of industrial output is considerably smaller than their share of energy consumption. In 1990, energy consumption in six industries accounted for 67 per cent of manufacturing energy demand while producing only 23 per cent of total manufacturing GDP (Statistics Canada 1991; 1992). The highly energy-intensive industries (Table 1) are led by pulp and paper, with 21.8 per cent of consumption in 1990, followed by iron and steel, chemicals (excluding petrochemical feedstocks), and smelting and refining. Alberta, British Columbia, and the Atlantic provinces have the highest intensities, largely because of the relative dominance there of pulp and paper and of mining and smelting.

Table 1

Intensity of manufacturing energy use and share of manufacturing GDP and energy demand, 1970–90

Manufacturing sector	Intensity (megajoules per 1981 dollar of industrial output)	Share (%) of Manufacturing GDP, (1986$)			Share (%) of manufacturing demand for primary and secondary energy, 1990
		1970	1980	1990	
Pulp and paper	115	10.1	7.9	7.0	21.8
Iron and steel	91	5.8	4.1	3.0	21.7
Smelting and refining	62	2.4	1.7	2.6	10.6
Cement	35	0.8	0.7	0.4	3.4
Petroleum refining	107	2.2	2.2	2.1	5.0
Chemicals	46	6.3	6.7	7.6	13.8
Other manufacturing	10	72.4	76.7	77.3	32.6

Sources: Statistics Canada (1991; 1992).

The most energy-intensive industries in Canada all produce "semi-finished," or intermediate goods. This stage of the production chain tends to be capital intensive, requiring large energy inputs and generating relatively low added value. In contrast, later stages, such as fabrication and assembly, use less energy and add much more value while earlier stages need moderate energy inputs and add the least value (Hamilton 1988).

Though energy-intensive industries still retain a dominant share of manufacturing demand for primary and secondary energy, their share of manufacturing GDP has been in decline since 1970. Most of this decline is attributable to the pulp and paper industry, whose share has decreased from 10 per cent to 7 per cent, and iron and steel, which has fallen from just under 6 per cent to 3 per cent. These structural changes contributed to a 16 per cent decline in overall manufacturing energy intensity between 1973 and 1987. Marbek Resource Consultants (1989) estimate that over half of the decline was the result of structural change arising from an increase in the share of manufacturing GDP held by less energy-intensive industries.

A subsequent study (National Energy Board 1991) identified two distinct periods of trends in energy intensity. Between 1973 and 1980, intensity increased at a fairly steady rate, but it declined equally steadily thereafter. Most of the improvements between 1980 and 1989 resulted from adoption of new production processes and were motivated more by concerns for competitiveness and increased productivity than by a need to reduce energy costs. Goodman (1986) notes that some of the decline can also be attributed to a general trend toward more knowledge-intensive production of value added. In pulp and paper, for example, the most rapidly growing sectors are not in the basic commodities of wood pulp and newsprint, but in the knowledge-intensive, high value-added sectors of fine specialty papers, printing-writing papers, tissues, and sanitary products.

Marbek Resource Consultants (1989) also calculated changes for mining and manufacturing, which together account for about 97 per cent of industrial output in

Canada. Improvements in energy intensity between 1973 and 1987 were not as significant in these activities, because mining has a relatively low energy intensity and its share of industrial output fell during that period. This structural change outweighed improvements in mining, resulting in a combined 5 per cent decrease, as opposed to the 16 per cent decrease in manufacturing alone.

Industrial improvements in energy efficiency can be divided into four general types: housekeeping improvements, retrofits, process changes, and product changes. Housekeeping improvements, which involve no capital cost and include such actions as reduced heating temperatures or improved maintenance of heating equipment, are usually the most popular because of their low cost. Between 1973 and 1985, about 40 per cent of improvements are estimated to have involved housekeeping (Energy, Mines and Resources Canada 1987). Retrofits accounted for 30 per cent of the remaining improvements, process changes 18 per cent, and product changes 12 per cent. The fact that improvements in industry have been dominated by the lower-cost alternatives suggests considerable potential for additional improvements at higher levels of investment.

Energy Prices

One of the most important factors affecting the energy intensiveness of Canadian industry has been historically low energy prices. Canadian industry enjoys the lowest average prices among all OECD countries for electricity and natural gas, and the second lowest prices, after the United States, for heavy and light fuel oil (OECD/IEA 1990). The average price of electricity per kilowatt-hour for industry is considerably lower in Canada than it is in the United States. In Toronto and Montreal, for example, it is about 60 per cent lower than it is in New York City, and between 15 and 20 per cent lower than in Chicago (Energy, Mines and Resources 1991).

Though low energy prices make Canadian industry more competitive, they encourage wastefulness rather than efficiency (Science Council of Canada 1992a). Even when efficiency measures are implemented, low energy prices limit industry's likely investment in such measures.

Cheap electricity has been a particularly controversial issue in Quebec, which has used it to attract energy-intensive industry. In particular, Quebec offers several primary metals fabricators a preferential electricity rate structure tied to spot prices in the metals markets (Science Council of Canada 1992a). As metal prices decrease, so do the fabricators' electricity costs. The U.S. government has challenged this rate structure as an unfair subsidy, and it is questionable whether Quebec will be able to offer such an advantage in future.

ENERGY TRADE AND SUPPLY SHIFTS

Canadian energy trade has changed during the last decade, especially in oil imports and natural gas exports. Oil imports decreased in volume during the mid-1980s because of the impact of the National Energy Program (NEP) and the economic slow-down resulting from the 1982 recession. Oil imports increased again after 1984 in concert with the dismantling of the NEP, lower world oil prices, and a healthier economy.

Natural gas exports were lower during the first half of the 1980s than in the 1970s because of the high export prices imposed by Canadian energy policy and the devel-opment of a U.S. surplus in natural gas. In the late 1970s and early 1980s rulings by Canada's National Energy Board restricted exports because Canadian reserves fell below the required 15-year reserve level. This surplus requirement has since been removed. In the late 1980s, market controls imposed by the U.S. Federal Energy Reg-ulatory Commission hampered the ability of Canadian exporters to compete in the United States (National Energy Board 1988; Verleger 1988). In 1987, for example, the commission ruled that Canadian natural gas producers could not pass on to their American customers the costs of transport through Canadian pipelines.

Despite U.S. regulatory constraints, Canadian exports since deregulation in the mid-1980s have been strong. Exports to the United States have increased by 80 per cent since 1986. The National Energy Board foresees a continued upward trend in gas exports, though at a slower rate of increase. The bulk of this increase is expected to come from exports to the growing northeastern U.S. market. This market still relies heavily on imported oil, and, even with current low international prices for oil, natural gas is a competitive alternative. In addition, electric utilities and industry have to comply with a revised and tougher U.S. Clean Air Act.

The future health of the Canadian oil and gas industry will depend to a large extent on its ability to exploit existing reserves and identify a new resource base. Currently, 18 per cent of Canada's remaining established reserves of conventional light crude oil lie on Canada Lands, while the bulk of the rest can be found in the Western Canada Sedimentary Basin (National Energy Board 1991). Under anticipated economic conditions, established reserves are economically recoverable with use of known technology. Only those at Norman Wells, in Northwest Territories, are currently in production. As the price of oil rises in future, we can expect to see a production shift from Alberta and Saskatchewan to Canada Lands, which have a substantial portion (63 per cent) of Canada's other discovered resources (National Energy Board 1991). Other recoverable resources are not yet economically viable. The National Energy Board projects that 27 per cent of Canada's light crude oil will come from frontier production at Hibernia and the Mackenzie Delta–Beaufort Sea region by 2010. By that time, productive capacity for light crude will have declined from 70 per cent of total crude oil and equivalent production[6] to around 60 per cent. Gains in productive capacity[7] will occur for heavy oil[8] and will benefit both Alberta and Saskatchewan, which have all of Canada's established reserves.

The relative distribution of established reserves and other discovered resources of natural gas is similar. The shift to frontier production will not happen as quickly, how-ever, because, in terms of energy equivalence, the western Canadian resource base is greater. Just over 15 per cent of productive capacity is expected to come from the Mackenzie Delta–Beaufort Sea region and the east coast off-shore fields by 2010 (National Energy Board 1991).

Exploration in Canada Lands has gone through two phases. The early 1970s saw an initial spurt of activity in northern Canada Lands as a result of promising oil discoveries on the neighbouring North Slope of Alaska. Exploration declined through the mid-1970s in response to disappointing results. However, in the latter third of the decade, concern over security of energy supply caused the federal government to introduce new exploration incentives aimed particularly at Canada Lands. Exploration there jumped significantly, though at a rather steep price. Plourde (1989) notes that, between 1977 and 1986, an average well drilled in northern Canada cost the same as 86 wells in western Canada, while the average well drilled off the east coast cost the equivalent of 90. Because of federal grants and generous depletion allowances, corporate contributions to exploration expenditures in Canada Lands during this time were only about 19 per cent of actual costs.

Fiscal incentives and oil-exploration expenditures in Canada Lands have declined noticeably since then as the world price of oil has dropped and as concern over security of energy supply has faded. Regional development has emerged as a new motivating factor for subsidizing energy exploration and development in Canada Lands (Plourde 1989) – for example, the $2.7 billion in federal money committed to development of the oil-fields at Hibernia off Newfoundland's coast intended to help, revitalize Newfoundland's depressed economy.

The fortunes of Canada's oil-refining industry during the past two decades have followed those of the upstream industry fairly closely. Oil refinery capacity rose through the 1970s but fell off in the 1980s as demand for refined petroleum products declined. While two refineries were opened between 1974 and 1985, twelve closed and others were reduced in size due to declining demand and rationalization (Southam Energy Group 1991). Refineries have also had to adapt to the demand for altered product specifications arising from environmental goals that seek to reduce the atmospheric pollution from automotive emissions.

Refinery capacity is fairly evenly divided among western Canada, Ontario, and the Atlantic provinces/Quebec, each with one-third (National Energy Board 1991). Though demand has stabilized, the remaining 28 refineries are experiencing new supply and demand pressures to modify production. The Science Council of Canada (1992b) predicts that declining oil reserves will require oil refineries to deal with increasing competition from the emergence of alternative fuels that have feedstocks other than crude oil, such as methanol, ethanol, propane, natural gas, hydrogen, and certain synthetic fuels. Environmental considerations will continue to influence production activities in the refining industry. Advances in reducing the lead content of gasoline were made during the 1970s, but further reductions are anticipated. Canadian and u.s. demand for "cleaner" oil-based fuels will require new refining capabilities for lead-free, reformulated gasoline.

CURRENT ENERGY POLICY AND FUTURE
TRENDS

In an era of relatively stable energy prices, two developments are shaping energy policy in Canada. One stems from growing public concern with environmental issues
and has strengthened opposition to construction of energy megaprojects, such as nuclear power plants and hydro-electric dams, because of their potential for harming the
natural environment, the socio-cultural environment, and human health. The other is
the threat of global warming and the need to reduce reliance on fossil fuels. A related
concern arises from recognition that fossil-fuel combustion worsens many other types
of local and regional air-quality problems, such as ground-level ozone and acid rain.
Taken together, these issues have had a powerful influence in shaping certain aspects
of energy policy.

Energy and the Environment

The National Energy Board regularly publishes a report on present and future energy
supply and demand in Canada. One measure of the growing importance of environmental considerations in energy policy making is the appearance in its 1991 report of
a chapter on the environmental implications of energy supply and demand. The chapter estimates future energy-related emission levels for a number of gaseous pollutants,
including carbon dioxide (CO_2), nitrogen oxides and nitrogen dioxide (NO_x), volatile
organic compounds (VOC), methane (CH_4) and sulphur dioxide (SO_2). These pollutants exacerbate global warming (CO_2, NO_x, VOC, CH_4), acid rain (NO_x, SO_2), and
ground-level ozone (NO_x, VOC).

Much recent concern has focused on global warming. About 55 per cent of the past
decade's increase in global warming potential has been attributed to emissions of carbon dioxide from human activity (Hengeveld 1991). Most emissions in Canada are
generated in the transportation sector, by power plants, and by residential, commercial,
and industrial heating equipment (Environment Canada 1991). Thus energy supply and
demand are a key consideration in developing policies on global warming. A joint fed-
eral-provincial-territorial task force has reviewed energy-efficiency and fuel-switching
options for reducing emissions to 20 per cent of 1988 levels by 2005, as called for by
the International Conference on the Changing Atmosphere in Toronto in 1988. The
task force claimed that cost-effective measures would achieve no more than 65 per
cent of the Toronto target, while technically achievable measures would achieve 88 per
cent of the target at best (Task Force on Energy and the Environment 1989).[9] Canada
took a compromise position and agreed to stabilize its emissions at 1990 levels by the
year 2000 (Environment Canada 1991). Even this target will be achieved only through
active intervention, since the National Energy Board (1991) forecasts an increase of
14 per cent by the year 2000 for its business-as-usual scenario.

While Ottawa is aiming for stabilization of carbon dioxide rather than reduction,
local response has focused on reduction. On the basis of reports by community-based

advisory committees, Toronto (City of Toronto Special Advisory Committee on the Environment 1989; 1991) and Vancouver (City of Vancouver Task Force on Atmospheric Change 1990) formally adopted as policy a commitment to reducing emissions of carbon dioxide by 20 per cent by 2005. Two other cities have recently adopted similar targets (Maclaren 1992). Ottawa aims to reduce emissions associated with municipal activities by 50 per cent by 2005. Regina plans to reduce municipal emissions by 20 per cent by 1998 and emissions within the city by 20 per cent by 2005.

Toronto has probably made the most progress. It has established a $23-million Atmospheric Fund to support a number of proposed reduction initiatives, including tree planting in Canada and Central America, public education, research and development, and energy-efficiency projects. It has already implemented a remote tailpipe-sensing demonstration project, created a municipal Energy Efficiency Office, and enacted a number of measures designed to reduce reliance on the car and improve heating and lighting efficiency (Harvey 1993).

Some of the latter measures are demand-management strategies. The underlying philosophy is to reduce demand for energy rather than perpetuate traditional management strategies, which have emphasized growth in the supply of energy. The electrical energy sector and, to a lesser extent, the transportation sector have been the target of most such initiatives to date. In energy-related emissions, these two sectors combined produce one-half of all energy-related CO_2 and SO_2 in Canada as well as 85 per cent of NO_x emissions, about 70 per cent of VOC emissions, and close to 15 per cent of methane emissions (Jaques 1990; Kosteltz and Deslauriers 1990).

Transportation demand management aims to encourage use of transportation modes other than the private automobile (for example, through construction and promotion of bicycle paths), encourage efficiency in transportation systems (by, for example, establishing ride-sharing programs and high-occupancy vehicle lanes), and discourage automobile use (for instance, by raising parking rates). Its effectiveness can vary widely, depending on existing transportation and urban spatial structure, the price of gasoline, the cost of public transit, and the combination of initiatives included in the program.

Evidence from the United States indicates that transportation-demand management can reduce reliance on the car by 2 per cent to 18 per cent (Mierzejewski 1991). In Canada, a number of municipalities have implemented initiatives. One of the most innovative is Toronto's requirement that certain new development proposals include a transportation demand-management plan that indicates how the employer intends to minimize related automobile use. Calgary's Air Improvement Resolution (AIR Calgary) encourages residents to use alternative transportation and leave their cars at home twice a month on weekdays (Maclaren 1992). A more radical proposal for Calgary, which is under review, calls for a fuel tax, an increase in downtown parking rates, and a decrease in free parking for downtown employees.

Electric utilities in several provinces are developing energy supply plans that will emphasize demand management. Such programs in Canada's three largest utilities

(Hydro Québec, Ontario Hydro, and B.C. Hydro) are expected to save energy amounting to approximately 16 per cent of 1990 generating capacity (Energy, Mines and Resources 1991).

It is less costly to develop policies that save a unit of electricity demand than it is to build a new generating facility to satisfy an additional unit of demand. Demand management includes all actions taken by a utility to influence electricity use. The three main types of strategy are improving energy efficiency, shifting load, and interrupting load (Energy, Mines and Resources 1991). Load shifting involves reducing peak demand during daylight hours and shifting demand it to non-peak periods. In interruptible load strategies, utilities offer customers contracts that promise lower electricity rates in return for the risk of supply curtailments during periods of high demand. Improvements to energy efficiency form the cornerstone of many plans. They include encouragement of greater appliance efficiency, increased energy-efficiency in new buildings, building retrofits, and energy efficient lighting. Robinson (1987) estimates that the technical potential for improved energy efficiency in Canada could be as much as 51 per cent in the residential sector and 42 per cent in industry. Though only 3 per cent of savings in electrical utility capacity from demand-side management were attributable to such improvements in 1990, their contribution is expected to reach 50 per cent by the year 2000 (Energy, Mines and Resources 1991).

Environmental considerations have also caused electrical utilities in Canada to modify their energy supply strategies. Perhaps the best-known example was cancellation in 1994 of Quebec's 3,000-megawatt, $13-billion, Great Whale hydro-electric complex. The two forces behind cancellation were an opposition campaign by environmentalists, led by the Cree, and decreasing demand in the U.S. northeast, where the electricity was to have been sold.

Another environmental consideration has been concern over acid rain. As a direct result, the provinces east of Saskatchewan reached an agreement with the federal government in 1985 to reduce overall emissions of sulphur dioxide by 50 per cent of 1980 emission levels by 1994 (Environment Canada 1991). Though the bulk of the reduction was to be achieved by the non-ferrous-metal-smelting industry, power plants burning fossil fuel were also affected. Ontario, New Brunswick, and Nova Scotia have significant thermal electric–generation capacities and were made subject to utility emission caps. The 1994 caps for the three provinces were set at roughly equivalent levels, which, for New Brunswick and Nova Scotia, were actually above their 1980 emission levels. However, since Ontario's 1980 emissions were so much higher, it was required to reduce emissions by 60 per cent by 1994.

Ontario was able to meet this target by placing heavier reliance on nuclear and hydro rather than coal-fired electricity and increasing its use of low-sulphur coal. In the last few years, however, operational problems with its nuclear plants and the constraints of its reduction goal have forced Ontario Hydro to import more U.S. electricity (National Energy Board 1991). Ironically, most of these imports are generated by fossil-fuel power plants, and the net effect has been to transfer production of emissions from Ontario to the U.S. Northeast.

Free Trade

The FTA removed a number of barriers to energy trade between Canada and the United States and reduced Canadian control over domestic energy supplies and energy exports, while NAFTA confirmed the provisions of the FTA but did not introduce any major new policy initiatives.

The impact of the FTA on the Canadian energy sector has been widely debated. Though Canada-U.S. energy trade has increased considerably since 1989, it is unclear, what proportion, if any, is attributable to the FTA. A situation close to free trade in energy had already existed since 1985, when Canada deregulated oil and gas prices. Carmichael (1988) suggests that short-term benefits of the FTA for the Canadian energy sector will probably be small, but in the long term the confidence created by a stable energy policy and a large, secure market will encourage investment in expensive oil and gas megaprojects in the Canadian north and the off-shore Atlantic region. Historically, export to the United States justified construction of major interprovincial transmission systems and helped reduce the unit cost of delivering oil and gas to eastern Canada (McDougall 1991).

At the time of the agreement, all U.S. imports of natural gas and electricity were from Canada, as well as more than half of its uranium imports, about 20 per cent of its coal imports, and 13 per cent of its oil imports (U.S. Energy Information Administration 1986). A much smaller proportion of Canadian oil and gas imports originated in the United States. All of Canada's coal and electricity imports came from the United States, but Canada did not import any U.S. oil or uranium and imported less than 5 per cent of its natural gas (National Energy Board 1991). Canada is much more important to the United States as a source of imports than the United States has been to Canada.

If another oil crisis materializes, the FTA will restrict Canadian energy policy initiatives. Under its provisions, Canada will never again be able to set prices for energy exported to the United States above Canadian prices, as it did during the 1970s and in 1980 with the NEP. Nor can it limit energy exports to the United States, as it did during the 1970s, unless it limits Canadian production at the same time. The FTA's proportional sharing agreement specifies that if Canada wishes to restrict its energy exports to the United States, the amount of the cut cannot exceed the average share of total domestic supply that the United States received during the preceding 36 months.

Most analysts agree that the FTA represents an extension and clarification of Canada's trade obligations for energy under the General Agreement on Tariffs and Trade (GATT). One exception is inclusion of electricity in the definition of energy products, since the status of electricity under GATT was unclear; the provision eliminates the threat of future U.S. tariffs on Canadian electricity exports. American analysts saw the clause as a victory for Canada (Calzonetti 1990), while some Canadian critics have argued that it will reduce the provinces' ability to control electricity exports and to offer lower rates to local industry in order to promote industrial development (Jackson 1988; Ryan 1991).

As with many other aspects of the FTA, debate over the energy provisions split along regional lines. Provincial governments that most strongly supported the energy provisions were those that had been most strongly opposed to the NEP. Uslaner (1989) reports that public opinion in favour of the FTA was strongest in Alberta and weakest in Ontario and the Atlantic region. But Quebec, an oil-consuming province that had favoured the NEP, also supported the FTA. Uslaner (1989) attributes at least part of Quebec's support to the potential beneficial effect that the FTA was expected to have on sales of electricity to the United States from new hydro-electric developments in James Bay.

McDougall (1991) argues that the residual regulatory powers of Canada's National Energy Board and the U.S. Federal Energy Regulatory Commission (FERC) may act as a non-tariff barrier to Canadian-U.S. energy trade. Though the FTA calls for consultation between the National Energy Board and the FERC in the event that either party objects to regulatory actions by the other, the agreement lacks an unambiguous statement regarding the limits of either party to regulate the construction of new energy transmission systems, such as international pipelines and electric transmission lines. It is also vague about what Canadian gas exporters can do in response to discriminatory measures imposed by U.S. regulatory agencies. In response, NAFTA has introduced new restrictions on the ability of U.S. state and federal energy regulators to interfere in Canadian exports of natural gas to the United States (Loizides and Rheaume 1993; Watkins, DataMetrics Ltd., and University of Calgary 1993).

Some analysts suggest that Ottawa should still seek to soften the impact on consumers of upward oil price shocks by means of an energy tax credit scheme (Carmichael 1988). Plourde (1991) notes that the FTA does not preclude differential pricing for domestic and foreign consumers, as long as the discrimination can be attributed to market forces rather than government price controls. In addition, price controls are permissible if they do not deliberately favour domestic over foreign users. Plourde (1991) observes that an apparent loophole in the agreement allows the Canadian government to impose price controls similar to those in place during the NEP. He suggests that if price controls were to be imposed on the wellhead price of domestic oil and then the provisions of the proportional sharing agreement were invoked to restrict U.S. access to Canadian oil, market forces might be expected to bid up the price of Canadian crude for U.S. consumers so that a differential pricing regime results.

The energy-related provisions of NAFTA are almost identical to those of the FTA. Though Canada has little energy trade with Mexico, NAFTA may affect competition between Canada and Mexico for energy exports to the United States in the long term. The Mexican government has generally discouraged natural gas exports to the United States, and NAFTA will probably lead to a change in this policy. However, the transportation infrastructure for such exports is not yet in place (Energy, Mines and Resources Canada 1991).

NAFTA will probably affect most the trade of energy-related equipment and services (Loizides and Rheaume 1993; Watkins, DataMetrics, and University of Calgary 1993). NAFTA reduces tariffs on imports to Mexico and liberalizes government procurement policies, both of which should increase opportunities for Canadian suppliers in the Mexican energy sector. Mexico is planning a major upgrading and refurbishment of its petroleum facilities, so the benefits of NAFTA to Canadian suppliers may become apparent fairly quickly.

CONCLUSION

The last two decades have seen a number of changes in the Canadian energy map. The types and amount of energy consumed by Canadians have changed considerably, as have the sources, types, and quantities of energy imported and exported. Most significant, oil's role as an energy source has declined, as has growth in overall demand for energy. These changes can be traced directly to policy initiatives driven by both internal and external forces. I have identified key forces, including fluctuations in the domestic and international price of oil, declining reserves of conventional oil, growing concern about energy-related environmental effects, and the FTA.

Because of abundant energy resources and government controls on the domestic price of oil, Canada has for almost two decades had lower energy prices than most other industrialized nations. Some have argued that low energy prices have given industry a competitive advantage, while others conclude that they have discouraged energy efficiency and left Canada at a comparative disadvantage. This debate has yet to be resolved.

Declining reserves of conventional oil have pushed energy exploration northward toward the Arctic and eastward to the Atlantic off-shore. Since 1985, exploration activity has decreased in these regions because of low international oil prices, but in the long term they will probably contribute a major portion of the country's oil and gas supplies.

The policy response to problems of energy-related pollution, such as global warming and acid rain, has been implementation of programs to reduce demand for energy, increase energy efficiency, and switch to less polluting fuels. Advances in ameliorating the environmental harm caused by energy consumption are being made by federal, provincial, and local government throughout Canada. Despite policy commitments to achieving additional improvements, progress will depend to a large extent on society's continued willingness to pay for an environmentally sustainable future.

Liberalization of trade in North America is a major initiative whose full effect on the energy sector has yet to be felt. Critics of the FTA and NAFTA have concentrated on the limitation on Canada's ability to impose export restrictions and differential oil export prices. Canada has not yet experienced any damage as a result but may if there is another international oil crisis. Though the benefits of the agreements have been minor to date, supporters suggest that in the long term benefits will most probably occur through increased investment in the Canadian energy sector. NAFTA may also open the Mexican market to Canadian suppliers of energy services and equipment.

NOTES

1 Primary energy requirements include domestic production plus imports minus exports plus/minus stock changes.
2 The National Oil Policy of 1961 guaranteed western Canadian producers the oil market east of the Ottawa Valley and allowed imports to service the rest of the country east of this boundary. The policy protected Canadian oil producers against competition from cheaper imported oil in the large Ontario market and stimulated growth of the refining and petrochemical industries in Ontario (Economic Council of Canada 1985).
3 Canada Lands include Yukon, Northwest Territories, and all off-shore resources.
4 End use demand for energy is energy used by final consumers.
5 Industrial energy intensity is measured here as the ratio of industrial energy use to industrial GDP.
6 Crude oil and equivalent production is production of light oil, heavy oil, and bitumen.
7 Productive capacity refers to the maximum rate at which crude oil or bitumen can be produced from a well or deposit.
8 Heavy crude oil includes both conventional heavy crude oil and crude bitumen. Most of Canada's bitumen is located in oil sands deposits in Alberta.
9 The Changing Atmosphere Conference recommended that about half of the 50 per cent reduction be achieved through improvements in energy efficiency and the remainder through supply modifications (World Meteorological Association 1989). In contrast to the pessimistic view of the Task Force on Energy and Environment, Robinson (1990) calculated that recommended levels of improvements in energy efficiency were both technically and economically achievable for Canada. He did not investigate the potential of supply modifications to meet the remainder of the recommended reduction goal.

REFERENCES

Calzonetti, F. 1990. "Canada-U.S. Electricity Trade and the Free Trade Agreement: Perspectives from Appalachia." *Canadian Journal of Regional Science*, 13, 171–77.
Carmichael, E.A. 1988. "Energy." In J. Crispo, ed., *Free Trade: The Real Story*, 66–76. Toronto: Gage.
Chapman, J.D. 1989. *Geography and Energy: Commercial Energy Systems and National Policies*. London: Longman.
CIPECC (Canadian Industry Program for Energy Conservation Council). 1988. *Canadian Industry Program for Energy Conservation: Energy, Efficiency, and the Environment*. Ottawa: Energy, Mines and Resources Canada.
City of Toronto. Special Advisory Committee on the Environment. 1989. *The Changing Atmosphere: A Call to Action*. Toronto: City of Toronto.
– 1991. *The Changing Atmosphere: Strategies for Reducing CO_2 Emissions*. Toronto: City of Toronto.
City of Vancouver. Task Force on Atmospheric Change. 1990. *Clouds of Change: Final Report of the City of Vancouver Task Force on Atmospheric Change*. Vancouver: City of Vancouver.

Doern, G. Bruce, and Toner, Glen. 1985. *The Politics of Energy: the Development and Imple-mentation of the NEP.* Toronto: Methuen.

Economic Council of Canada. 1985. *Connections: An Energy Strategy for the Future.* Ottawa: Supply and Services Canada.

Energy, Mines and Resources Canada. 1980. *The National Energy Program: 1980.* Ottawa: Supply and Services Canada.

– 1987. Canadian Industry Program for Energy Conservation. Internal memorandum, Ottawa.

– 1991a. *Electric Power in Canada 1990.* Ottawa: Energy, Mines and Resources Canada.

– 1991b. *North American Free Trade Agreement (NAFTA): Potential Impact on the Canadian Energy and Mineral Sectors.* Ottawa: Energy, Mines and Resources, Mineral Policy Sector.

Environment Canada. 1991. *The State of Canada's Environment.* Ottawa: Supply and Services Canada.

Flavin, C., and Durning, A.B. 1988. *Building on Success: The Age of Energy Efficiency.* World-watch Institute, Paper No. 82. Washington, DC.

Goodman, R.J. 1986. "Industrial Restructuring: Energy Demand under Siege." *Energy News-letter,* 7 no. 3, 3–9.

Hamilton, K. 1988. "Energy Intensiveness and Economic Performance since 1971." *Canadian Economic Observer,* 12, 4.1–4.16.

Harvey, L.D.D. 1993. "Tackling Urban CO_2 Emissions in Toronto." *Environment,* 16–20, 38–44.

Hengeveld, H. 1991. *Understanding Atmospheric Change: A Survey of the Background Science and Implications of Climate Change and Ozone Depletion.* Environment Canada, Atmo-spheric Environment Service, State of the Environment Report No. 91–2. Ottawa.

Jackson, A. 1988. "The Trade Deal and the Resource Sector." In D. Cameron, ed., *The Free Trade Deal,* 91–103. Toronto: Lorimer.

Jaques, A.P. 1990. *National Inventory of Sources and Emissions of Carbon Dioxide (1987).* Ottawa: Environment Canada, Environmental Protection Service.

Kosteltz, A., and Deslauriers, M. 1990. *Canadian Emissions Inventory of Common Air Con-taminants (1985).* Ottawa: Environment Canada, Environmental Protection Service.

Loizides, S., and Rhéaume, G. 1993. *The North American Free Trade Agreement: Implications for Canada.* Ottawa: Conference Board of Canada.

McDougall, J.N. 1991. "The Canada-U.S. Free Trade Agreement and Canada's Energy Trade." *Canadian Public Policy,* 17, 1–13.

McLachlan, Michele, and Itani, Imad. 1991. *International Comparisons: Interpreting the En-ergy/GDP Ratio.* Calgary: Canadian Energy Research Institute.

Maclaren, V.W. 1992. *Sustainable Urban Development in Canada: From Concept to Practice,* 3 vols. Toronto: Intergovernmental Committee on Urban and Regional Research.

Marbek Resource Consultants. 1989. *Energy Demand in Canada, 1977–1987: A Retrospective Analysis.* Ottawa: Energy, Mines and Resources Canada.

Mierzejewski, Edward A. 1991. "Transportation Demand Management for Quality Develop-ment." *Journal of Urban Planning and Development,* 117, 77–84.

National Energy Board. 1988. *Canadian Energy Supply and Demand 1987–2005.* Ottawa: National Energy Board.

– 1991. *Canadian Energy Supply and Demand 1990–2010*. Ottawa: National Energy Board.

OECD/IEA (Organization for Economic Cooperation and Development/International Energy Agency). 1988. *Energy Policies and Programmes of OECD Countries, 1987 Review*. Paris: OECD/IEA.

– 1990. "Energy End-Use Prices, 4th Quarter 1989." *CADDET Newsletter*, 2, 18–19.

Plourde, A. 1989. "Canadian Fiscal Systems for Oil and Gas: An Overview of the Last Two Decades." *Energy Studies Review*, 1, 1–15.

– 1991. "The NEP Meets the FTA." *Canadian Public Policy*, 17, 14–24.

Robinson, J.B. 1987. "Insurmountable Opportunities? Canada's Energy Efficiency Resources." *Energy*, 12, 403–17.

– 1990. "Decarbonating Energy Systems: The Potential for Reducing CO_2 Emissions through Reduced Energy Intensity in Canada." *Energy Studies Review*, 2, 1–17.

Ryan, J. 1991. "The Effect of the Free Trade Agreement on Canada's Energy Resources." *Canadian Geographer*, 35, 70–82.

Science Council of Canada. 1992a. *The Canadian Electric Power Sector*. Science Council of Canada, Sectoral Technology Strategy Series No. 12. Ottawa.

– 1992b. *The Canadian Oil and Gas Sector*. Science Council of Canada, Sectoral Technology Strategy Series No. 5. Ottawa.

Southam Energy Group. 1991. *Energy in Canada, 1989–90*. Calgary: Southam Business Information and Communications Group Inc.

Statistics Canada. 1991. *Quarterly Report on Energy Supply-Demand in Canada, 1990–IV*. Ottawa: Statistics Canada.

– 1992. *Energy Statistics Handbook*. Ottawa: Statistics Canada.

– 1994. *Quarterly Report on Energy Supply-Demand in Canada, 1993–IV*. Ottawa: Statistics Canada.

Task Force on Energy and the Environment. 1989. *Report on Reducing Greenhouse Gas Emissions*. Ottawa: Task Force on Energy and the Environment.

United States. Energy Information Administration. 1986. *Energy Information Administration Monthly Review, December 1986*. Washington, DC: Energy Information Administration.

Uslaner, E.M. 1989. "Energy Policy and Free Trade in Canada." *Energy Policy*, 17, 323–30.

Venleger, P.K., Jr. 1988. "Implications of the Energy Provisions." In J.J. Schott and M.G. Smith, eds., *Free Trade Agreement: The Global Impact*, 117–35. Washington, DC: Institute for International Economies.

Watkins, G.C., DataMetrics Ltd., and University of Calgary. 1993. "NAFTA and Energy." In S. Globerman and M. Walker, eds., *Assessing NAFTA: A Trinational Analysis*. Vancouver: Fraser Institute.

World Meteorological Organization. 1989. *The Changing Atmosphere: Implications for Global Security*. Geneva: World Meteorological Organization.

CHAPTER NINE

Agriculture in a World of Subsidies

WILLIAM C. FOUND

Canadian agriculture is in a precarious state. While it is a major contributor to Canada's wealth (about 9 per cent GDP) and accounts for a significant share of Canada's international-trade surplus (about 7 per cent), its future is uncertain (Agriculture Canada 1988). Farm incomes and returns on investment are very low. Much of the land area under cultivation is marginal in terms of climate and soils, less productive than prime areas of cultivation south of the U.S. border. The historical development of farm organizations and government assistance has enabled agriculture to thrive much of the time, but recent low prices for farm products, particularly in international markets, have placed Canadian agriculture in jeopardy. Fierce international competition has been the result, in large part, of agricultural subsidies provided to farmers in other countries by their governments.

The recent strategy of the Canadian government to deal with the nation's agricultural problems has been promotion of liberalized international trade, focusing first on the Free Trade Agreement FTA with the United States and then as the North American Free Trade Agreement (NAFTA). This strategy is based on the assumption that Canada can benefit from a system that emphasizes free competition, natural comparative advantage, and removal of various forms of governmental intervention. It is also intended to help persuade the governments of other countries to reduce their subsidies for their own producers. At best, the strategy was an optimistic gamble, based more on faith in a liberal economy than on much substantive analysis. At worst, it may lead to the dismantling of Canadian institutions that have helped the nation's agriculture to develop, leaving the sector more open to internal and external erosion. It can also lead the government to underestimate the importance of other factors, such as the fiscal policy that determines the value of the Canadian dollar, which in turn determines the price of agricultural exports, which have more affect on Canada's export economy than most of the articles in NAFTA.

The following consideration of agricultural land use, management, production, trade, government and non-government agencies, and international agreements serves to illustrate and explain the precarious nature of Canadian agriculture and the dubious wisdom of pursuing a strategy of trade liberalization.

AGRICULTURAL LAND USE
AND MANAGEMENT

The most comprehensive evidence concerning agricultural land use in Canada, and which antidates trade liberalization with the United States, comes from the 1986 Census of Agriculture. Figure 1, which is based on maps produced from the census, indicates the spatial extent of agriculture. Farmland is concentrated in the south, reflecting severe limitations of climate and soils further north. Most productive land is located within 200 km. of the U.S. border, in the agricultural heartlands of Ontario and Quebec, extending considerably farther north in the wheat lands of Saskatchewan and adjacent parts of Alberta and Manitoba. In 1986 the total farm area was 67.8 million ha, only 7 per cent of the total land area. Of this total, 68 per cent was "improved farmland" (i.e., its management involved some significant application of labour and capital). Only 33 million ha (49 per cent) was actually cultivated cropland. The total number of farms was 293,089, with an average size of 231 ha. The proportion of the population employed directly in agriculture was 3.7 per cent (Statistics Canada 1989).

Types of agricultural land use vary greatly by region. Agriculture in Ontario, Quebec, and the Maritime provinces is generally intensive, with a mixed farm economy involving livestock and a number of crop types. The predominant land uses are hay, corn, small grains, soybeans, and specialized vegetables and fruits. Farms are relatively small, with most under 100 ha. In contrast, farms in the prairie provinces tend to specialize in wheat or canola, with most in excess of 200 ha. Areas with moisture and/or soil limitations in western Canada, northern Ontario and Quebec, eastern Quebec, and the Maritimes contain large proportions of unimproved land, used mostly for grazing cattle.

Canadian agriculture is highly mechanized and generally highly capitalized. In 1986 the average value of farm capital per farm was $374,000, with 73 per cent attributable to land and buildings and the remainder to livestock, machinery, and other capital inputs. In 1985, 50.3 per cent of the total improved farmland was fertilized, and 49.9 per cent was sprayed for weeds and brush. Ten percent was sprayed for insects and disease.

Average sales of agricultural products per farm in 1985 were $70,917, with average farm business expenses of $60,000. Twenty-nine per cent of business expenses were for livestock (such as feed, livestock, and poultry purchases, and veterinary services), 14 per cent for crop expenses, 14 per cent for machinery, 11 per cent for interest payments, and 9 per cent for wages. The average net family income of $11,000 is very low by Canadian standards. The return on total capital investment (about 3 per cent, assuming no return for farm-operator labour) is minuscule. Small economic returns help to explain why only two-thirds of farmers in the 1986 census reported farming as their principal occupation and why farmers reported an average of 172.9 days for off-farm work during 1985.

Considerable regional variation in farm expense, farm income, and off-farm employment is evident for 1986 but is much less than variation in type of land use and

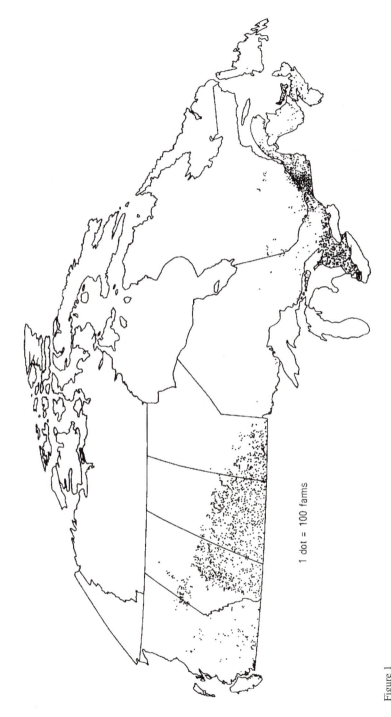

Figure 1
Number of farms in Canada, 1986. *Source*: Statistics Canada (1989, Map 4).

1 dot = 100 farms

farm size. Higher productivity per ha in eastern Canada and in other core agricultural areas is compensated for, in part, by larger farms in the west and in areas of more marginal agriculture.

Certain trends during the past 30 years help to reveal the character of Canada's agricultural land use. Between 1956 and 1986 Canada's total farm area declined by 3.5 per cent. During the same period the number of farms was reduced by half, and the average farm size doubled. The farm population declined from 2.75 million to just under one million (from 17 per cent of Canada's population to 3.7 per cent). The total improved farm area and area under crops increased 13 per cent and 29 per cent, respectively. Labour inputs declined and capital inputs increased greatly, with overall increases in most types of production. Livestock production became more efficient (for example, the number of dairy cows fell by half, but total production of milk increased). In general, more intense land use, through greater capitalization on better-quality land, led to overall increases in production.

Two other land-use trends have been a source of major concern: the loss of agricultural land to urbanization and a decline in environmental conditions (as in soil erosion and accumulations of chemicals). A combination of city growth and liberal land-use planning regulations has led to an estimated loss of 174,790 ha of farmland due to urban expansion during 1966–86 (Environment Canada 1989). Much of this loss has occurred in areas of prime agricultural land, particularly in southern Ontario. Greater cultivation of open-row crops (such as corn) and increased uses of heavy machinery for cultivation have led, many believe, to dangerous levels of soil erosion (Manning 1986). At the same time, growing use of artificial fertilizers and other chemical inputs has helped to maintain higher yields but masked the harm caused by erosion and exacerbated the environmental risk to surface and ground-water quality.

An analysis of census and other data reveals that Canadian farming, despite considerable rationalization, is economically marginal and subject to environmental deterioration because of unchecked urbanization and capitalization. Its precarious nature is easily identified.

PRODUCTION AND TRADE

The distribution of farm cash receipts ($20.2 billion) varies greatly among crops (45 per cent) and livestock (50 per cent) and between regions. By far the greatest share of income from cereals and oil seeds is in the west, whereas income from potatoes tends to be concentrated in the Atlantic provinces, and from fruits and vegetables in Ontario (Table 1). Receipts from livestock, dairying, and poultry and eggs are more evenly distributed among the west, Ontario, and Quebec. Given the variety of agricultural output, Canada is more than self-sufficient in the production of most farm products and produces sizeable surpluses in cereals, flax seed, and canola (Table 2). Canadian production is significantly deficient in soybeans, fruits and vegetables, and

Table 1
Distribution of farm cash receipts by commodity and region in Canada (regional shares %), 1986

Commodity group	Value ($million)	Distribution (%)	Atlantic Provinces	Quebec	Ontario	Western Canada
Cereals*	5,677	28.1	2.2	3.8	7.8	*86.2*
Oilseeds	1,061	5.2	–	–	21.4	*78.6*
Potatoes	272	1.3	*36.8*	15.0	14.2	34.0
Fruits	267	1.3	9.6	14.9	*39.6*	35.9
Vegetables	573	2.8	4.6	18.1	*57.2*	20.1
Other crops†	1,224	6.1	3.1	7.9	*63.3*	25.7
Total crops	9,076	44.9	3.5	5.4	21.2	*69.9*
Cattle and calves	3,553	17.6	2.4	8.2	32.9	*56.5*
Hogs	2,095	10.4	4.0	*31.9*	*31.8*	23.3
Dairy	2,779	13.7	6.2	*36.3*	*34.5*	13.0
Poultry and eggs	1,435	7.1	8.2	24.7	*37.0*	31.1
Other livestock	215	1.1	12.9	16.1	29.8	*41.2*
Total livestock	10,078	49.9	4.8	23.4	33.7	*38.1*
Other cash receipts‡	1,060	5.2	2.7	33.5	15.7	48.1
Grand total	20,215	100	3.5	15.9	27.1	53.5

Sources: Coffin (1988, 47); Agriculture Canada (1986).

Note: Figures in italics indicate leading region (or regions) for each commodity.

* Includes Western Grain Stabilization and crop insurance payments.

† Includes tobacco, sugar beets, floriculture, and nursery.

‡ Includes receipts from the sale of forest and maple products, dairy supplementary payments, deficiency payments, provincial stabilization payments, and other supplementary payments.

mutton and lamb but is at least equal to local demands for most other products. The pattern of self-sufficiency can be explained in part by natural environmental conditions (for example, Canadian climate and soils are conducive to producing cereals but are inadequate for year-round production of fruits and vegetables). The pattern is also a function, however, of explicit policies toward certain types of production. Some agricultural marketing boards (such as milk boards), with the support of federal and provincial governments, control production levels so that supplies generally equal local demands, but the prices resulting from supply management are not competitive in international markets. Consequently, surplus production is restricted.

Canadian agricultural exports normally exceed 40 per cent of the total value of farm receipts. The typical trading pattern is a sizeable agricultural trade surplus (over $2 billion), as shown in Table 3. Canada exports a large range of agricultural products, but cereals and oil seeds account for over half of the total value (see also Norcliffe, this volume, chap. 2). About half of the value of imports is represented by fruits, nuts, vegetables, and tropical plantation crops.

Table 2
Canadian self-sufficiency ratios for selected agricultural products, 1981–82

Product	Self-sufficiency ratio	Year
Wheat	479	1981–82
Barley	203	1981–82
Oats	104	1981–82
Rye	337	1981–82
Corn	126	1981–82
Flaxseed	426	1981–82
Canola	162	1981–82
Soybeans	63	1981–82
Fresh fruit	36	1981
Canned fruit	29	1981
Fresh vegetables	72	1981
Canned vegetables	75	1981
Frozen vegetables	103	1981
Beef	99	1982
Pork	121	1982
Mutton and lamb	43	1982
Turkey meat	94	1981
Chicken meat	96	1981
Milk	103	1981
Whole milk cheese	90	1982
Skim milk powder	280	1981
Creamery butter	105	1981
Cheddar cheese	143	1981
Other cheese	81	1981

Source: Agriculture Canada (1985).

China, Japan, the United States, and the USSR/CIS have been major purchasers of Canadian products, with the United States alone accounting for over 30 per cent of exports. The United States is, however, even more important as a supplier of Canadian imports, accounting for over 57 per cent of the value. Consequently, Canada has a $1-billion trade deficit in agricultural products with its neighbour.

Agricultural trade is central to Canada's economy, and it is characterized by a high degree of international market integration. Canada's agricultural trade is not a major factor, however, within the aggregate pattern of international trade. Even in wheat and flour, which account for much of Canada's exports, Canada's share in international trade is only 17 per cent (Warley 1986). And while the United States is crucial to Canadian agricultural trade as both a purchaser and seller, Canadian trade is significantly less important to the United States, which has much higher total agricultural production and a more diversified pattern of international trade. So agricultural trade though important to Canada, is not large enough to determine international prices. Canada is very much a "price taker" rather than a "price setter," which makes it very vulnerable to actions and policies beyond its control.

Table 3
Value of Canadian imports and exports, by agricultural product, for the 1987 calendar year

Commodity	Imports		Exports	
	$000	(%)	$000	(%)
All agricultural products	6,767,170	(100)	8,886,318	(100)
Grains	91,332	(1.3)	3,779,268	(42.5)
Grain products	209,369	(3.1)	265,799	(3.0)
Animal feeds	133,537	(2.0)	225,214	(2.5)
Oilseeds	143,626	(2.1)	734,611	(8.3)
Animals (live)	120,569	(1.8)	326,336	(3.7)
Red meats	487,665	(7.2)	1,064,943	(12.0)
Other animal products	436,952	(6.5)	587,536	(6.6)
Dairy products	142,981	(2.1)	145,436	(1.6)
Poultry and eggs	121,780	(1.8)	55,863	(0.6)
Fruit and nuts	1,647,409	(24.3)	156,110	(1.8)
Vegetables	882,789	(13.0)	432,101	(4.9)
Sugar	233,633	(3.5)	35,513	(0.4)
Plantation crops (tropical)	838,503	(12.4)	122,637	(1.4)
Other	1,116,491	(16.5)	954,951	(10.7)

Source: Agriculture Canada (1988b).

ROLES OF GOVERNMENT

Federal, provincial, and, to a lesser degree, local governments all affect Canadian agriculture, and all have played a creative role in its development and support. Historically, all three have reflected a society that tends to honour democratic action,
freedom of individual choice on use of land, and protection of the family farm. At the
same time, all three have been expected to provide the infrastructure and social services to support viable agriculture for a dispersed rural population. They have also
been expected to intervene in the free-enterprise economy, especially during difficult
economic times, to bring stability and financial relief to farmers. Recently governments have been emphasizing the "free-choice" aspect of Canada's rural tradition. In
1986 the federal minister of agriculture and his ten provincial counterparts all signed
a National Agriculture Strategy supporting government action that is "equitable and
sensitive to regional economies" and that supports market orientation, competitiveness, importance of the private sector, and support for the family farm (Coffin 1988).
These same ministers have faced renewed calls from farm organizations to intervene
in the free-market economy to provide financial support in the face of depressed farm
incomes.

The British North America Act, 1867, gave the provinces jurisdiction over land-
ownership and other property rights. Consequently, the provinces have had the right
to immediate control over many factors that influence agriculture, including land

surveys, road construction, property taxation, and marketing within provincial boundaries. The federal government has traditionally had jurisdiction over interprovincial matters, such as inter-regional and international trade, and interprovincial transportation. Two examples of federal action were the building of the railway system, which, particularly in western Canada, has been essential for settlement and the development of export agriculture (see also Lea and Waters, this volume, chap. 16), and creation of the Canada Wheat Board, which has been responsible for most of Canada's international wheat sales. Ottawa has also been a coordinator, seeking to ensure that provincial and federal policies and programs work together for the common good of Canadian farmers, facilitating interprovincial agricultural trade, and sometimes intervening to prevent trade wars between provinces. It has also aided inter-regional equity through provision of special funds, programs, and services for poorer regions. In the twentieth century, particularly since the economic and environmental disasters of the early 1930s, Ottawa has worked increasingly with the provinces in shared-funding programs to support agriculture. The historical division of responsibility has gradually blurred, programs in support of agriculture have become more complex, and the effects of government intervention are now more difficult to trace.

The most important, immediate impact of government on Canadian agriculture is through direct expenditures in support of farmers (see also Davis, this volume, chap. 20). In 1985–87 the average annual value of this support was $4.69 billion (Coffin 1988), 23 per cent of the value of total farm receipts. So despite a tradition of free enterprise and limited intervention in the economy, government financial assistance to agriculture is very significant. Direct subsidy occurs at about the same rate as in the United States but at a much lower rate than in European countries. Federal expenditures, which amounted to 63 per cent of all government support during 1985–87, were for, in declining order of importance, transportation subsidies (particularly through the Western Grain Transportation Act), stabilization programs, marketing, research, crop insurance, federal-provincial agreements, production assistance, farm credit, and miscellaneous other programs. Provincial expenditures supported production assistance (including extension services), farm credit, stabilization payments, marketing, research, crop insurance, and other programs. The single most important form of government assistance was for transportation subsidies, which amounted to $1 billion – 21.5 per cent of annual government direct expenditures – during 1985–87.

Governments at the local level (county, township, municipality, and so on) make a number of decisions that affect farming at the micro-level (such as granting building permits and improving local roads), and they often implement provincial government policy or programs (as with property tax rebates to farmers and land drainage subsidies). They also prepare official plans, which often include zoning by-laws delineating the location of farmland. All provinces have some form of policy, guidelines, or legislation that encourage or require local municipalities to consider the preservation of good farmland in the preparation of plans for land-use change. In some cases, the legislation is quite specific and requires that good farmland be protected (Audet and Le Henaff 1984). Examples include British Columbia's Agricultural Land Commission Act, Quebec's Agricultural Land Protection Act, and Newfoundland's Development Areas (Land) Act.

Planning guidelines in some other provinces have been weak statements in support of the preservation of farmland or have been implemented in a casual manner, so that they have had little effect. Strict government plans or procedures to protect farmland have met with much public criticism in all provinces – from farmers, who often stand to make substantial capital gains through upgrading of their land to "higher," urban land uses; from developers, whose business it is to convert land to more intensive urban uses; and from local politicians. who seek to please local interest groups and who traditionally strive to reduce local taxes by increasing the total property assessment in the region.

Despite recent government emphasis on the free-enterprize tradition in Canadian agriculture, all levels of government continue to have a major intervening role in creating a nurturing environment for Canadian farming. Without this intervention Canadian agriculture could not have reached current levels of production and stability, precarious though they be. Removal of many of the programs would be detrimental, though the complex roles of various levels of government make it difficult to predict the full impact of such removal.

OTHER INSTITUTIONS

A great many non-governmental organizations concerned with various aspects of agriculture have developed at the federal, provincial, and local levels. Some have been established by government to administer programs or to address questions identified by government. At the federal level these include the Canada Wheat Board, the Canadian Grain Commission, the Canadian Dairy Commission, the Livestock Feed Board of Canada, the Western Grain Transportation Authority, the Agricultural Stabilization Board, and the National Farm Products Marketing Council (Coffin 1988). At the provincial level the most influential bodies, created as a result of facilitating provincial legislation, are agricultural-products marketing boards.

Approximately 100 marketing boards across Canada have been created during the past 70 years. Usually they began as grass-roots organizations of producers of a particular product, brought together out of a common concern for their economic welfare. Common problems have included price instability, lack of infrastructural support, and a sense of disadvantage when dealing with processors or with the wholesalers who buy their products. Canadian farmers have suffered the economic disadvantages of disaggregation into large numbers of producers, spread across vast areas of countryside, often selling their products to a handful of processors whose market behaviour tends to be oligopolistic. The responses of marketing boards have ranged from the simple sharing of market information, through advertising campaigns for their product, establishment of "fair" marketing procedures (as through centrally located public auctions), pooling, price negotiation with buyers, to aggressive pursuit of foreign export markets. A few marketing boards have established systems of supply management, designed to avoid creation of price-depressing surpluses and to create greater price stability.

The major products under marketing-board supply management are milk, eggs, chicken, and turkey. Provincial boards operate producer quotas, and Canada-wide co-ordination is achieved through the actions of national marketing agencies. The boards have achieved what many believe to be reasonable market stability, with decent re-turns on investment. The producers are not permitted to operate as unconstrained mo-nopolies, which might take market advantage of their ability to control levels of production. Under the supervision of federal and provincial bodies, the marketing boards must ensure provision of adequate supplies, and their prices must reflect costs of production, or represent the results of committee decision. To protect the stability of the supply-management systems, and to maintain prices that are usually higher than those in the United States, Ottawa matches the actions of the marketing boards with import controls.

Important to farmers are a number of local institutions. These include agricultural societies, clubs (such as 4-H), and local agricultural cooperatives. The cooperatives have long provided access to farmer-owned market outlets and relatively inexpensive capital inputs.

Despite creation of numerous farm-related organizations, and many efforts to mo-bilize political action by farmers, no single organization represents a strong force in Canadian political circles. The two major national organizations, the Canadian Feder-ation of Agriculture and the National Farmers Union, have a history of competition, which has weakened their aggregate effect. None of the country's major political par-ties has the general backing of Canadian farmers. Only in Quebec, where the Union des Producteurs agricoles du Québec has government authorization to represent farm-ers and to require membership of all farmers, can any institution represent agricultural interests in a truly effective way.

INTERNATIONAL AGREEMENTS

At the international level Canadian agriculture faces trade restricions, oversupplied markets, subsidized competition, and markets too large to be affected much by varia-tions in Canadian production. Consequently international agreements, particularly those that seek to reduce subsidies, are of great interest to the Canadian government. It, together with the U.S. government, has sought general international agreement to move toward more liberalized trade in the belief that it would best serve the interests of Canadian agribusiness.

The current international pattern of subsidies and other government interventions into the farm economy is the product of historical events. Government programs to stimulate agriculture, and to insulate it from direct competition in international mar-kets, began in Europe, Japan, and elsewhere earlier in this century, when those countries were major importers of food and were taking extreme steps to build up their internal agricultural economies. This seemed particularly appropriate in the face of severe shortages of food. Over the years these programs, in combination with modern methods of farming, have led to great increases in production. The European Union (EU) is now a major exporter of agricultural products, and the total cost of farm-support programs in the EU and Japan is in the order of U.S. $100 billion per year (Winglee 1989). Meanwhile, some of the recent government support for agriculture in Canada and the United States, countries that have traditionally subsidized farmers to a much lesser extent, has been provided to help protect farmers from the ill effects of subsidized competition in international markets. The net result has been a kind of international trade warfare in agricultural products, which many participants would like to control or stop through new international agreements.

The main forum where Canada has sought reform in international agricultural trade is the General Agreement on Tariffs and Trade (GATT). Canada has long argued there for trade liberalization and has taken some satisfaction from recent agreements among GATT members. The Uruguay Round of negotiations sought multilateral agreements on agriculture in order to improve market access through reduction of trade barriers; increase competition by reducing subsidies, which limit agricultural trade; and minimize the adverse effects of sanitary regulations and barriers on trade (Warley 1987). Similarly, the Council of Ministers of the Organization for Economic Co-operation and Development (OECD) has been seeking reform based on the following principles: agricultural policies should provide less support; guaranteed prices should be reduced for commodities in excess supply, or output should be subject to production restrictions; market signals and mechanisms should be allowed a role even when supply is controlled by administrative decisions; and public assistance should support farmers' incomes directly rather than indirectly – for example, through support of commodity prices and farm production activities (Warley 1987). While Canada spoke strongly in favour of trade liberalization in both GATT and the OECD, it has probably exhibited its commitment most convincingly through the relevant sections of the FTA and NAFTA. Some might argue that the major function of the Canadian-U.S. agreements on agricultural trade is to facilitate similar, multilateral agreements, which could help both countries more than their bilateral agreements.

Since bilateral agricultural trade was fairly liberal before the signing of the FTA many believe that the agreement will not have a large, aggregate, direct effect (Farrell 1988). The FTA calls for elimination of almost all tariffs within 10 years, but these tariffs averaged only 6 per cent to begin with. It also specifies a range of adjustments designed to eliminate or reduce government subsidies on items traded between the two countries. Each nation will consider the export interests of the other in any use of export subsidies to third countries. Canada and the United States are often in serious competition for export markets (for wheat, for example), and some believe that significant government intervention to improve either country's relative competitiveness in international markets could have a far greater effect than all of the agriculture-related items in NAFTA (Schmitz 1988). Ottawa, and many professional agricultural economists, believe that NAFTA will be advantageous for Canada. At the same time, some regions of Canada (such as Ontario) expect local economic losses, as they become subject to unprotected competition from the United States that they can not match (Ministry of Agriculture and Food 1988).

The actual long-term effects of NAFTA are not clear, in part because of lack of substantive research at the regional level. The extent to which a combination of NAFTA and GATT rulings may erode vital Canadian programs and institutions is similarly unknown. For example, while supply-management marketing systems are to be protected, some observers believe that they will be ruled an invalid form of intervention (Troughton 1990). NAFTA, and Canada's general strategy on trade liberalization, represent real risks to Canadian farmers.

CONCLUSIONS

I have traced the status of Canadian agriculture, its supporting programs and institutions, and both its domestic and international problems. While it is a significant component of the country's economy, it is a precarious enterprise, restricted to a small proportion of the land base and severely limited by environmental constraints. Decades of modernization, modest government support, and development of influential agricultural institutions have helped farmers in Canada overcome some of the disadvantages associated with long distances from markets and poor market structures. Still, Canadian agriculture is so highly integrated within international trade that its institutions can provide only limited protection from worldwide trends in agricultural economy. Recent global surpluses in many agricultural products, for example, have contributed to low prices and declining incomes for Canadian farmers, reflected in declining values for farmland (Statistics Canada 1989).

Ottawa's major response to these problems is to promote greater liberalization of international trade. The outcome of this strategy is uncertain. Supporters, who include a number of the country's economists (Miner and Hathaway 1988), foresee greater sensitivity to "market signals," disappearance of detrimental government interference, greater access to international markets, rationalization within agriculture to make it more competitive, and development in accordance with the principle of comparative advantage to the ultimate benefit of Canadian farmers.

This optimism, however, seems to represent ideological faith in neoclassical economic principles rather than substantive analysis. The strategy has a number of potential flaws. Trade liberalization, and the corresponding transformation of agriculture in accordance with comparative advantage, would probably lead to greater specialization in a limited number of products (such as wheat) in a limited number of regions (such as the west), with an increase in economic and ecological vulnerability resulting from over-specialization and greater integration within international markets. Canada has a significant comparative disadvantage in some forms of production (Brinkman

1987), and more liberalized trade may lead to economic depression in some rural regions. Just as agriculture in some parts of Canada (the Maritimes, for instance) has become marginalized as land use has become more capital intensive and productive in the more advantaged parts of the country, so some regions of Canada may become marginalized as production within North America becomes more concentrated within the U.S. agricultural heartlands.

Canada is not an "equal" partner with the United States, and there is no reason to believe that it would fare well in serious disputes. Recent U.S. subsidies in support of American grain exports (after signing of the FTA and NAFTA) should raise questions about the value of trade agrements with United States, since subsidies of this sort probably have far greater direct impact on the health of the Canadian farm economy than the entire NAFTA. Similarly, recent movements in the value of the Canadian dollar, which reflect domestic fiscal policy, have had much more influence on the saleability of agricultural exports than any article in NAFTA.

Another problem with trade liberalization is its potential to damage the rural environment. Programs to protect soil and water and to preserve land for agriculture may be incompatible with an allocative system that is sensitive merely to short-term economic signals. The development of Canadian agriculture has depended heavily on a balance between free enterprise and the creation of programs and institutions that support Canadian farmers. Recent movement toward trade liberalization places these programs and institutions in serious jeopardy, without any firm indication that the risk will lead to useful reforms of agricultural subsidies in other countries. Canadian agriculture, already in a precarious state, may be further weakened.

ACKNOWLEDGMENTS

I wish to thank the Faculty of Environmental Studies, York University, for providing a grant in support of the research for this chapter. Thanks also to Michael Troughton, who provided many useful suggestions during preparation of the text.

REFERENCES

Agriculture Canada. 1985. *Self-Sufficiency Trends in Canadian Agriculture: 1960–1980 and Future Prospects.* Ottawa.

– 1986. *Handbook of Selected Agricultural Statistics 1986.* Ottawa.

– 1988a. *The Canada-U.S. Free Trade Agreement: An Assessment.* Ottawa: Government of Canada.

– 1988b. *Canada's Trade in Agricultural Products 1985, 1986 and 1987.* Ottawa.

Audet, R., and Le Henaff, A. 1984. *Land Planning Framework of Canada: An Overview.* Lands Directorate, Environment Canada, Working Paper No. 28, Ministry of Supply and Services Canada.

Brinkman, G.L. 1987. "The Competitive Position of Canadian Agriculture." *Canadian Journal of Agricultural Economics*, 35, 263–88.

Coffin, H.G. 1988. "Driving Forces, Instruments, and Institutions in Canadian Agricultural Policies." In R. Allen and K. Macmillan, eds., *US-Canadian Agricultural Trade Challenges: Developing Common Approaches*, 57. Washington, DC: Resources for the Future.

Environment Canada. 1989. *Urbanization of Rural Land in Canada. 1981–86*. SOE Fact Sheet No. 89–1, Minister of Supply and Services Canada, Ottawa.

Farrell, R.R. 1988. "Epilogue." In R. Allen and K. Macmillan, eds., *US-Canadian Agricultural Trade Challenges: Developing Common Approaches*, 197–202. Washington, DC: Resources for the Future.

Manning, E.W. 1986. "Planning Canada's Resource Base for Sustainable Production." In I.S. Knell and J.R. English, eds., *Canadian Agriculture in a Global Context: Opportunities and Obligations*, 53–68. Waterloo, Ont.: University of Waterloo Press.

Miner, W.M. and Hathaway, D.E., eds. 1988. *World Agricultural Trade: Building a Consensus*. Halifax: Institute for Research on Public Policy.

Ministry of Agriculture and Food. Government of Ontario. 1988. *Assessment of the Impacts of the Canada-U.S. Free Trade Agreement on the Ontario Agriculture and Food Sector*. Toronto.

Schmitz, A. 1988. "Summing Up: The Canadian Perspective." In K. Allen and R. Macmillan, eds., *U.S.-Canadian Aqricultural Trade Challenges: Developing Common Approaches*, 193–5. Washington, DC: Resources for the Future.

Statistics Canada. 1989. *A Profile of Canadian Agriculture*. Ottawa: Minister of Supply and Services Canada.

Troughton, M.J. 1990. "An Ill-Considered Pact: The Canada-U.S. Free Trade Aqreement and the Agricultural Geography of North America." Paper presented at the Annual Meeting, Association of American Geographers, Toronto, 20 April.

Warley, T.R. 1986. "Canadian Agriculture in a Global Context: An Overview." In I.S. Knell and J.R. English, eds., *Canadian Agriculture in a Global Context: Opportunities and Obligations*, 53–68. Waterloo, Ont.: University of Waterloo Press.

– 1987. "Issues Facing Agriculture in the GATT Negotiations." *Canadian Journal of Agricultural Economics*, 35, 517.

Winglee, P. 1989. "Agricultural Trade Policies of Industrial Countries." *Finance and Development* (International Monetary Fund and the World Bank), 26 no. 1, 9.

PART THREE

Manufacturing and the Impact of Technological Change

The locational pattern of secondary manufacturing is one of high concentration in and around the metropolitan centres of Toronto and Montreal (Figure 1), with the most important industrial region, the Golden Horseshoe of southern Ontario, stretching from 50 km east of Toronto to the U.S. border at Niagara. Historically, industrial investments by both Canadian- and foreign- (mainly U.S.-) owned corporations have been encouraged by a variety of means, including tariff protection. The result is a manufacturing sector serving a small market with a highly diversified range of products. A consequence has been short production runs. While this has been a disadvantage in some industries where economies of scale are possible and has resulted in lower productivity, increasingly manufacturing technology has favoured flexible production systems, small-batch production, and even customized production, and these modes actually and potentially have reduced the disadvantage of the small market. Some Canadian firms have successfully circumvented any problems caused by the limited domestic market by seeking foreign sales.

In aggregate, Canada experienced very slow growth of manufacturing productivity through the 1980s (average annual rate 1.3 per cent, 1979–91, compared with 2.4 per cent for the United States), and this fact has had a significant influence in its decline in international competitiveness; though productivity in manufacturing improved 1950–80, the 24 per cent gap in 1980 grew to 34 per cent by 1990 (Economic Council of Canada 1992). Partly as a consequence of slower productivity growth and a higher Canadian rate of inflation, markets and jobs were lost – the problem had become one of higher unit-labour costs in Canadian manufacturing. Economists have noted contributory influences to explain the productivity problem, including increases in the real price of energy, tight monetary policies in response to inflation that reduced use of installed capacity, and the exchange value of the Canadian dollar after 1986 (though the trend has not always been constant), while indicating that the core issue is Canada's poor record of industrial innovation (Economic Council of Canada, 1991, 1992). Internationally, the strongest response to industrial competition has been increased investment in innovation. This involves expenditure on

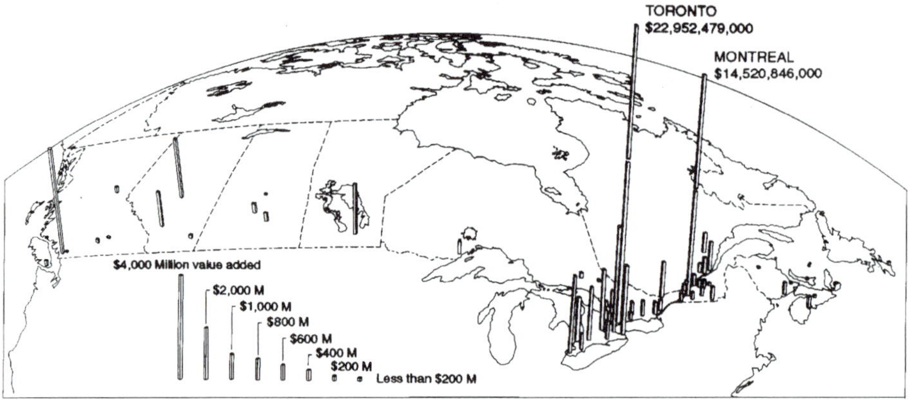

TORONTO
$22,952,479,000

MONTREAL
$14,520,846,000

$4,000 Million value added
$2,000 M
$1,000 M
$800 M
$600 M
$400 M
$200 M
Less than $200 M

Figure 1
Manufacturing output, Canada, 1986. *Source*: Statistics Canada (1988).

research and development (R&D) and on design and engineering inputs, use of new machines and management techniques, and new marketing ventures. But in these areas Canadian achievements are weak; for example, Canada has low levels of expenditure on R&D and lags behind U.S. and other economies in adoption of advanced technologies in manufacturing. The limited scale of high-technology industry in terms of production and jobs follows this, as does the deficit on high-technology trade.

In recent decades, the government of Canada has actively sought to increase the exposure of manufacturing to the competitive pressures which may induce it to become more specialized, innovative, and profitable and to grow. The resultant agreements under the General Agreement on Tarrifs and Trade (GATT), the Canada–United States Free Trade Agreement (FTA), and the North American Free Trade Agreement (NAFTA) have increased competition by eliminating residual tariff protection and beginning reduction of non-tariff impediments to imports and exports. In locational terms, the clearest implication of NAFTA is the integration of Mexico, where wage rates are at the lowest in North America – further incentive for Canadian industry to increase its technological intensity.

Attempts by Canadian firms to increase industrial productivity have already helped lower the proportion of the workforce employed in manufacturing: from about 27 per cent to about 15 per cent (1960–92). Increased international competition seems to have had an influence throughout the 1980s because Canadian manufacturing has secured a declining share of the domestic market, dropping from 73 per cent in 1980 to an estimated 53 per cent in 1993. Nevertheless, the relative fall in the contribution of manufacturing production to GDP, from 20 to 18 per cent (1971–92) is more modest. Some of the decline in the rate of increase in manufacturing output (real value added) derives from increasing use of service inputs such as product design, marketing intelligence, data processing, and industrial services. This change is part of a broader restructuring which occurred with intensity in the early 1980s and over the last three years. One legacy has been the rising proportion of manufacturing production that is exported. This figure has doubled since 1980, rising from less than 24 per cent to an estimated 48 per cent in 1993.

The reduction in the growth of industrial employment is explored by Ray (this volume, chap. 10) using a longitudinal data file of all Canadian firms for 1978–86; he finds that small firms grew larger and more numerous, while large firms declined in both respects. Size thus underpins industrial and regional changes in employment. The death rates of firms are highest among younger and smaller firms, but it is birth rates, influenced by location, industry, and the business cycle, that are of particular interest, since new (small) firms – the more competitive replacements – are primarily responsible for growth in industrial employment. Regional variations place Ontario in the lead, but Quebec, the next highly industrialized province, is in last place.

In contrast with retailing and commercial services, in manufacturing the size-mix of firms is unfavourable to employment growth, and many large firms have responded to more open competition by rationalizing product lines, specializing, and using equipment more efficiently. Given the prevailing wage level, some firms have abandoned products too costly to produce in Canada, while others have ceased production entirely or have shifted production sensitive to labour cost to a low-wage location elsewhere. For both domestic and foreign-owned firms, but for u.s. firms in particular, the last decade has seen many corporations calculate the merits of a Canadian manufacturing component in their North American operations. Taking industry, size, and other factors into account, Ray has discovered that foreign firms held employment better than domestic counterparts and are not the primary category of plant closures. Cycles of recession have generated peaks in closures, but MacLachlan (this volume, chap. 11) shows that the recent peak for Ontario is much larger than the peak in 1982. We can presume that international competition and market integration, and the restructuring of multinational corporations, are strong influences.

Restructuring of corporations and plant closures in Ontario do not appear to result in investment gains in other locations in Canada. The macro-economic evidence, in particular (reviewed by Gertler, this volume, chap. 15) indicates that there is little spatial redistribution of capital within Canadian manufacturing, and the strong regional-sectoral pattern of the Canadian space economy appears to be both cause and effect. As an alternative analysis of closure MacLachlan advances a business behaviour approach using work of Chandler (1962) and develops a corporate decision model to describe the relationship between subsidiaries and parents, thus providing insights into the strategic decision to close plants. His case study traces the process whereby management by both the parent and the subsidiary fails to establish a rationalized position within the corporation for a subsidiary that has been a miniature replica of its u.s. parent.

Despite general weaknesses in productivity and innovation, some Canadian manufacturing has clear technological strengths. Using mainly directory and proprietary data sources, Britton (this volume, chap. 14) identifies the main clusters in Canadian technology-intensive activity. Canada has developed international strengths by innovation in new industrial activities, especially in telecommunications systems, including communications satellites, and in computer software, and firms serve international product niches. Other new technology-based activity reflects advances made by strong, older Canadian companies, which have moved with the times in established product fields such as transportation, where significant innovation has been achieved. Despite the innovative weakness of many firms in resource industries (discussed in this volume, part II) some companies produce equipment for energy production and have developed advanced applications in environmental monitoring equipment and in services relying on resource and other industries for their market. High-technology industry is highly concentrated in and around Toronto, Montreal, and Ottawa, but there is limited potential for new locations to develop elsewhere. With little secondary manufacturing, the Atlantic region and western Canada are hoping to share in the growth of R&D and technology-based industrialization, but investment has not yet led to significant alternative locational concentrations.

Innovation, of course, is not a restricted preserve of the new industries, and across the industrial spectrum firms' ability to implement new product designs and to customize output is associated with fast and effective acquisition of new production technology. Yet, as noted above, Canada has low rates of adoption of new technology, other than the commonplace use of personal computers and computer workstations in office, inventory, and communications tasks. Presumably the low levels of R&D and innovation in most industries depress demand for other new machines, though duty remission on machinery imports has long assisted Canadian firms in acquiring new equipment. Certainly the Canadian machinery industry is small and weak, and the failure of the market to stimulate local equipment suppliers is a general problem of Canadian industry, especially as proximity of equipment producers to user firms is understood to affect the quality of their relationship. Gertler (this volume, chap. 15) explores this proposition and concludes that Canadian firms often have to acquire new equipment from distant suppliers and have been further thwarted by the absence of knowledgeable operatives, a problem that reflects the failure of Canadian industry to invest in the training of workers. Gertler's chapter provides insights into the constraints on the acquisition of production technology that limit the competitiveness of Canadian firms.

Perhaps the greatest changes in work practices and production techniques have been those adopted by the auto industry, which directly employs about 14 per cent of manufacturing and which, as a source of demand, has significant connections with other industries (such as steel, rubber, and plastics). Under the regime of the Auto-Pact (see Holmes, this volume, chap. 13) the Canadian segment of the assembly industry has increased its productivity and efficiency, and a few innovative parts makers have also prospered. The lower labour costs and exchange value of the Canadian dollar have helped to maintain Canadian jobs and production. Nevertheless, major shifts are occurring in the way production is organized.

The recent adoption of new production systems by North American-owned firms will allow efficient design, production, inventory, and delivery relations between parts and assembly plants. Modern flexible machines are used where possible, and innovations in labour relations are being introduced so that well-trained workers can realize a greater proportion of the potential productivity of these investments. Some plants of the "Big 3" auto manufacturers now match the productivity levels of Japanese transplants, but the full integration of the Mexican industry promised by NAFTA means that U.S. auto assembly firms will have much greater freedom to choose their locations within North America, and this places Canadian component suppliers at a new disadvantage, since only a minority are innovative competitors.

Conventional economic expectations are that trade liberalization will induce long-term decreases in employment and production for mature industries such as clothing, whose products have relatively low technological content and which continue to be made using little advanced production equipment. The structural and productivity weaknesses of this industry, as argued by Cannon (this volume, chap. 12) are attributable to the small scale and poor capitalization of firms, which limit their ability to change their methods of production. The complex web of national and international policy conventions has also inhibited the response of the industry to international competition, while the high proportion of immigrant women in less skilled jobs has allowed it to perpetuate its low wage mode of adjustment. This has spawned a substantial level of subcontracting, including homeworking, where low wages and limited benefits prevail. In terms of modernization the Canadian clothing industry trails that in other industrial countries, but technological restructuring has become more feasible as equipment costs fall, and Canadian expenditures have risen since FTA, though the scale of the industry has declined. Duty-free access to the U.S. market, however, is limited under FTA by U.S. constraints on imports of clothing made of fabric from outside North America, and under NAFTA the situation is worse.

Though industrial decline and restructuring are the negative results of trade liberalization and recession on industries slow to adopt technological change, these activities have the potential to restructure in other ways. For clothing, as for most other industries, the key to survival in domestic and export markets is innovation that relies on the work of both designers and marketing experts functioning in conjunction with new equipment and skilled operatives. They have positive effects on each industry's value added, especially when production is shifted to higher market segments, where performance characteristics of new designs and the income elasticity of the market have greater importance and where fast changes in consumer tastes can be accommodated. This proposition for restructuring and retaining or regaining industrial strength, based on the need for higher-valued products and increased productivity, applies to industries with varying design, technological, and market characteristics, and it is a remarkably generic prescription for Canadian industry – both those that rely on the traditional resource base for their basic inputs and those industries – secondary manufacturing – where human capital is a more evident basis for economic success.

REFERENCES

Chandler, Alfred D. 1962. *Strategy and Structure: Chapters in the History of Industrial Enterprise*. Cambridge, Mass.: MIT Press.
Economic Council of Canada. 1991. "Canadian Productivity." *Au Courant*, 11 no. 3, 12–13.
– 1992. "Productivity Growth." *Au Courant*, 13 no. 1, 5–6.
Statistics Canada. 1988. *Manufacturing Industries of Canada: Sub-provincial Areas*. Cat. No. 31-209. Ottawa: Minister of Supply and Services.

CHAPTER TEN

Employment Creation by Small Firms

D. MICHAEL RAY

SMALL FIRMS AND NET EMPLOYMENT GROWTH

The contribution of small firms to net employment creation in Canada, which has been substantial and growing since the 1960s, is now pre-eminent (Rothwell and Zegveld 1982, 125–8; CFIB 1983; Thompson 1986; Cléroux 1988; Statistics Canada 1988; and Laroche 1989). Small firms – those employing fewer than 20 in the base year – generated over one million net new jobs 1978–86 (Figure 1), while large firms (over 100 employees) declined, offsetting comparatively small growth by medium firms. It was once thought that large firms grow larger and small firms grow more numerous. During 1978–86, however, small firms grew larger and more numerous. It was the large firms that declined in size as well as in numbers (Berry, Conkling, and Ray 1993).[1]

Canadian interest in the contribution of small firms to employment growth has been heightened by U.S. work in which David Birch (1979) argued that small firms make three principal contributions. First, they produce most of the net growth in employment. Second, regional differences in employment growth can be attributed largely to regional differences in the birth rates of new small firms rather than to job losses from cut-backs and closures. Third, about half of the jobs in new firms are created by independent entrepreneurs. Birch also claimed that Canadian data pointed to similar conclusions.

Employment creation by small firms may have important policy implications. Canada has had the fastest-growing labour force in the Western world, coupled with a secular increase in unemployment rates (Osberg 1988; Gera and Rahman 1989), which have increased at corresponding points in successive business cycles from 3.4 per cent (1966) to 6.9 per cent (1975) to 8.3 per cent (1978) to 9.6 per cent (1986). Moreover, of those employed, an increasing proportion are involuntarily in part-time work (Cameron and Sharpe 1988, 6). There are large regional variations in unemployment rates: in Ontario it was 7 per cent in both 1978 and 1986, but it increased in the Atlantic

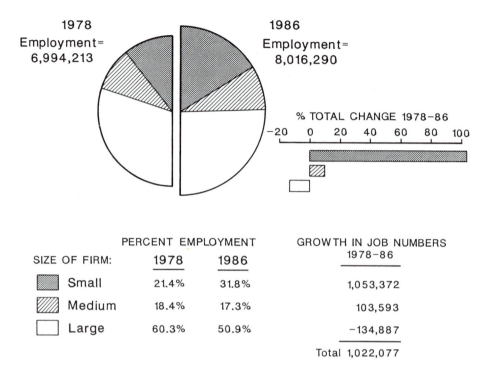

Figure 1
Employment growth and distribution by size of firm, Canada, 1978–86. *Source*: Special tabulations prepared by Small Business and Special Surveys Division, Statistics Canada. *Note*: Data are for the private sector and exclude community service (SIC 801–31), public administration (902–91), and unclassified (000). Jobs are in average labour units.

provinces from 12 to 15 per cent, and in British Columbia from 8 per cent to 13 per cent. Thus regional disparities in unemployment rates have been increasing at a time when geographic mobility of our ageing labour force can be expected to diminish (Grey 1985; Osberg 1988, 9). It is not surprising then that unemployment has often topped the Gallup Poll as Canada's most important policy issue.

Detailed regional studies of net employment growth by British geographers raise concerns of significance for Canada about the potential contribution of small firms to reducing unemployment. First, most of the new jobs created by small firms are in the service sector, not manufacturing (Fothergill and Gudgin 1982; Bannock 1986; Horne et al 1989). These jobs tend to be lower paid and therefore have lower multiplier effects. Second, gross changes underlying net employment growth are striking. For instance, about half of the firms in Canada in 1978 failed to survive to 1986; almost half of those that did survive had reduced employment. Of the seven million jobs in the commercial (non-public) sector in Canada in 1978, two and a half million had disappeared by 1986, so a net gain of one million jobs required a gross gain of three and a half million. But how are these job gains and losses distributed by firm size? Did large firms create more gross jobs than small firms? Is the churning concentrated among small firms creating two classes of firms and of jobs – temporary jobs in unstable small firms, and lasting jobs in stable large firms?

Third, should small firms be a substantial component of regional policy (Mason and Harrison 1985)? Job losses have been greatest in large firms. Hence regional job losses are greatest in regions with a concentration of large firms. Small-firm birth rates are highest in areas where small firms predominate. Thus policies to encourage small-firm birth rates and growth are likely to increase regional disparities, not decrease them (Fothergill and Gudgin 1982, 113; Storey 1986). Too little is known about the spatial pattern of gross job loss and gain at different geographic scales: at the state level in the United States death rates are relatively constant but birth rates vary (Birch 1979), while Lloyd and Dicken (1979) found that death rates in the Manchester and Liverpool conurbations were generally high and varied spatially, whereas birth rates were lower and less varied.

Fourth, are there limits to the growth of small firms? Their current resurgence may simply mark the end of the fourth Kondratieff long wave of economic growth (Shutt and Whittington 1984, 11), and in this view the resurgence of small firms may be both short-lived and at the expense of jobs in large firms (Fothergill and Gudgin 1982, 131). Small firms tend to focus on supplying specialized products with limited demand (Bolton 1971, 90), suggesting that their growth is necessarily limited (Storey 1986). The motivational and social attitudes of their owner-managers (Stanworth and Curran 1986) also limit growth. Underlying this set of issues is the belief that size class may be an inadequate basis for classifying firms (Horne et al 1989, 5).

Finally, there is growing recognition of the need for a multi-factor explanation of growth. Fothergill and Gudgin (1982) identified four main factors influencing regional growth rates: industrial structure, urban size, firm size-mix, and regional policy. Traditionally, industrial structure has been regarded as the key to explaining differences in regional growth, with a residual regional effect measured by shift-share analysis (Jones 1940, 249–80; Watts 1987), but this view has been weakened by new research. The factors may be different in Canada, where manufacturing decline in the 1981–82 recession may have been greatest in the non-metropolitan regions (Kale and Lonsdale 1987). And the migration of Canadian workers who have been laid off twice or more seems to suggest that they find better re-employment opportunities in the larger metropolitan areas (Grey 1985). More important than urban size structure is probably country of control because of the extent of foreign ownership in Canada, its regional and sectoral concentration, and the distinctive industrial behaviour of foreign firms (Watkins et al., 1968; Ray 1971; Britton and Gilmour 1978).

It is now possible to examine a number of these issues for Canada using special tabulations of the Small Business and Special Surveys Division of Statistics Canada. These tabulations link files on firms and employment at Statistics Canada, at Revenue Canada, and at Employment and Immigration (McVey 1987; Statistics Canada 1988). The number of firms, and their employment (measured in average labour units, which converts full time, part-time, and temporary workers to a common base), are cross-tabulated by province, two-digit SIC, three size classes of firm, and Canadian and foreign control for the period 1978–86. These two years are identified by the Economic Council of Canada as corresponding points on the business cycle: unemployment rates were moderately low after having peaked in 1981 to 1983 (Gera and Rahman 1989). Additional tabulations for the period classify firms and gross employment changes by the components of change (continuously identified firms increasing or decreasing employment 1978–86, births after 1978, and deaths before 1986).

EMPLOYMENT GROWTH BY ECONOMIC SECTOR

If net employment growth is tabulated by economic sector instead of by size class, then just three sectors – retail, finance and commercial services – added 954,999 jobs from 1978 to 1986. Small-firm employment was concentrated in these three fast-growth sectors (58 per cent of their employment, compared with 41 per cent of the total labour force), while they were under-represented in the slow-growth manufacturing sector (10 per cent of their employment, compared with 29 per cent of the total labour force).

There is concern about the quality of these new service jobs because they are generally for fewer hours and lower pay (Canadian Labour Market and Productivity Centre 1988). Laroche (1989) has noted that not only do smaller firms pay lower wages, both for the total of all sectors and for manufacturing, but also the gap has been widening. So if small-firm growth is sector-driven, policies to encourage small firms are using the wrong lever to generate the wrong sorts of jobs.

In fact, small firms grew faster than large in every industry at the two-digit SIC level and produced more than half the net growth in every sector of the economy (Figure 2). In transportation, almost the entire labour force is in large firms; these declined, resulting in a net reduction for the sector, even though small and medium-sized firms expanded their employment. In several other sectors – for example, primary, construction, and wholesale trade – small firms offset net job losses by medium-sized and large firms to produce net sector growth.

The consistently good performance of small firms in every economic sector contrasts with the decline of large-firm employment in all sectors except retailing, finance, and commercial services. Hence industry mix cannot explain size-class performance. On the contrary, size mix has become more important to the growth performance of industry than has industry mix to size performance. Consider employment growth in forestry, wood industries, and paper and allied industries by size class. The paper industry did best, followed by wood, with the forest industry last in each of the three size classes (Table 1). But their ranking on the predominance of small firms is exactly the reverse: the paper industry is dominated by large firms and the forestry industry by small. And it is this size ranking that determines industry growth ranking. So forestry

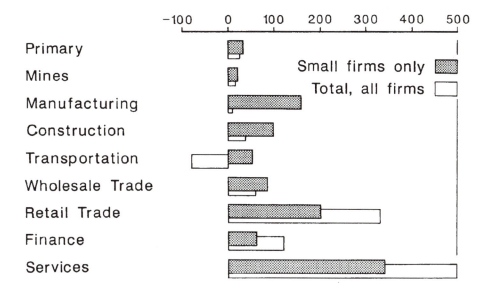

Figure 2
Net growth in jobs (000s), by economic sector, Canada, 1978–86. In all sections but retail trade, finance, and services, total employment growth is less than the subtotal for small firms because of net job losses in medium and/or large firms.

scored the highest growth rate notwithstanding its lowest ranking in every size class. This is an example of Simpson's paradox (Simpson 1951; Blyth 1972; Paik 1985; Cohen 1986). The lesson is clear: size effects have become so important that failure to correct employment growth rates in different industries and regions to take account of firm size-mix differences confounds industry and regional effects with size effects.

A method is needed to partition out industry, region, and firm-size effects on net employment growth: shift-share analysis (Watts 1987, 232–6) is limited to two effects, however, and multi-factor partitioning extends shift-share to the multivariate case and corrects its mathematical flaws (Ray 1990). Partitioning industry growth rates for size mix, regional mix, country-of-control mix, and various interaction effects shows that in retailing and commercial services, a favourable size mix reinforced a positive industry effect to drive growth rates higher still. Equally, an unfavourable size-mix effect reinforces a negative industry effect in manufacturing and transportation to slow growth even more. Conversely, finance did well despite an unfavourable size mix, and primary industries did well only because of a favourable size mix. The most important finding is, however, that compositional differences in industry mix narrowed the growth differences between small and large firms. The intrinsic size effect of small firms was slightly higher than the 70 per cent growth that they actually achieved. The pre-eminence of small firms in employment creation is a true attribute of their size, not of their composition.

Table 1

Employment growth in the forestry, wood, and paper industries by size class, 1978–86

	Forestry		Wood		Paper	
Size class	%	Rank	%	Rank	%	Rank
Small (0–19)	52	3	80	2	179	1
Medium (20–100)	−28	3	−6	2	29	1
Large (100+)	−40	3	−19	2	−14	1
Total	−3	1	−4	2	−9	3
% small	36	1	12	2	1	3
% large	35	3	65	2	94	1

Source: Special Tabulation, Small Business and Special Surveys, Statistics Canada.

Note: Employment growth is net change in average labour units 1978–86 as a proportion of 1978 employment. Forestry is division 2, SIC 031 and 039. Wood is manufacturing group 8, SIC 251–9. Paper and Allied are group 10, SIC 271–4. The percentage of labour force in small and large firms for these industries are for the base year, 1978.

MANUFACTURING EMPLOYMENT
GROWTH AND FOREIGN OWNERSHIP

The contrast between small and large firms is particularly marked in manufacturing. Small firms more than doubled employment 1978–86, whereas large firms decreased theirs by more than 10 per cent. Small firms grew in every two-digit industry, whereas large firms declined in all but two, rubber and plastics, and printing and publishing.

Attention inevitably focuses on the role of the large foreign-controlled firms that account for more than a third of manufacturing employment. They shed 10.4 per cent of their labour force 1978–86, while Canadian-owned firms increased theirs by 6.7 per cent. Are the multinational corporation (MNC) "snatchers" investing in host countries when good profits are to be made but quick to leave in a recession (McAleese and Counahan 1979)? Certainly the exit barriers – that is, the costs of closing down – are lower for MNCs than for domestic firms (Harrington 1981; Harrington and Porter 1983; MacLachlan 1986). They can transfer equipment to another country rather than liquidating. Intangible assets, such as goodwill and proprietary knowledge, which can be of considerable value, need not be lost in an MNC transfer of production but are usually lost for Canadian non-multinationals that close. Furthermore, MNCs may be under political pressure to maintain employment at home during a recession, whatever the employment cost to branch plants in other countries. Or they may use the economic reality of recession as an excuse to move production to lower-cost areas, such as export-processing zones in Asia (International Labour Office [ILO] 1988).

Finally, Canadian branch plants tend to show up badly in the portfolio planning now used by MNCs (Davidson and McFetridge 1984). Portfolio planning weighs the value of branch plants by their competitive position and business attractiveness. Competitive position, measured in market share, is reduced in Canada by the sheer number of U.S. companies with subsidiaries and the resulting miniature-replica effect. Business attractiveness is reduced by Canada's relatively slow economic growth.

Surprisingly, in view of the actual decline in employment in foreign-controlled manufacturing firms and the theories to explain it, the longitudinal data indicate that when industry-mix, size-mix, and other effects are taken into account, foreign firms did better than Canadian (Ray 1990). Thus the poor performance of large firms, and the pre-eminence of small firms in employment creation, cannot be explained by the concentration of foreign ownership among large firms and in manufacturing industry. Indeed, Canadian firms have been increasing their foreign investment rapidly and the ILO now assumes that "the employment growth abroad of Canadian MNE's has probably been greater than in the home country" (ILO 1981, 19). Indeed large firms, foreign or domestic, have done worse than small firms, foreign or domestic, in job creation both in manufacturing and in the economy generally.

TURNOVER IN FIRMS, JOBS, AND WORKERS

Though net employment trends are better documented, easier to understand, and attract more public and policy attention, it is the underlying components of gross employment growth that measure the processes of sectoral churning, employment growth, and worker mobility (Dunne, Roberts, and Samuelson 1989). This churning, at whatever level, is not cost-free, and so it is important to know how it is distributed by size class.

Firm churning is clearly related to size-class (Figure 3). The death rates of firms decrease with age and with size of firm. Typically, 35 to 40 per cent of all firms opened between 1979 and 1984 had closed within three years. The rate for new large firms depended on the stage in the business cycle: 20 per cent opened at the beginning or end of the business cycle closed within three years, but the rate peaked at 38 per cent for firms opened in the recession year of 1981. The consistency of failure rates for small firms suggests that internal factors are responsible (Cross 1981). Moreover, death rates vary more among size classes in Canada than they do among sectors or regions. Thus sectoral and regional death rates are affected by their size mix. Mining and manufacturing, with the largest average firm sizes, had the lowest death rates. Construction and retailing, with the smallest average firms sizes (apart from primary), had the highest.

Birth rates, too, are related to size class: the proportion of births drops with each increase in firm size from very small to very large. But birth rates are much more responsive to the stage in the business cycle, the general growth rate in each industry, and general regional factors. It is births, then, rather than deaths that reallocate resources among firms, sectors, and regions, and the evidence suggests that the new firms are more specialized, productive, and profitable than the ones that they replace (Baldwin and Gorecki 1990).

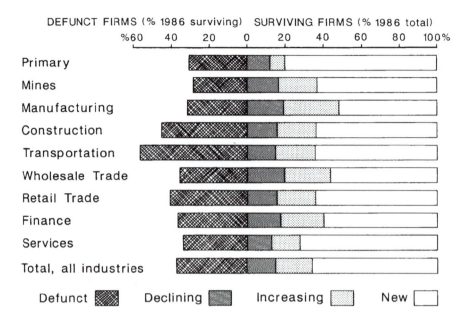

Figure 3
Churning of firms by economic sector, Canada, 1978–86.

Over the longer run, some of the employment fluctuations within each continuing firm cancel out, so that births and deaths do become increasingly prominent components of net job creation. Total job losses from declines and closures combined occur at about the same rate in all size classes, while job gains from births and increases combined drop rapidly with size class. Thus the rapid rate of net growth of small firms is due primarily to higher gross gains, and the poor performance of large firms to lower gross gains. Even new firms, established after 1978, showed the same size-related dynamic: small new firms in manufacturing had doubled employment by 1986, medium new firms had not grown, and large new firms had declined (Baldwin and Gorecki 1990).

One measure of the efficiency of job creation is the ratio of net employment increase to gross job turnover. Turnover is simply the sum of the absolute gains and losses expressed as a percentage of base-year employment. The gains for small firms were 39 per cent (firms increasing employment) plus 85 per cent (births), and losses were 13 per cent (firms decreasing employment) plus 40 per cent (deaths), for a gross job turnover of 177 per cent. The rate of turnover decreased with size class. It was 101 per cent for medium and 52 per cent for large. But net employment increase dropped more rapidly than job turnover. Small firms had to generate 1.8 jobs per net job retained. Medium-sized firms needed 6.8 jobs. Large firms were losing jobs faster

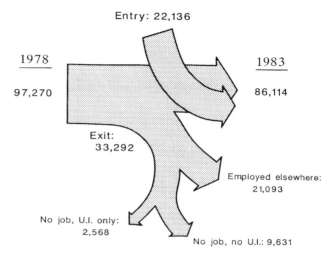

Entry: 22,136

1978

97,270

1983

86,114

Exit:
33,292

Employed elsewhere:
21,093

No job, U.I. only:
2,568

No job, no U.I.: 9,631

Figure 4
Changes in the labour force in the steel industry, Canada, 1978–83.

than they could generate them. On this basis the small firms were the most efficient net employment generators.

The steel industry exemplifies worker turnover in an industry dominated by large firms. Employment expanded 1978–80 and then declined substantially 1981–83 (Allen 1985), the number of workers falling from 97,270 in 1978 to 86,114 in 1983. The net decrease of 11,156 workers comprised 33,292 who left the industry and 22,136 who were recruited (Figure 4). Of those who left, 9,631 obtained no known job or unemployment insurance. About half were 55 or older and may have retired, but some three thousand were under 45. A small number obtained unemployment insurance, but most leavers found work elsewhere but generally are poorer paid than previously.

Steel workers who left for other industries moved in almost equal numbers to other manufacturing jobs, to service industries, and to other activities (primary, construction, and transportation). This group and those moving into manufacturing had generally earned above-average wages in steel and did about as well in their new job, except in construction, where incomes were about 10 per cent lower (adjusted for inflation). Workers moving into retail trade, personal services, finance, insurance, and real estate dropped in earnings by 20 per cent or more.

Another study of seventeen thousand workers in all industries who had two or more lay-offs between 1978 and 1982, but who were working in both those years, shows that 55 per cent of those in primary metals (which includes steel) had returned to work with the same employer (Grey 1985). Only transportation-equipment manufacturing had a higher employer-attachment level. The average for all industries was 31 per cent and was lowest for wholesales and retail trade (18 per cent) and finance (20 per cent). And in general, attachment levels were highest in manufacturing, construction, and mining. Grey suggests three explanatory factors. First, as noted, a move out of manufacturing to services incurs a drop in income. Second, in mining particularly, employment is concentrated in smaller, one-industry towns, reducing alternative work opportunities. Third, in both mining and manufacturing average firms are very large, so that when they begin laying off workers, the local labour market is likely to become congested. Nevertheless, most workers who are laid off end up with a different employer in a different sector (Economic Council of Canada 1988, 30). Workers in goods-producing industries (primary, mining, manufacturing, and construction) who changed their sector after a lay-off were twice as likely to move to the services-producing industries as laid-off services workers were to switch to goods producing. Thus lay-offs are a powerful stimulus to the shift of employment from manufacturing to services.

Rates of worker turnover are very high. On average, 15 per cent of the workforce was laid off each year 1974 to 1983, though a third are temporary lay-offs. Attritions equal another 19 per cent a year. But job turnover for whatever reason is disproportionately concentrated among workers under 35 years of age. And the proportion of workers returning to the same employer after a lay-off is disproportionately low for this age group (Baldwin and Gorecki 1990). It may be for this reason that the underlying tenure rates have actually been increasing, notwithstanding heavy job turnover and factors such as the 1981–83 recession. Thus the high rate of churning among small firms has not reduced job retention or tenure prospects of workers.

THE REGIONAL EQUATION OF EMPLOYMENT GROWTH

Regional differences in the individual components of change are quite small. But they are consistent, and so they grow into substantial aggregate differences in job turnover rates and net employment growth. The biggest contrast is between Ontario and Quebec. Ontario did better than Quebec in three of the growth components, with bigger contributions from growing firms and smaller losses from deaths and declining firms. So it stood at the head of the regional-growth league, and Quebec at the bottom. Quebec had a higher birth rate of firms than Ontario in every size class. But Ontario had a higher rate of employment growth in all two-digit manufacturing industries and in all two-digit service industries but communications and personal services. However, Quebec did better in all primary and mining industries but non-metal mining. Ontario also did better in size-specific growth; it had the smallest rate of decline of large firms in manufacturing and the highest aggregate manufacturing growth rate. Quebec had the second-highest rate of decline of large firms and the second-lowest manufacturing growth rate. On job turnover rates Ontario was lowest with 86 per cent for 1978–86, and Quebec was highest with 125 per cent. Ontario had to generate 2.8 new jobs for each net addition (the lowest regional figure) and Quebec 4.6 (the highest). And these very consistent contrasts are echoed by similar contrasts between the western regions and Atlantic Canada.

Table 2
Partitioned rates of employment growth by region, 1978–86

Growth effect	Actual employment growth	Region-mix effect	Industry- mix effect	Size-mix effect	Control effect	Other effects
Canada	14.6	0	0	0	0	14.6
Atlantic	10.9	−7.1	−0.2	2.6	−1.3	16.9
Quebec	7.9	−9.5	0	0	−0.1	17.5
Ontario	18.6	8.6	−0.2	−2.8	1.1	11.9
Prairies and NWT	16.0	−3.1	0.6	3.2	−0.9	16.2
BC and Yukon	15.1	−1.5	−0.1	4.4	−0.6	12.9

Source: A multifactor partitioning of special tabulations on employment growth by five regions, nine industry sectors, three size classes, and two country-of-control classes, provided by Small Business and Special Surveys, Statistics Canada (Ray 1990).

Note: Other effects include national growth rate (+14.61), allocation (7.81 per cent), and ten interaction effects (six two-way, three three-way, and one four-way). The aggregate value of the interaction effects varies by region.

Systematic regional differences in employment growth indicate intrinsic regional growth effects. These effects can be measured using multifactor partitioning (Table 2). One effect common to all regions, industries, and size groups is the national growth-rate effect – in this case 14.61 per cent. Another is the allocation effect (Esteban-Marguillas 1972; Ray 1990) – the extent to which location of economic activity enhances national growth rates. It is the difference between the actual national growth rate and what it would be if all activity were distributed in perfect proportion to total employment. It equals 7.81 per cent. The specific effects of industry-mix, size-mix, country-of-control-mix, and region effects are all measured in a way similar to industry mix in shift-share analysis. Finally, interaction effects, such as industry-size interaction, measure industry-specific economies of scale.

Employment growth in each region, and the nation, is equal to the sum of the individual effects. The effects measure how favourable the mix of industries, size classes of firms, and the country-of-control mix is in each region compared to the national mix. A region with a concentration of small firms (British Columbia) has a better size-mix effect than one with a concentration of large firms (Ontario). There can be no size-mix effect for the nation as a whole because it sets the norm by which the regions are judged. A similar argument applies to the other specific effects – including the region effect itself.

The region effect is a comparative measure of how much faster or slower firms of a given size, industry, and country of control tend to grow in that region than in the nation. It is an intrinsic effect that operates consistently equally on all firms, whatever their size, industry, or country of control. Therefore, just as a region can have a favourable (or unfavourable) size mix, so can a size class have a favourable (or unfavourable) region mix. And just as we can tabulate the partitioned employment effects by region to identify how their composition affects their growth, so also we can determine the effects by size class to identify how its composition has affected its growth – as we see below.

The region effect quantifies the pervasive inter-regional differences in employment growth that affect each specific industry, size, country-of-control combination. The general pattern of Ontario-Quebec and west-east contrasts is as expected, but the scale of the region effect is surprising. Firms in Ontario grew 18 percentage points faster than counterparts in Quebec. The analysis does not offer any explanation, though the political climate could well have contributed. And given that Ontario alone accounted for over half of the national increase in employment, the other regions all have negative regional effects. These results emphasize the importance of multifactor explanations of regional growth and the danger of attributing regional differences in employment growth to a single principal factor such as birth rates of new firms, as in the work of Birch.

The industry-mix effect almost vanishes at the highly aggregated region and industry scale used. Inter-regional variations in industry mix begin to emerge only at the two-digit, ten-province level of disaggregation. A rerun of the multifactor partitioning at this level picks out the contribution of mineral fuels (petroleum and natural gas) and mining services (drilling) in Alberta to the aggregate industry-mix effect in the prairies and the drag that the forestry industry has been on BC employment growth (Hayter 1988).

The size-mix effect too is very muted, even though small firms created most of the net employment growth in every province in Canada (Figure 5). This is because inter-regional disparities in the size distribution of the labour force are small. Inter-regional differences in the growth rates of firms in the same industry-size-control class are assigned to regional effects, not size effects. So the inter-regional differences in the performance of large manufacturing firms, for instance, which account for three-quarters of total manufacturing employment and therefore determine the aggregate employment performance of the industry in each province, are quite correctly counted as part of the regional effect, not the size effect. And indeed, the three provinces where large manufacturing firms declined at less than their national average rate – Nova Scotia, Ontario, and Alberta – were precisely those in which all firms tended to perform above the national average for their industry-size-control class. Hence these three provinces, the only ones in which manufacturing employment increased 1978–86, are the only three with a positive regional effect (Figure 6).

Similar considerations apply to the influence of country of control. The difference in the intrinsic growth rate of Canadian and foreign firms is considerable, giving Canadian firms a −3.6 per cent control effect and foreign +15.1 per cent. Ontario was the only region where foreign-controlled employment was disproportionately high – 53.6 per cent of the nation's foreign-controlled employment, compared to 41.0 per cent of all employment. So the control-mix effect by region, which measures how differences in the distribution of foreign-controlled employment affected growth, is quite small. It ranges from +1.1 for Ontario to −1.3 for the Atlantic provinces.

The net effects on regional growth of the mix of industries, sizes, and control are further reduced because they tend to offset one another. Ontario suffers from a concentration of manufacturing and of large firms but benefits from the concentration of

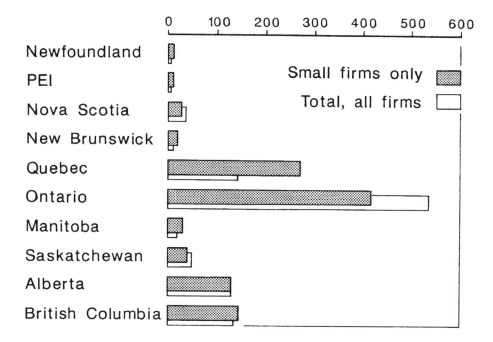

Figure 5
Net growth in jobs (000s), by province, 1978–86.

foreign ownership. The Atlantic provinces had a larger concentration of small firms than Ontario but lost on foreign ownership. Their industry-mix effect was the same as Ontario's: what it gained from a lower concentration of manufacturing it lost from a high concentration of transportation (nationally a slow-growth industry) and a low concentration in primary industries (nationally a fast-growth industry). Nevertheless the aggregate differences in employment growth were considerable. These differences can be summarized as a growth quotient: proportion of the nation's growth divided by proportion of its employment. The highest quotient is Ontario's (1.27), followed by 1.09 in the prairies and 1.04 in British Columbia. The losers were Quebec (0.55) and the Atlantic provinces (0.74).

Such large regional differences imply large inter-regional movements in the labour force. However, there has been little tabulation or analysis of workers' mobility, and the conclusion from existing research is that regional mobility is low (Grey 1985; Baldwin and Gorecki 1990): only 11.6 per cent of Grey's sample were in a different region five years later.

The standard deviation was, however, quite large: 5.0 per cent. It seems that the size of the local urban centre was more important than the rate of employment growth

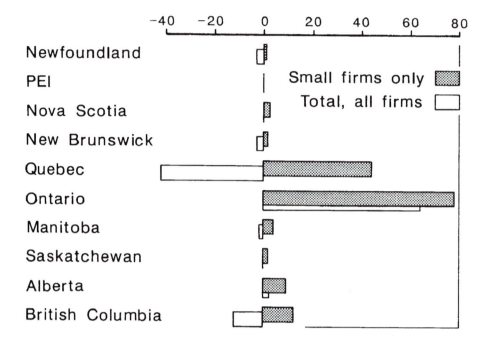

Figure 6
Net growth in jobs (000s) in manufacturing, by province, 1978–86.

in determining whether laid-off workers stayed or moved. For instance, the percent-age of laid-off workers who left was 3.1 for Montreal, 5.0 for Toronto, and 3.9 for Vancouver. By contrast the figure for southern Manitoba was 15.5 per cent. Data on Canadian workers in census metropolitan areas (CMAs) who had a permanent lay-off between 1974 and 1981 shows that 18.2 per cent had moved (Baldwin and Gorecki 1990). Of those who moved, 84 per cent went to another CMA. The proportions were much higher in heartland CMAs (between 85 and 100 per cent) and much lower in the Atlantic provinces (60 to 69 per cent) and the prairies (60 to 77 per cent). These re-gional differences may, however, only reflect the closer spacing of CMAs in Ontario and Quebec.

SUMMARY AND CONCLUSION

Small firms have become pre-eminent in employment creation in Canada, responsible for most of the net addition to jobs in every sector, in every province and at each stage of the 1978–86 business cycle. The proportion of Canadian workers employed in small firms (those with under 20 employees in 1978) grew from one-fifth in 1978 to one-third in 1986. This pre-eminence was in no way due to sectoral, regional, or country-of-control mix of small firms. The employment growth of small firms was in fact reduced, not boosted, by their composition.

The growth performance of small firms can be decomposed by multifactor partitioning and flow-charted to show the effects at work (Figure 7). Small firms employed 1,496,747 people in 1978. The national growth rate (14.6 per cent) could have been expected to add 211,969 jobs by 1986. But against this, the industry mix of small firms cost 13,633 jobs, regional mix 7,739, and country-of-control mix another 118,235. The biggest single negative influence, however, consisted of internal diseconomies of scale (measured by the industry-size interaction), which lopped off a further 510,823. The pure size effect, however, generated 1,491,841 jobs, offsetting all the losses, to give small firms a net growth of 70 per cent and bringing their 1986 employment up to 2,550,118.

One result of substantial size-class differences in growth rate in every industry is that the firm-size composition of industry has become a major factor in determining aggregate industry growth. Failure to adjust industry growth rates to take account of their firm size mix confounds industry effects with size effects and runs the risk of Simpson's paradox. In the example of the forestry, wood and paper industries paper did best in every size class but worst in aggregate because of an unfavourable size mix.

Net employment growth is bought at the price of considerable churning of firms and of jobs. The closure rates of small firms are particularly high. However, the very fact that these firms are small cushions the impact of their closure rates on employment loss. Moreover, when job losses from cut-backs are added to losses from closures, then the rates of gross job losses are about the same for all size classes.

Firm births, too, are concentrated in the small size class. However, the jobs created by these births are not matched by the employment growth of large firms. So large firms, at least in Canada 1978–86, did not generate more gross jobs than did small firms. The difference between small and large firms in net employment generation is the result primarily of differences in gross job creation and is not achieved at the cost of higher job churning among small firms. Indeed, small firms are more efficient job generators as measured by the ratio of net job gain to gross job turnover.

Given the substantial size differences in the components of employment, it is clear that the performance of economic sectors and regions needs to be standardized for difference in size mix. The multifactor partitioning should therefore be repeated for each component of employment change to provide a standardized comparison of growth dynamics.

Moreover, given the very high rates of job churning in all size classes, it would be clearly wrong to think of the labour force as divided into two classes – a small-firm workforce limited to temporary jobs and a separate large-firm workforce locked into tenured positions. For example, one-third of the workers in Canada's steel industry, which is dominated by large firms, had left between 1978 and 1983, and more than a third of those who did so accepted lower-paid jobs in the service sector. If typical,

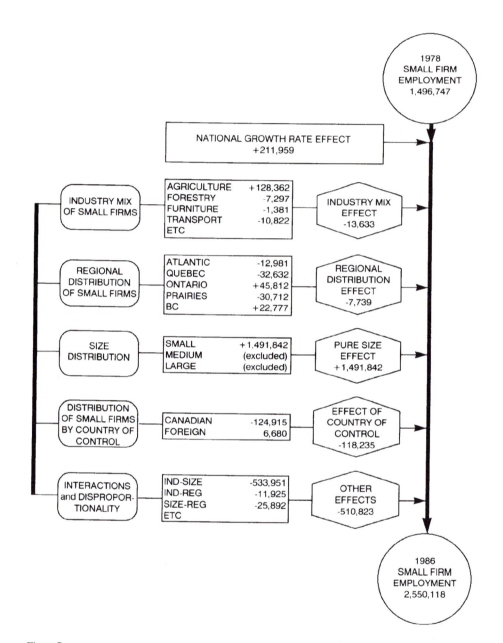

Figure 7
Partitioned effects on the growth of small firms, Canada, 1978–86.

these figures imply that permanent lay-offs from large manufacturing firms could account for a large proportion of the net employment growth in the service sector. Additional tabulations are needed to track longitudinally the distribution of workers by industry, size class, and region to identify how far employment shifts are fed by worker mobility versus new entrants to the labour force. What data are available suggest that age of worker is more important than size of firm in influencing rates of employer attachment and of sectoral and regional mobility and that length of job tenure is not being adversely affected by the relative growth of small firms. The data also indicate that employment shifts are being accommodated by low mobility rates. Thus high job churning is more than enough to facilitate regional economic adjustment.

Though small firms have dominated employment creation in every region of Canada, they have not created broad inter-regional differences in growth, because inter-regional differences in firm-size composition are very limited and regional differences in the growth rates of small firms are a symptom of systematic differences that affect all size classes and sectors and are therefore regional, not size effects. Regional differences in employment growth in Canada emerge as a multifactor phenomenon, in which the pure region effect plays a principal role. The sharpest regional contrast is between Ontario, at the top of the growth league, and Quebec, at the bottom. Western Canada and the Atlantic provinces offer a muted echo of the Ontario-Quebec contrasts. The details of this broad picture need to be filled in with tabulations of the longitudinal data for small areas and for selected industries. The distribution of small firms within CMAs and at different levels of the urban hierarchy may well be producing shifts in employment growth that are averaged out at the provincial level.

Available data do not yet help to pick the potential winners among each new crop of small firms. The distribution of employment growth within each size class needs to be cross-tabulated by other attributes such as age of firm and type of entrepreneurship. The evidence at present is that most employment growth is generated by a minority of small firms, two to five years old, in manufacturing rather than retailing, and located in a large city rather than a rural area (Swift, no date). Nor do the tabulations available explain why larger firms have done so much worse than small firms in employment creation. The theories put forward that large foreign-controlled MNCs can be expected to divest at a higher rate than Canadian firms are contradicted by the data, which show higher rates of divestment for Canadian firms. Only further cross-tabulations can assess differences in the employment growth rates of large Canadian firms with and without foreign operations.

The tracking of individual firms and workers made possible by the new longitudinal business microdata at Statistics Canada marks a turning point for regional economic analysis. It provides for the first time an effective mechanism to monitor and document the patterns and processes of employment change, to develop and test methods to analyse these changes, and to assist in the establishment of policies to realize more fully the undoubted contribution that small firms have been making to employment growth in Canada.

EDITOR'S NOTE

I received the original version of this chapter in March 1990; the author revised it in September 1992.

REFERENCES

Allen, Don. 1985. "Analysis of the Steel Industry Labour Market and the Adjustments Facing its Work Force." Paper for Canadian Steel Trade Conference, at Sault Ste Marie, May.

Baldwin, J.R., and Gorecki, Paul K. 1990. *Structural Change and the Adjustment Process: Perspectives on Firm Growth and Worker Turnover*. Ottawa: Economic Council of Canada.

Bannock, Graham. 1986. "The Economic Role of the Small Firm in Contemporary Industrial Society." In James Curran, John Stanworth, and David Walker, eds., *The Survival of the Small Firm*, 8–18. Hants: Gower.

Berry, Brian J.L., Conkling, Edgar C., and Ray, D. Michael. 1993. *Economic Geography*. Englewood Cliffs, NJ. Prentice Hall.

Birch, D.L. 1979. *The Job Generation Process*. Cambridge, Mass.: MIT Program on Neighborhood and Regional Change.

Blyth, Colin D. 1972. "On Simpson's Paradox and the Surething Principle." *Journal of the American Statistical Association*, 67, 364–6.

Bolton, J.E. 1971. *Report of the Committee of Enquiry on Small Firms*. HMSO Report Command 4811. London: HMSO.

Britton, John N.H., and Gilmour, James M. 1978. *The Weakest Link*. Ottawa: Science Council of Canada.

Cameron, Duncan, and Sharpe, Andrew, eds. 1988. *Policies for Full Employment*. Ottawa: Canadian Council on Social Development.

Canada. Department of Regional Industrial Expansion. 1986. A Study of Job Creation in Canada 1976–1984. Internal working document. Ottawa.

– 1984. An Analysis of Job Creation Based on the T.4 Employment Estimates Data Base 1978–82. Internal working document. Ottawa.

Canadian Federation of Independent Business (CFIB). 1983. *A Study of Job Creation 1975 to 1982 and Forecasts to 1990*. Toronto: CFIB.

Canadian Labour Market and Productivity Centre (CLMPC). 1988. "The Quality of the New Service Jobs." *Labour Research Notes*, No. 2. Ottawa: CLMPC.

Cléroux, Pierre. 1988. *Job Creation in Canada – 1978 to 1986 – Industrial and Provincial Employment Statistics by Size of Firm*. Toronto: CFIB.

Cohen, Joel E. 1986. "An Uncertainty Principle in Demography and the Unisex Issue." *American Statistician*, 40, 32–3.

Cross, M. 1981. *New Firm Formation and Regional Development*. Hants: Gower.

Davidson, W.H., and McFetridge, D.G. 1984. "Recent Directions in International Strategies: Product Rationalization or Portfolio Adjustment?" *Columbia Journal of World Business*, 19, 95–101.

Dunne, Timothy, Roberts, Mark J., and Samuelson, Larry. 1989. "Plant Turnover and Gross Employment Flows in the U.S. Manufacturing Sector." *Journal of Labor Economics*, 7 no. 1, 48–71.

Economic Council of Canada. 1988. "Adjustment and the Labour Market." In *Managing Adjustment: Policies for Trade Sensitive Industries*, 21–35. Ottawa: Ministry of Supply and Services.

Esteban-Marguillas, J.M. 1972. "A Reinterpretation of Shift-Share Analysis." *Regional and Urban Economics*, 2, 249–55.

Fothergill, S., and Gudgin, G. 1982. *Unequal Growth: Urban and Regional Employment Change in the UK*. London: Heinemann.

Gera, Surendra, and Rahman, Syed Sajjadur. 1989. "Sectoral Shifts and Canadian Unemployment: Evidence from the Microdata." Paper presented at Canadian Economic Association Meeting, University of Laval, Quebec.

Grey, Alex. 1985. Aspects of Labour Flexibility in Canada: Patterns of Regional, Inter-Firm and Occupational Mobility. Manuscript. Department of Employment and Immigration.

Harrigan, Kathryn Rudie. 1981. "Deterrents to Divestiture." *Academy of Management Journal*, 24, 306–23.

Harrigan, Kathryn Rudie, and Porter, Michael E. 1983. "End Game Strategies for Declining Industries." *Harvard Business Review*, 61, 111–20.

Hayter, Roger. 1988. *Technology and the Canadian Forest-Product Industries: A Policy Perspective*. Ottawa: Science Council of Canada.

Horne, M.R., Lloyd, P.E., Pay, J.L., and Roe, P. 1989. "Structuring Knowledge of the Small Firm: A Framework for Informing Local Authority Intervention within the Private Sector." Working Paper No. 20, Northwest Industry Research Unit, Manchester University.

International Labour Office (ILO). 1981, 1985. *Employment Effects of Multinational Enterprises in Industrialized Countries*. Geneva: ILO.

– 1988. *Economic and Social Effects of Multinational Enterprises in Export Processing Zones*. Geneva: ILO.

Jones, J. Harry. 1940. "A Memorandum on the Location of Industry." Appendix II to the Royal Commission on the Distribution of the Industrial Population (Barlow Report). HMSO Command 6153, reprinted 1963, 249–80. London: HMSO.

Kale, Steven K., and Lonsdale, Richard E. 1987. "Recent Trends in US and Canadian Non-Metropolitan Manufacturing." *Journal of Regional Studies*, 3 no. 1, 1–13.

Keeble, David. 1976. "The Impact of Industrial Structure." In *Industrial Location and Planning in the United Kingdom*, 31–45. London: Methuen.

– 1980. "Industrial Decline, Regional Policy and the Urban-Rural Manufacturing Shift in the U.K." *Environment and Planning A*, 12, 945–62.

Laroche, Gabriel. 1989. *Petites et moyennes enterprises au Québec: Organisation économique, croissance de l'emploi et qualité du travail*. Seneva: Institut International d'Études sociales, Organisation International du Travail.

Lloyd, P.E., and Dicken, P. 1979. "New Firms, Small Firms and Job Generation: The Experience of Manchester and Merseyside 1966–1975." Working Paper No. 9, Northwest Industry Research Unit, Manchester University.

McAleese, Dermont, and Counahan, Michael. 1979. " 'Stickers' or 'Snatchers'? Employment in Multinational Corporations during the Recession." *Oxford Bulletin of Economics and Statistics,* 41, 345–58.

MacLachlan, Ian. 1986. "Foreign Direct Disinvestment in Ontario Industrial Plant Closure as a Divestment Instrument." Paper read at Association of American Geographers Conference in Minneapolis.

McVey, J.S. 1987. "Development of New Longitudinal Business Microdata at Statistics Canada." Working paper, Business Microdata Integration and Analysis, Statistics Canada, Ottawa.

Mason, Colin M., and Harrison, Richard T. 1985. "The Geography of Small Firms in the U.K.: Towards a Research Agenda." *Progress in Human Geography,* 9, 1–37.

Osberg, Lars. 1988. "The Future of Work in Canada: Trends, Issues and Forces for Change." Working Paper No. 11, Canadian Council on Social Development, Ottawa.

Paik, Minja. 1985. "A Graphic Representation of a Three-Way Contingency Table: Simpson's Paradox and Correlation." *American Statistician,* 39, 53–4.

Ray, D. Michael. 1971. "The Location of United States Subsidiaries in Canada." *Economic Geography,* 47, 389–400.

– 1990. "Standardizing Employment Growth Rates of Foreign Multinationals and Domestic Firms in Canada: From Shift-Share to Multifactor Partitioning." Working Paper No. 62. Geneva: ILO.

Rothwell, R., and Zegveld, W. 1982. *Innovation and the Small and Medium Sized Firm.* London: Francis Pinter.

Shutt, J., and Whittington, R. 1984. "Large Firm Strategies and the Rise of Small Units: The Illusion of Small Firm Job Generation." Working Paper No. 15, Northwest Industry Research Unit, Manchester University.

Simpson, E.H. 1951. "The Interpretation of Interaction in Contingency Tables." *Journal of the Royal Statistical Society,* Series 8, 13, 238–41.

Stanworth, John, and Curran, James. 1986. "Growth and the Small Firm." In James Curran, John Stanworth, and David Walker, eds., *The Survival of the Small Firm 2,* 81–99. Hants: Gower.

Statistics Canada. 1988. *Developing a Longitudinal Database on Businesses in the Canadian Economy: An Approach to the Study of Employment.* Ottawa: Supply and Services.

Storey, David. 1986. "New Firm Formation, Employment Change and the Small Firm: The Case of Cleveland County." In James Curran, John Stanworth, and David Walker, eds., *The Survival of the Small Firm 2,* 8–24. Hants: Gower.

Swift, Catherine, no date. Financing the Rapidly Growing Firm: Recent Canadian Experience. Manuscript, Canadian Federation of Independent Business, Toronto.

Thompson, Pat. 1986. "Small Business: Canada's Engine of Economic Change and Growth." Toronto: CFIB.

Watkins, M.H., et al. 1968. *Foreign Ownership and the Structure of Canadian Industry.* Ottawa: Privy Council Office.

Watts, H.D. 1987. *Industrial Geography.* London: Longman Scientific and Technical.

CHAPTER ELEVEN

Organizational Restructuring of U.S.-based Manufacturing Subsidiaries and Plant Closure

IAN MACLACHLAN

One of the unifying themes in Canadian economic geography is the "restructuring" of nearly every aspect of the Canadian space economy. The advent of the North American Free Trade Agreement (NAFTA) will be a significant factor in the industrial restructuring of Ontario's heartland economy. The full effect of the agreement on individual regions and labour markets is still conjectural. It appears that southern Ontario stands the greatest risk of employment losses because of its large secondary manufacturing sector and continued dependence on foreign-owned subsidiaries for employment in key manufacturing industries. However, Watson (1987, 251) argues that tariff protection may actually have discouraged manufacturing and that realization of economies of scale would lead to an increase in manufacturing value-added in Ontario.

This chapter offers a model of organizational restructuring to explain how foreign-owned subsidiaries in the secondary manufacturing sector may acquire new roles within their parent firms and how such restructuring may sow the seeds for closure of a typical Ontario branch plant. I illustrate this model with a case study of a branch plant that went through multiple phases of strategic re-evaluation and adjustment before its eventual closure in 1983. The adaptations that this firm made in response to changing corporate, market, and technological circumstances are relevant to a more general understanding of disinvestment as a possible outcome of trade liberalization.

FREE TRADE AND PLANT CLOSINGS IN ONTARIO

The steady liberalization of trade that began with the advent of the General Agreement on Tariffs and Trade (GATT) has prompted restructuring of foreign firms in Canada for several decades. With over 40 per cent of its manufacturing assets, output, and employment controlled from outside the country, Ontario's space economy has been especially vulnerable to the effects of industrial restructuring. The most notable events on the road to a more liberal trade environment include the Canada-U.S.

Automotive Products Agreement (Auto Pact), which permitted duty-free trade of automobiles and parts starting in 1965, the Kennedy Round of GATT (1963–67), which lowered tariff barriers, and the Tokyo Round (1973–79), which further lowered tariffs and regulated many forms of non-tariff barriers. By 1987, when the last of the Tokyo Round's tariff reductions had been phased in, two-thirds of Canada's imports from the United States entered free of duty and the Canadian tariff rate (averaged over all imports from the United States) had fallen to less than 4 per cent (Economic Council of Canada 1987, 26–7). Thus trade liberalization had all but negated the original rationale for many of Ontario's American-controlled branch plants before the Canada–United States Free Trade Agreement (FTA) began to take effect in 1989.

Seen against this background, NAFTA codifies and continues a trend toward more liberal trade between Canada and the United States that began in the 1940s. Branch plants have been adjusting to freer trade for some time. The crucial question for Ontario continues to be how many foreign subsidiaries will respond to the freer trade by closing plants and how many will adopt new strategies to adjust and benefit from new trade opportunities.

Notwithstanding the potential for adjustment and growth of manufacturing enterprises that can benefit from trade liberalization, many establishments will be unable to survive in markets dominated by intense foreign competition. Predictions of net employment creation tend to divert attention away from the painful reality of localized unemployment resulting from the closure of branch plants and the organizational and strategic realities of branch plant operations in a more competitive environment. Critics of liberalization of trade have warned that foreign-owned manufacturing subsidiaries are more likely to close when trade barriers fall for three principal reasons: higher production costs in Canada, lower barriers to exit for subsidiary firms, and negative perceptions of the Canadian business climate.

High-cost Canadian subsidiaries are more vulnerable to foreign competition than larger-scale U.S. plants and will be the first victims of intensified global competition. Some survey evidence has suggested that U.S. firms would close down Canadian subsidiaries in the event of freer trade because they could supply the Canadian market more cheaply from south of the border (Matthews 1971, 63). Baranson (1985, 16) argues that U.S.-owned plants will be withdrawn from Ontario when manufacturing in Canada is no longer a condition for serving the Canadian market. Britton (1977, 368) applies locational principles to conclude that rationalization would tend to favour plants in more centrally located North American cities and that Canadian subsidiaries would close faster than U.S. branch plants.

Foreign-owned firms have relatively low barriers to exit. Just as multinational corporations are better equipped than domestic firms to surmount barriers to entry, they are less encumbered by the barriers to exit (Shapiro 1983). Since foreign companies are able to deploy assets at relatively low cost, there is a smaller investment to recover in the event of liquidation. Subsidiaries that are not closely integrated with home-country operations are relatively easy, managerially, to open and equally easy to close (Townroe 1975, 48). Foreign-owned parent firms owe no loyalty to the Canadian labour force or business community. When American corporations contract they will choose to sacrifice jobs in Canada before those at home (Britton 1977; Carr 1988).

The nationalist industrial policies of the Trudeau era gave Canada's business climate a poor reputation. Shapiro (1983) argues that foreign firms are likely to be more sensitive to such policies than domestic ones and more likely to scrap plants and leave the market. Canada's plant-closing legislation is much stronger than most U.S. state legislation, and American policy analysts have warned that plant-closing laws encourage firms to "vote with their feet" and seek out more amenable industrial climates (Batt 1983; McKenzie 1982).

While these factors seem to imply that Canada's trade agreements will reduce industrial employment in Ontario, some analysts have been much more sanguine. An econometric simulation by the Economic Council of Canada (1988, 24) suggests that Ontario would gain a net 95,000 jobs (7,000 in manufacturing) as a result of the productivity increases and investment induced by the FTA. MacCharles's (1984) investigation of the effect of falling trade barriers showed that foreign-owned firms invested in Canada for the long term and resolved to improve their positions by adaptive strategies to reduce unit costs. Rugman's (1988) survey of ten U.S. subsidiaries in Canada and 16 Canadian-based multinationals did not detect any significant differences in the intentions of the two groups. On balance both groups anticipated growing investment in Canada and increased exports from Canada. Rugman (1988, 7.21) found "no evidence ... that U.S. subsidiaries in Canada will close plants and create job losses." Lazar's (1988) survey of employment intentions indicated that foreign-owned corporations would increase employment by 26 per cent between 1985 and 1995 in the event of a favourable free trade agreement being negotiated. This compares to a 22 per cent increase in employment if there were no changes in Canadian-U.S. trade relations between 1985 and 1995. This survey evidence of stated intentions is corroborated by recent trends in plant closings.

The Employment Adjustment Branch of Ontario's Ministry of Labour has kept detailed records on plant closings affecting more than 50 employees since 1981. Both the number of manufacturing plants closed and associated employment loss decreased from 1982 until 1986 and has increased ever since (Figure 1). Through the mid-1980s, when the Tokyo Round of tariff cuts were being phased in, plant closings decreased. They peaked in 1982 and reached an all-time high in 1990; these maxima coincide with recession years. From 1989 to 1990 plant closings grew by 44 per cent while employment losses increased 53 per cent.

Critics of freer trade have attributed these peak levels of dislocation to the out-migration of U.S.-owned branch plants. However, closure of foreign-owned manufacturing plants accounted for 39 per cent of both the total number of plant closings in 1990 and of the total number of employees laid off by such shutdowns. The proportion of foreign plant closings was actually down marginally from previous years. Since employment in foreign-owned manufacturing accounts for 42.5 per cent of Ontario's manufacturing labour force, the size of employment losses caused by plant closures suggests that foreign-owned branch plants are, if anything, more robust than domestic

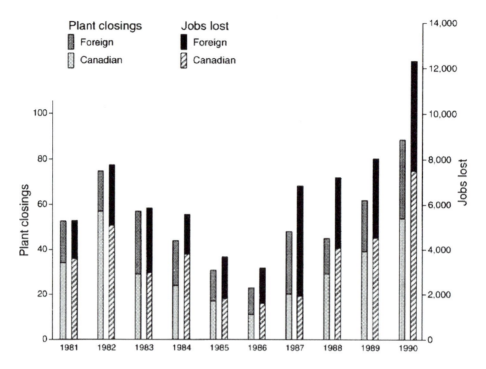

Figure 1
Foreign and domestic manufacturing plant closings and employment losses in Ontario, 1981–90. *Note*: Includes only manufacturing plant closings that result in lay-off of 50 workers or more. *Source*: Raw data from Ontario Ministry of Labour, Employment Adjustment Branch.

firms. At this early stage of implementation of the FTA there is no evidence of an exodus of foreign-owned branch plants; the level of foreign plant closings has remained approximately proportionate to their numbers.

Nevertheless, foreign-owned branch plants will continue to experience competitive pressures as a result of trade liberalization; some will adapt and some will close. To gain a better grasp of the actual adjustment mechanisms and locational behaviour within foreign-owned manufacturing it is necessary to understand how remote production facilities are linked to parent firms and how those links may change as trade barriers fall.

To that end, I propose a model of organizational restructuring to explain a particular form of structural change that takes place within the intangible hierarchy of multinational firms when trade barriers decrease, foreign competition increases, and the original rationale for foreign direct investment gradually erodes. Organizational restructuring often has a visible expression on the economic landscape in the form of technological, employment, and product change in existing plants, shifts in head office functions, and altered patterns of international trade between subsidiary and parent firms. I test the model with a case study of a typical foreign subsidiary to show its relevance to structural change and plant closure in a liberalized trade regime.

CORPORATE ORGANIZATION STRUCTURE

The strategy and the bureaucratic organization structure of firms are closely related (Chandler 1962). Structure constrains the strategic autonomy of business units within the firm and channels their interaction. The key characteristic of organization structures is the criterion by which the corporation is hierarchically divided. When firms adopt new strategies, they come under pressure to design new organizational structures to complement the strategy. Restructuring changes the organizing criterion and redefines strategic business units, presenting new opportunities to rationalize operations. Thus a new organizational structure often sets the stage for investment and expansion or disinvestment and plant closure.

   Corporate organizations may be structured according to three criteria: function, region, and product. Small enterprises and business units with narrow product lines tend to organize their activities along functional lines (manufacturing, marketing, finance). A regional organization structure may be selected to facilitate entry into foreign markets and to adapt to conditions in the host country. The regional structure is especially appropriate for firms that customize marketing activities to suit local cultures and conditions in foreign countries and whose operations do not require close coordination with domestic counterparts. Diversified firms tend to apportion responsibility among strategic business units according to product groups. This has become the most common form of organization structure for large firms in Canada. Lecraw and Thompson (1978, 82) found that 42 per cent of the largest corporations operating in Canada had product division structures. The spatial span of control of the product division depends on the firm's regional orientation. Companies that are developing a global orientation may switch from regionally defined business units in each of their foreign markets to a global product division structure.

Strategy and Structure

In reality these organizational criteria may be hybridized. Some functions may be controlled on a regional basis, while others are product specific. Manufacturing subsidiaries may produce at a global-output scale while marketing is decentralized in regional units. Product and regional structures may be combined in the same firm. As U.S.-based multinationals began to expand after 1945 they were typically configured as product division organizations at home, with international operations organized along regional lines (Davis 1976). As corporations diversify, their foreign subsidiaries may begin to operate as corporations in their own right with product divisions paralleling the parent's home organization structure. This has been termed the "mirror effect" (Brooke and Remmers 1978, 36–8). The emulation of the parent's structure in foreign subsidiaries is typical of firms practising a localized or multidomestic strategy, which takes a unique approach to each national market's idiosyncrasies.

The opposite of a localized strategy is a global strategy, in which firms ignore regional differences and attempt to manipulate global preferences through cosmopolitan marketing and aggressive pricing (Levitt 1983). They produce standardized "world products" in huge volumes for sale in global markets. A global product division structure is the appropriate organizational vehicle for firms that have elected this strategy.

Regional Mandates and Strategic Autonomy

Though most foreign-owned subsidiaries are established as legally incorporated entities within the host country, they do not necessarily function as corporations. The subsidiary may comprise a diverse mixture of more or less unrelated business units, each responsible to different parent business units in the home country. In other instances the Canadian subsidiaries of transnational firms may be integrated, self-contained business units in their own right. To understand the motivations for the restructuring of a subsidiary it is necessary to appreciate its strategic role within the multinational parent. Figure 2 adapts a construct developed by White and Poynter (1984) to show the four principal roles for manufacturing subsidiaries. The role is a function of both the strategy and structure of the corporate parent and the level of autonomy and integration in the subsidiary. A fully integrated subsidiary is responsible for its own research and development, technology, production, marketing, distribution, and product service. Many subsidiaries are "truncated" and fulfil only some of these activities.

The value added in market satellites is generally confined to the packaging and final assembly of products to reduce shipping costs and to tailor products to the needs and conditions found in foreign markets. A firm of this type differentiates its imported product by establishing a locally based distribution and service network to maintain a brand image in the foreign market and allow it to compete with local products and other imports.

The miniature-replica branch plant is the most widespread role for foreign-owned subsidiaries. Branch plants were typically established as import substituters to avoid the tariff on finished goods; hence they are sometimes called tariff factories. It has long been acknowledged that tariff factories in Canada operate at less than optimal production volumes (Eastman and Stykolt 1967), since their mandate is limited to a small market. To spread overhead costs, these plants tend to produce a wider product line with greater vertical integration than is the case in American plants of equivalent size (Caves 1975). Thus the unit costs of tariff factories run from 20 to 40 per cent higher than their American counterparts (MacCharles 1984, 55). A gradual decrease in trade protection has encouraged subsidiaries to "reverse old strategies of product diversity in a single market in favour of single products in a diversity of markets" (Crookell 1983, 25).

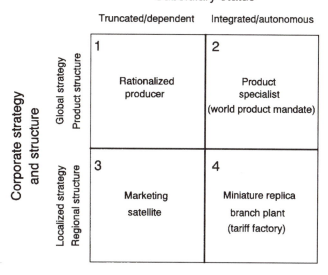

Figure 2
Strategic roles of manufacturing subsidiaries in Canada.

A rationalized producer is specialized and dedicated to production of a narrow range of goods. As a working part of an integrated product division controlled from the parent's home base, its responsibilities are confined to routine, day-to-day operations. The parent allocates products to the rationalized subsidiary, which is dependent on the parent business unit for centralized design, purchasing, and marketing. The many U.S.-based auto parts plants in southern Ontario set up under the Auto Pact exemplify rationalized manufacturing (Holmes 1983; see also Holmes, this volume, chap. 13).

A product specialist has the exclusive rights to manufacture and market a specific product. It develops, produces, and markets a limited product line with a mandate to serve an extensive market area. A subsidiary with a "world product mandate" is a product specialist authorized to sell in global markets. It has near-complete strategic autonomy within the constraints of the product charter. But it is not free to diversify into new products and needs parental approval to undertake major capital expenditures (Science Council of Canada 1980).

A Model of Organizational Restructuring in a Foreign Subsidiary

Firms operating in dynamic technological and competitive environments are constantly under pressure to adopt new strategies which in turn call for new organizational structures, especially when several business units are involved in different regional markets. While the precise form of restructuring varies, firms with domestic product divisions and regionalized international operations most commonly transform themselves into a global product structure, which is consonant with global strategies. Levitt (1983, 83) argues: "The world's needs and desires have been irrevocably homogenized. This makes the multinational corporation obsolete and the global corporation absolute."

New world product division structures change the reporting lines of regional production units and centralize control over individual product lines. Centralized control and a truly global orientation reveal new ways to rationalize production not possible under a multidomestic strategy and a regional organization structure. Thus the subsidiaries of firms shifting from regional to global product structures come under pressure to adopt new missions. Figure 3 presents a model of this transformation by comparing a schema of branch plants in a regional structure with the pattern of plants after adoption of a continental product structure and consequent rationalization of branch plant production.

Figure 3a models a typical spatial allocation of production in a diversified firm organized along regional lines for its single foreign subsidiary. In this scenario the national subsidiary is relatively autonomous within its national market territory, and its corporate structure mirrors the parent's. Its mandate does not extend outside this territory, and it is not equipped to respond to growing foreign competition across the wide range of its diversified product line. Thus the regionally decentralized management is likely to look for locally based solutions to global challenges that affect the entire firm (Doz 1978, 83). As a product division grows, a new continental product market strategy may become vital to effective competition within each of its markets. A more spatially extensive strategy calls for organizational restructuring that links products in both home and host countries.

If the firm adopts a continental product strategy without a change in structure there would be conflict between its diversified regional subsidiaries and the plethora of product division managers in the parent's home country. Staff functions in the diversified branch become subject to multiple lines of communication on production requirements, technical developments, competitive opportunities, and distribution channels. Thus firms restructure, placing subsidiary product divisions under the authority of their respective parent divisions. The regional head office becomes redundant for all functions except to provide a corporate shell for legal and tax purposes. Continental product divisions permit rationalization of production at a continental scale (Figure 3b). Organizational restructuring at the corporate level, particularly a shift from a regional structure to an international product structure, may be the catalyst for rationalization of subsidiaries that makes production facilities redundant and hence candidates for closure. Thus the strategy of the firm, the orientation of its corporate organization structure, and the ability of subsidiaries to adjust to trade liberalization are closely related.

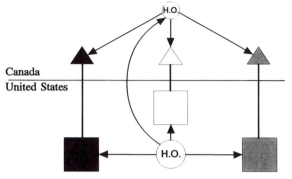

a. Regional organization structure

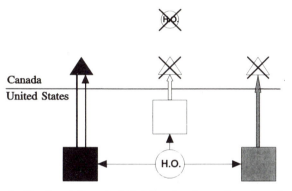

b. Continental product division structure after rationalization

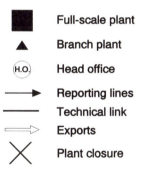

◼	Full-scale plant
▲	Branch plant
Ⓗ.ᴏ.	Head office
⟶	Reporting lines
—	Technical link
⟹	Exports
✕	Plant closure

Figure 3
Spatial model of organizational restructuring in a U.S.-based manufacturing subsidiary.

The Research Process

The strategic model outlined above implicitly requires time during which decisions are made to identify and resolve problems of business performance that derive from either external or corporate factors of economic change. Therefore using the model as a research tool requires that instances of corporate restructuring be viewed in their full historical and corporate contexts. For this reason, a case study design has been implemented rather than a cross-sectional survey, and intensive research has been implemented by means of semi-structured interviews with the key actors (Sayer and Morgan 1985; Schoenberger 1991).

The case study of McGraw-Edison reported here has been extracted from a group of retrospective inquiries of corporate closures (MacLachlan 1990). The initial step in data collection was to conduct a thorough search of the business and trade periodicals and local and national newspapers and to obtain annual reports covering the last 30 years. Respondents were identified through personnel listings in past issues of *Scott's Industrial Directory*. These contacts led to referrals to other respondents able to answer specific questions. While some respondents held lower-level positions, such as foreman or sales representative, most interviews were with former presidents, vice-presidents, and plant managers. Most people were still working in the industry but not in the same firm or any of its divested businesses. All respondents were assured of complete anonymity, and most were willing to be interviewed.

For each interview an agenda in the form of questions was prepared based specifically on the respondent's position and period of tenure in the firm. This was used to begin the meeting, though discussion often moved to other issues. Documentary information was sought, and some respondents tendered boxes of corporate plans, correspondence, and accounts from their period in the firm. A transcript prepared from interview notes and the documentary sources was the basis for a draft case study. A copy of the transcript was sent to all respondents for comment. Respondents agreed on the main facts, but there were some differences in opinion on key issues. These were resolved through further discussion, which occasionally generated the names of additional sources of information.

CORPORATE STRATEGY AND ORGANIZATIONAL CHANGE: McGRAW-EDISON, CANADIAN POWER SYSTEMS DIVISION

This case study of a large U.S. electrical products conglomerate, its Canadian subsidiary, and its Canadian Power Systems Division (CPS) supports the model outlined in Figure 3. The reorganization of strategic business units created the opportunity to divest some business units and to assign new missions to its foreign subsidiaries. Though established as a typical tariff factory, CPS attempted to adapt to trade liberalization by specializing first as a marketing satellite and later as a product specialist.

Despite these attempts to adapt, the plant's obsolete physical layout, falling profitability, and adverse market conditions dogged it for years before it finally succumbed and was closed.

Corporate Policy

McGraw-Edison was a highly diversified manufacturer of electrical goods, ranging from capital equipment for utilities and manufacturing plants to consumer goods such as appliances. Despite its reliance on a relatively technical and capital-intensive industry McGraw was philosophically opposed to the risks of developing innovative products. Its growth strategy was to acquire other firms that had developed new products and proven themselves in the marketplace. The firm had assessed and rejected the option of internal growth. "We believed a safer method was to consolidate with sound companies in the electrical manufacturing field with proven products and proven distribution" (McGraw-Edison 1958). McGraw's conservative growth strategy was justified by the uncertainty attendant on the development of new products. Since only sound companies were acquired, and it was considered imprudent to meddle with success, the acquired firms were permitted to remain autonomous in almost all respects.

Though the firm focused its acquisitions on the electrical product industry it behaved like a conglomerate and had a conglomerate style of organization structure in the 1960s (Figure 4). It had one small head office and a single subordinate tier of unrelated divisions. The firm was a loosely knit confederation of formerly independent companies. It had the very shallow organizational structure characteristic of the conglomerate style of organization formed during the merger and diversification wave of the 1960s.

After three years of failing profitability, a major restructuring program was begun in 1970. New technology and corporate planning programs were initiated to hasten growth and integrate some of the diverse and redundant activities in the decentralized divisions (Figure 5). A more centralized, hierarchical structure was imposed to strengthen existing operations and concentrate on areas with the greatest growth potential. Product groups were created to add depth to the hierarchy and centralize control of 26 divisions into six product-related groups and one foreign subsidiary, McGraw-Edison of Canada. The new structure was planned to create larger divisions, which would benefit from coordination and staff support: "It will reduce the fractionation of divisions having common markets, help to achieve greater operating efficiencies by insuring closer coordination of manufacturing, marketing, and administration activities and allow the firm to capitalize on its combined strengths in such important areas as long range planning, market research, product design, engineering, manpower development, and facility planning" (McGraw-Edison 1973, 4). The groups were organized along market lines to create integrated business units with the potential to be industry leaders. Small divisions were amalgamated, and production facilities were consolidated.

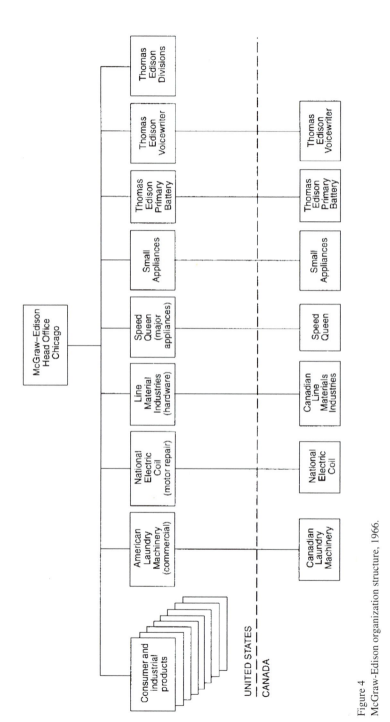

Figure 4
McGraw-Edison organization structure, 1966.

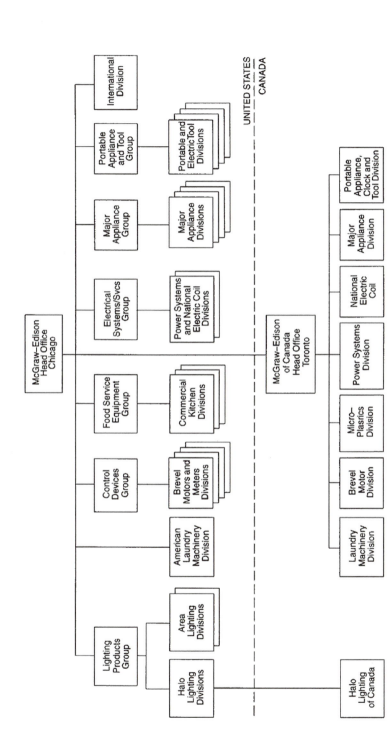

Figure 5
McGraw-Edison organization structure, 1976.

McGraw-Edison of Canada

The new strategy and centralized structure included an entirely new approach to the Canadian market. A functional subsidiary headquarters was created in 1970, with a Canadian president (Figure 5). The new structure amalgamated the diverse Canadian branches into a miniature conglomerate with a product-based organization similar to its parent – the mirror effect.

Like its parent, McGraw-Edison of Canada did not foster links between its constituent divisions. While there were many opportunities for intrafirm trade, there was little incentive for such transactions. Interdivisional sales were priced according to a "standard factory cost" formula, and divisions could earn more from selling their products to unrelated buyers than from internal markets. "Nobody wanted to sell to a sister plant – there was no money in it" (interview).

Since there was virtually no interaction between McGraw's Canadian divisions, there was little to be gained by grouping them together. However, strong links between parent and subsidiary divisions were required to facilitate imports of high-value components for assembly in Canada. In 1974, for example, subsidiary divisions purchased $15 million worth of goods from the parent, but sales to American divisions amounted to only $580,000. Under these circumstances the Canadian head office was an unwelcome intermediary between the north-south links of product divisions. Accordingly, the Canadian head office was eliminated from the corporate hierarchy in 1977; Canadian plants were integrated with their U.S. parent divisions, and the firm adopted a continental product structure (Figure 6).

Within three years some 14 of the firm's former Canadian plants were closed or divested. Closures included Power Systems plants in Toronto and Port Elgin; one Laundry Machinery and three Portable Appliances plants in Metropolitan Toronto; and two recently acquired businesses in Guelph. The Major Appliances division closed plants in Toronto and Cambridge before it was finally divested following a leveraged buyout. McGraw's Quebec City transformer plant was sold to the Canadian subsidiary of another electrical products manufacturer, and an electric motor plant was divested. Thus organizational restructuring was the precursor to withdrawal from the Canadian market, the loss of thousands of manufacturing jobs, and contraction of manufacturing in Canada.

While organizational restructuring was an important contributor to the exit from Canada, each individual establishment was subject to different market pressures. The closure or divestment of each requires an interpretation of competitive market conditions at the time of closure as well as the parent firm's strategy for that particular product line and business unit. The market for the Laundry Machinery Division was decimated by the advent of permanent-press fabric technology. McGraw-Edison closed five of its eight American plants, and so the Canadian plant closing was part of division-wide contraction. The CPS plant had been allowed to become increasingly obsolete as the subsidiary head office channelled funds to other parts of the Canadian conglomerate. Reduced markets for electrical distribution equipment and an inability

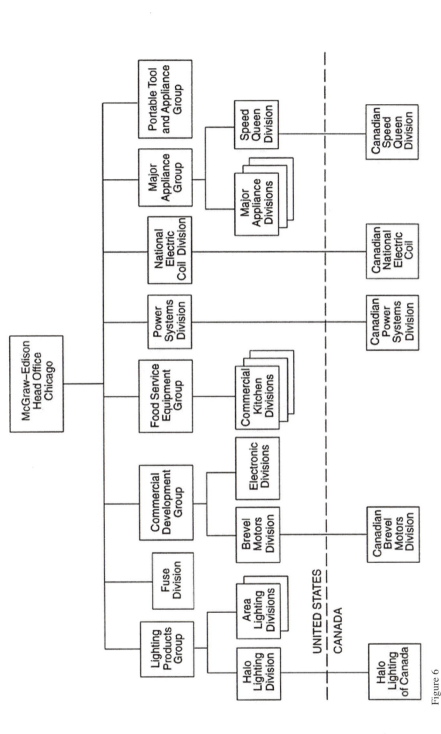

Figure 6
McGraw-Edison organization structure, 1978.

to compete with lower-cost manufacturers were influential factors in its closure. Divestment of the Canadian Major Appliances operation was undertaken at the same time as the parent divested all its white-goods divisions in the United States. This was undertaken as part of a strategy to abandon consumer products and concentrate on capital and intermediate goods.

The dramatic contraction of McGraw-Edison's Canadian operations is explicable in part by strategy and organizational restructuring internal to the firm and in part by market forces outside the firm's control. However, to fully appreciate the impact of these forces we should consider more closely the missions and mandate of individual subsidiary business units.

Role and Mandate of Subsidiary Business Units

All of McGraw-Edison's Canadian operations were configured as typical miniature-replica branch plants (Figure 2, cell 4). Tariff barriers were the prime rationale for locating these subsidiaries in Canada: "Transfer of completed products across the border attracts higher duties than parts. These duties are a dead additional cost that benefits no one at M-E" (McGraw-Edison of Canada, 1976). To reduce their exposure to these tariffs the branches imported the capital-intensive, high-technology components from U.S. divisions and did the simple fabricating and final assembling in Canada. Non-tariff barriers were also significant because of domestic procurement policies by provincial utilities and Canadian-U.S. differences in technical standards and electrical codes. As well, McGraw-Edison needed an active presence in the Canadian market to protect its flank. To prevent competitors from gaining an advantage it had to defend its Canadian market: "In our kind of business Canada cannot be successfully dealt with as though it were a 51st state. What we do in this market *does* affect what happens to markets in the U.S. We could, for instance, inadvertently give up enough of a market share for manufacturers to develop sufficient skills to successfully attack us in the U.S." (McGraw-Edison of Canada, 1978).

Canadian Power Systems Division

CPS was originally set up to manufacture a narrow line of electrical distribution equipment, but buoyant markets encouraged addition of a wide range of new utility products (such as substation switchgear, utility pole hardware, and street lights). The American parent division had diversified by acquiring new firms and adding their specialized plants to its growing production system. The Canadian branch kept up by adding extensions to its single, multipurpose plant in Scarborough. Under one roof CPS produced the same line of products that its parent division manufactured in 11 specialized plants scattered across the United States.

As in so many other sectors, the tariff barriers that had encouraged creation of CPS were gradually reduced and competition from American imports began to stiffen. The Canadian market became more heterogeneous as each of the ten provincial electrical utilities began to set its own standards and quality-assurance programs. In time there were nearly as many trade barriers between provinces as between Canada and the United States. Thus declining trade barriers and the splintered Canadian market began to erode the rationale for the Canadian branch.

Growth and Expansion

CPS had experienced rapid growth in both sales and profitability in the 1960s and early 1970s. In 1970, the midpoint of a period of steady growth, the Scarborough plant became the Power Systems division of McGraw-Edison of Canada. Sales doubled between 1970 and 1975, when returns on capital peaked at 30.5 per cent. CPS was the second-largest generator of funds in McGraw-Edison of Canada; in 1973 it was responsible for 24 per cent of sales and 44 per cent of total Canadian profits. However, these funds were used to finance acquisition and construction of a major appliance plant in Cambridge, the "glamour division" of the growing subsidiary conglomerate. Of the $2.7-million cash flow generated by CPS operations in 1975, a net of $3.2 million went to the Canadian corporate head office (McGraw-Edison of Canada 1976).

Starved of the resources required to modernize capital equipment and rationalize material flow, CPS became increasingly obsolete and uncompetitive. Yet there was no explicit or deliberate policy to run the plant down or "harvest" the operation in preparation for closure. It was assumed that the growth of the early 1970s would resume and that the plant could still turn a profit in spite of its antiquated condition. Rising revenue driven by market growth and inflation had contributed to a complacent attitude, and little attention was given to manufacturing investment, technological innovation, or cost control.

Decline and Retrenchment

Following this growth phase CPS entered into a protracted period of nearly continuous decline in both sales and profits. Stagnant markets for electric distribution equipment, aggressive competition from domestic and American manufacturers, and the plant's high cost structure all played a role in the slide. As performance deteriorated, production capacity was reduced and product lines that had been manufactured in Canada began to be sourced from more cost-competitive U.S. branches. Revenues and profits were buoyed up by resale of imported products; however, Canadian value added and production employment fell by some 40 per cent between 1975 and 1978.

The performance decline came at an inopportune time, since the parent firm had by then adopted a product-based corporate structure that placed CPS directly under the U.S.-based Power Systems Division (Figure 6). Without a Canadian head office there was no longer a commitment to build and maintain a Canadian conglomerate. CPS could not draw on Canadian cash flows to invest in new products and upgrade its obsolete facilities. It joined the American Power Systems Group as a losing operation when U.S. utilities markets were also becoming weak. The coincidence of organizational restructuring and market decline made the Scarborough plant an unwanted orphan, which resulted in further capital starvation.

Yet the plant was not closed. It scaled back on production operations and concentrated on resale of u.s. manufactured products. As the mission of tariff factory had been made obsolete by falling tariffs, cps began to adopt a marketing-satellite mission (shifting from cell 4 to cell 3 of Figure 2) and exports from the United States replaced Canadian-built goods. The manufacturing operation could survive only as a product specialist; however, none of the plant's existing product lines had sufficient competitive advantage to be the basis for a specialty. Consistent with McGraw-Edison's growth strategy, an entirely new transformer business was purchased in 1978, adding two new establishments to the Canadian operation – a small plant in Port Elgin and a larger facility in Quebec City. cps gradually specialized in this new product, and the remaining lines were gradually eliminated. Thus cps moved from cell 3 to cell 2 in Figure 2.

However the Scarborough plant was designed as a "job shop" and was poorly configured for assembly-line production of standardized transformers. Compounding its difficulties, the market for transformers went flat during the recession years of the early 1980s. As financial performance worsened, the small Port Elgin plant was closed in 1981. Continuing losses prompted the decision to divest both the Scarborough and Quebec City plants. The Quebec City operation was divested, but there were no buyers for the inefficient and unprofitable Scarborough plant, which was closed in 1983 at a net cost of $1.2 million.

CONCLUSION

Declining trade barriers had negated the role of the Scarborough plant as an import-substituting tariff factory. The need to become more competitive and the adoption of a new organizational architecture by the corporate parent precipitated two attempts at adjustment. First, cps shifted from its role as a tariff factory toward a market-satellite concept, and later it acquired an entirely new business and adopted a product-specialist mission. Both attempts to adjust to the more liberal trade environment failed.

As a member of McGraw-Edison of Canada, cps had been drained of capital when it was very profitable. Corporate restructuring removed cps operations from control by the Canadian head office, permitting continental rationalization. By then, however, the plant was obsolete and its physical layout was ill-suited to a product-specialist role. Financial performance was further compromised when markets contracted soon after acquisition. The events surrounding closure of McGraw-Edison's cps plant and elimination of the entire subsidiary organization accord well with the model presented in Figure 3.

The link between strategy and structural change in firms has been well established (Chandler 1962). The model developed here suggests that this link can be extended to the relationship between subsidiaries and parent firms. Adoption of new corporate structures has a direct effect on integration, autonomy, value added, and employment growth of individual manufacturing subsidiaries. And when subsidiaries are unable to adapt to restructuring, plant closure is the final alternative.

The prospect of an integrated North American market and heightened global competition will encourage reorganization by many transnational manufacturers. In some industries the prospects for adjustment are so constrained that plant closures seem inevitable. Intensifying off-shore competition and the technological obsolescence of existing plants makes reinvestment essential to survival in some industries. In other cases reinvestment will be drawn to locations outside Canada with lower cost structures, and Canadian plants will be retired. While these sombre prospects are plausible for some branch plants, many more can adjust to a more liberal trade environment. There is evidence that Canadian manufacturing subsidiaries are adopting new roles, but the final outcome will depend on highly individualized considerations: congruence of corporate strategy and structure; competitiveness and adaptability of individual establishments; and competitive market conditions in the industry. While the net effect of these changes on Ontario's space economy is still unclear, it is certain that organizational restructuring will continue to influence the role and survival of manufacturing subsidiaries in Ontario's industrial heartland.

REFERENCES

Baranson, Jack. 1985. Assessment of Likely Impact of a U.S.-Canada Free Trade Agreement upon Behaviour of U.S. Industrial Subsidiaries in Canada (Ontario). Unpublished, Ministry of Industry, Trade, and Tourism. Toronto.

Batt, William L., Jr. 1983. "Canada's Good Example with Displaced Workers." *Harvard Business Review*, 61 no. 4, 6–22.

Britton, John N.H. 1977. "Canada's Industrial Performance and Prospects under Free Trade." *Canadian Geographer*, 21, 351–71.

Brooke, Michael, and Remmers, H. Lee. 1978. *The Strategy of Multinational Enterprise*. 2nd edn. London: Pitman.

Carr, Shirley. 1988. "Jobs on the Line." In Ed Finn, ed., *The Facts on Free Trade*, 15–17. Toronto: Lorimer.

Caves, Richard E. 1975. *Diversification, Foreign Investment, and Scale in North American Manufacturing Industries*. Ottawa: Economic Council of Canada.

Chandler, Alfred D. 1962. *Strategy and Structure: Chapters in the History of Industrial Enterprise*. Cambridge, Mass.: MIT Press.

Crookell, Harold. 1983. "The Future of U.S. Direct Investment in Canada." *Business Quarterly*, 48 no. 3, 22–8.

Davis, Stanley M. 1976. "Trends in the Organization of Multinational Corporations." *Columbia Journal of World Business*, 9 no. 2, 24–34.

Doz, Yves L. 1978. "Managing Manufacturing Rationalization within Multinational Corporations." *Columbia Journal of World Business*, 13, 82–94.

Eastman, H.C., and Stykolt, S. 1967. *The Tariff and Competition in Canada*. Toronto: MacMillan.

Economic Council of Canada. 1987. *Reaching Outward: Twenty-fourth Annual Review.* Ottawa: Supply and Services Canada.

– 1988. *Venturing Forth: An Assessment of the Canada-U.S. Trade Agreement.* Ottawa: Supply and Services Canada.

Holmes, John. 1983. "Industrial Reorganization, Capital Restructuring and Locational Change: An Analysis of the Canadian Automobile Industry in the 1960s." *Economic Geography,* 59, 251–70.

Lazar, Fred. 1988. "Survey of Ontario Manufacturing." *Canadian Public Policy,* 14, 78–91.

LeCraw, Donald J., and Thompson, Donald N. 1978. *Conglomerate Mergers in Canada.* Royal Commission on Corporate Concentration Study No. 32. Ottawa: Supply and Services Canada.

Levitt, T. 1983. "The Globalization of Markets." *Harvard Business Review,* 61 no. 4, 91–102.

MacCharles, Donald C. 1984. "Do Foreign-Controlled Subsidiaries Have a Future?" *Canadian Business Review,* 11 no. 1, 18–24.

McGraw-Edison. 1958. *Annual Report.*

– 1973. *Annual Report.*

McGraw-Edison of Canada. 1967–74. Unpublished correspondence.

McKenzie, Richard B. 1982. "The Case for Business Mobility." In Richard B. McKenzie, ed., *Plant Closings: Public or Private Choices,* 125–33. Washington, DC: Cato Institute.

MacLachlan, Ian. 1990. "Industrial Plant Closure and Competitive Strategy in Ontario: 1981–1986." PhD thesis, University of Toronto.

Matthews, Roy A. 1971. *Industrial Viability in a Free Trade Economy: A Program of Adjustment Policies for Canada.* Toronto: University of Toronto Press.

Rugman, Alan. 1988. "Trade Liberalization and International Investment." Discussion Paper No. 347. Ottawa: Economic Council of Canada.

Sayer, Andrew, and Morgan, Kevin. 1985. "A Modern Industry in a Declining Region: Links between Method, Theory and Policy." In Doreen Massey and Richard Meegan, eds., *Politics and Method,* 147–68. London: Methuen.

Schoenberger, Erica. 1991. "The Corporate Interview as a Research Method in Economic Geography." *Professional Geographer,* 43, 180–9.

Science Council of Canada. 1980. *Multinationals and Industrial Strategy: The Role of World Product Mandates.* Ottawa: Minister of Supply and Services.

Shapiro, Daniel M. 1983. "Entry, Exit, and the Theory of the Multinational Corporation." In Charles P. Kindleberger and David B. Audretsch, eds., *The Multinational Corporation in the 1980s,* 103–22. Cambridge, Mass: MIT Press.

Townroe, P.M. 1975. "Branch Plants and Regional Development." *Town Planning Review,* 46, 47–62.

Watson, William G. 1987. "The Regional Consequences of Free(r) Trade with the United States." In William J. Coffey and Mario Polèse, eds., *Still Living Together: Recent Trends and Future Directions in Canadian Regional Development,* 237–73. Montreal: Institute for Research on Public Policy.

White, R.E., and Poynter, T.A. 1984. "Strategies for Foreign-Owned Subsidiaries in Canada." *Business Quarterly,* 49 no. 3, 59–69.

CHAPTER TWELVE

Restructuring in Mature Manufacturing Industries: The Case of Canadian Clothing

JAMES CANNON

The growth of manufacturing in low-wage, developing countries has presented a continuing competitive threat to labour-intensive industries in high-wage, industrialized nations. Clothing production represents a case in point. Throughout much of the period since the Second World War clothing producers in the high-wage economies have been viewed as an endangered species. While garment production has experienced dynamic change throughout this period, fears of extinction have not been realized. Innovative strategies of adaptation and accommodation have enabled producers to survive and at times thrive. However, the recent implementation of the North American Free Trade Agreement (NAFTA) poses challenges to Canadian garment producers. Recent history suggests not an inevitable total collapse of the Canadian industry, but a continuing fight for survival that will have dramatic effects on producers, workers, and communities.

This chapter examines the Canadian clothing industry's struggle to survive. The analysis begins by tracing changes in the size and significance of garment production in Canada over the past 30 years. Second, I examine distinctive structural characteristics of the industry, which have been vital to the sector's coping strategies. Third, I describe the organization of clothing production as a basis for discussing restructuring. Internationalization has been a leading cause of restructuring. Increased levels of import penetration have pushed such efforts along three distinct paths. First, producers in high-wage economies have appealed to governments for protection from imports and for business assistance programs to support industrial adjustment. Second, strategy has involved outward processing. Finally, producer-initiated forms of restructuring have included technological innovation, reorganization of the labour process, and revamping of enterprises. I consider a selection of these strategies and their implications for producers, workers, and communities. To conclude, I review prospects under NAFTA.

CLOTHING IN THE CANADIAN ECONOMY

Efforts to assess the effects of trade liberalization have universally concluded that Canadian clothing industries are likely to be hurt. Comprehensive analyses of actual effects are impeded by lags in the reporting of official statistics and by significant changes made in the classification of manufacturing and trade data. However, both the quantitative significance of clothing in the economy and its distinctive structural characteristics demand that industry performance be assessed.

In 1988, the Canadian clothing industries had 115,485 employees. Despite their image of decline, during the post-1982 economic recovery employment increased by 8 per cent (8,598 workers) and shipments by 24 per cent ($1.1 billion) in constant 1981 dollars. However, expansion during the recovery also needs to be seen in longer-term perspective. Clothing employment increased relatively slowly from 93,306 in 1961 to 104,300 in 1973 before declining to 91,306 in the early 1980s. In constant 1981 dollars, the value of shipments increased from $2.6 billion in 1961 to a peak of $4.7 billion in 1979 before declining to $4.0 billion in 1982. Thus the performance in the 1980s recovery must be seen in the context of employment decline in the previous 20 years and growth in the value of shipments, which was only 52 per cent of the all manufacturing rate.

Structural Properties of Clothing Production

Clothing production is characterized by internal diversity and heterogeneity. The sector consists of four three-digit industries, which are further disaggregated into 18 four-digit industries.

Three-digit	Industry name	No. of four-digit industries
243	Men's Clothing	5
244	Women's Clothing	5
245	Children's Clothing	1
249	Other Clothing	7

Differences among products, productions methods, markets served, and other features of particular industries create significant variations within and between industries.

Amidst this diversity, three structural characteristics are apparent. First, garment making is dominated by small production units. In 1988, 53 per cent of the 2,821 establishments ennumerated employed fewer than 20 employees, and 90 per cent employed fewer than 100. Moreover, the proportion of plants in the "smallest" category has increased by 13 per cent in the past decade. Only 3 per cent of all establishments had more than 200 employees.

The size structure of production units affects the way firms behave and the performance of the industry. The level and growth of productivity in clothing have been below that recorded in all manufacturing. Most firms are poorly capitalized, and size constrains both their willingness and their ability to undertake major investment.

Clothing production is geographically concentrated. Quebec and Ontario account for 86 per cent of industry employment and 90 per cent of shipments. Quebec, with 56 per cent of employment and 60 per cent of shipments, has the larger share of production. At a three-digit level, Quebec accounts for 73 per cent of children's clothing and is slighty over-represented in "other" clothing and women's clothing. Ontario has a relatively large share of men's clothing and is slightly over-represented in "other" clothing. While garment making is important to some small communities, the bulk of activity is found in larger centres. Clothing production is relatively more significant in the Montreal economy than it is in Toronto.

Quebec's share of industry employment fell from 65 per cent in 1961 to 56 per cent in 1988. Ontario's share rose from 25 per cent to 30 per cent over this period, with the remaining positive shift distributed throughout the rest of Canada. Geographic concentration underlines the economic importance of clothing production to particular areas and has increased the industry's political visibility.

The clothing sector's labour force has distinctive structural properties. It has a gendered division of labour in which most of the small number of higher-paid "skilled" positions are occupied by males. Numerically, however, the workforce is predominantly female (Seward 1990). In all manufacturing women account for one-third of employment, but over 75 per cent of clothing workers are female. A large proportion of female clothing workers are immigrants and have below-average educational achievement and low levels of competence in French or English. A high proportion of female clothing workers are married. Labour-force adjustment is compounded by these structural characteristics.

In summary, industry structure has profoundly affected restructuring and has led to strong public policy interest in the problems encountered by clothing capital, garment workers, and communities that rely on clothing industries.

Organization of Garment Making

Clothing production is structurally linked to a broader industrial complex. In an upstream direction, it is connected to the manufacture of yarns, fibres, and fabrics – the basic material inputs for garment-making. For clothing producers, availability of a wide range of fabrics, at internationally competitive prices, is essential to commercial success and often establishes a basis for conflict over trade policy with fabric producers and other interest groups in the clothing complex. Clothing production also involves downstream links with major buyers. The high level of buyer concentration among major retail clothing chains and department stores contrasts starkly with the fragmented nature of clothing production.

Four distinct stages of garment making can be identified – design, pre-assembly, assembly, and finishing (Rush and Soete 1984; Hoffman and Rush 1988). The details of production at each stage vary among industries according to such factors as product, scale of output, and production technology used. In general, recent technological developments have most affected design and pre-assembly.

Design, whether accomplished using traditional manual methods or electronically using computer assisted (CAD) technology, involves a combination of creative and engineering activity that results in production of a complete set of product specifications.

In the pre-assembly phase each constituent part of the design must be graded to facilitate production of a range of sizes. Next, pattern markers are produced as templates for cutting fabric. Dramatic increases in productivity can be realized where computer-automated grading and marking and cutting technology replace traditional manual methods. Components are then transferred to the sewing room for assembly. This involves the age-old task of sewing the pre-cut components to produce the garment. Efforts to increase productivity in the sewing room have focused on increasing the speed of sewing machines, developing sewing machines dedicated to specific tasks in an effort to speed throughput, and introducing automated transfer systems in an effort to reduce the time operators spend in handling, as opposed to sewing, garments. The final stage – finishing – involves tasks such as pressing and packaging garments to prepare them to meet the scrutiny of the buyer in the marketplace. Though new technologies are widely available, economic and cultural constraints have slowed adoption in many firms.

RESTRUCTURING

The structural properties and organizational characteristics of garment making described above have profoundly influenced the nature of restructuring. However, a more fundamental cause of restructuring has been post-1945 internationalization of clothing production. Canadian manufacturers have reacted to increased global competition by appealing to governments for protection and adjustment assistance, by engaging in outward processing, and by restructuring domestic production.

Government Intervention and Restructuring

An essential ingredient of post–Second World War economic reconstruction was trade liberalization. The GATT, negotiated under American leadership, was intended to encourage trade and international economic interdependence. Economic liberalism was viewed as the key to rebuilding the war-torn developed economies and to linking the developing economies into a growing international system. Because of its significance, the "textile" trade occupied an important place in this strategy.

In the developed economies during this period, textile and clothing production still accounted for 20 to 25 per cent of manufacturing employment. In developing countries, these sectors often became important instruments of industrialization. By the 1950s, European and American markets had become outlets for the products of the newly emerging "low-cost" producers. Increased import penetration in the developed economies demonstrated the domestic-adjustment implications of trade liberalization. The developed economies found the prospect untenable and sought "managed" trade, compatible with liberalization. Thus a sector-specific international trade regime, which was in clear violation of GATT principles, began to emerge (GATT 1984; Aggarwal 1985).

Canada's experience with the emerging textile trade regime highlighted two themes. First, as relatively minor players in the global textile trade, Canadian textile and clothing industries were subject to rather strong indirect effects of actions taken elsewhere. In particular, exporting nations whose products became subject to restraint in the relatively large U.S. and European markets were forced to seek alternative outlets. The resulting trade diversion meant that Canadian markets were subject to sudden import surges that directly affected domestic producers.

Second, development of the new system illustrated the incremental nature of policy evolution. Originally viewed as a temporary measure to facilitate "orderly development" for both industrialized and industrializing countries, regulation of the textile trade has expanded over the past 30 years to become, along with that of the agricultural sector, the most glaring exception in GATT's efforts to encourage trade liberalization (Trela and Whalley 1990).

By the late 1950s, ad hoc restraints on textile and clothing trade between industrialized and developing nations were proliferating. The Short Term Arrangement (STA) concluded under GATT auspices in 1961, followed by the Long Term Arrangement (LTA) a year later, permitted importing nations to negotiate trade restraint agreements with particular exporting nations on cotton textiles and clothing. In principle, these voluntary export agreements sought a balance that protected textile and clothing sectors in the developed economies against excessive short-term market disruption while facilitating industrialization and exports in the developing countries.

As time passed development of synthetic fabrics and an ever-widening set of industrializing countries meant that less and less of the international textile trade came under the LTA. In 1974 the Multi-Fibre Arrangement (MFA) was introduced as a comprehensive replacement of the LTA. The MFA was initially heralded as an effective liberalizing instrument. Restraint on imports had to be balanced by offers to exporters of annual growth guarantees and flexible access to developed countries markets. However, during successive renegotiations of the MFA, the rules have become more restrictive. Growth allowances have been reduced. The ability of exporters to shift quota among products (swing) and to employ unused portions of a previous year's quota (carry-over) or to borrow against a future year's quota (carry-forward) have all been progressively restricted, thus reducing flexibility for exporters. In short, as the Canadian case illustrates, the trade regime has become a serious irritant in North-South relations without easing the domestic issues that the policies were intended to address in the industrialized nations.

In Canada, import penetration gradually reduced the market serviced by domestic clothing producers. In the early 1960s imports accounted for less than 10 per cent of domestic sales, and the United States was the principal source of imports. By the mid-1970s import penetration had doubled, and low-cost producers located in developing countries had become the dominant suppliers of imports. This trend has persisted, with imports now accounting for over 30 per cent of domestic sales, 75 to 80 per cent of which originate with low-cost producers.

The Canadian government responded to industry appeals for assistance by establishing an Interdepartmental Committee on Low Cost Imports in 1960. Working within the framework established under the LTA, this group responded to industry complaints about surges in clothing imports, and by 1970 18 bilateral agreements specifying quantitative restraints on textile and clothing imports had been negotiated with low-cost supplying nations. However, it was readily apparent that this ad-hoc, reactive response was inadequate for dealing with dynamic developments in the international clothing trade.

In a effort to cope with the problem in a more systematic and comprehensive manner, the government introduced a sectoral policy for textiles and clothing in 1970, setting up the Textile and Clothing Board. A major function of the board was to investigate industry's claims that imports had "caused or threatened to cause serious injury to domestic production and employment" (Canada, Textile and Clothing Board, 1980). When claims were substantiated, the government undertook to introduce special measures of protection, provided that the industry presented a restructuring plan that would strengthen its competitive position. In principle, government was offering short-term protection on condition that the industry undertake positive adjustment (Mahon 1985b).

Introduction of the MFA in 1974 increased the scope of the international regulatory regime within which the Textile and Clothing Board operated. Imports from low-cost suppliers continued to increase, generating additional bilateral voluntary export-restraint agreements with exporting countries. Finally, in response to a 46 per cent increase in imports in 1976, the board launched an inquiry that resulted in Canada's unilaterally using emergency provisions of the MFA to impose global quotas that restricted future imports to 1975 levels and thereby provided breathing space for the industry. However, the longer-term objective was to return to a system of bilateral restraints, subject to regular review, under a renewed MFA.

By 1979, the system of bilateral restraints was back in place. In 1990, Canada was party to 29 bilateral restraint agreements (Canadian International Trade Tribunal 1990). The tribunal concluded that the system of bilateral agreements had proven to be a flexible policy tool from a Canadian perspective. Over the years the product emphasis has shifted, with clothing replacing textiles as the dominant focus of restraints. Moreover, the nations with which the restraints have been concluded have changed, reflecting developments in the international textile and clothing trade. However, it is not apparent that the contingent protection provided by restraints has stimulated the degree of restructuring that policy–makers had intended.

The second thrust of the federal government's response to import penetration involved general and sector-specific industrial policies intended to facilitate growth and adjustment. Textile and clothing firms have made modest use of programs open to manufacturing in general. The Regional Development Incentives Program (RDIP) attracted the most attention. Between 1969 and 1982, textile and clothing firms accounted for 7.1 per cent of the $3.2 billion invested in projects supported by the program. However, 75 per cent of the total went to the more capital-intensive textile sector. Moreover, RDIP's distributional and equity goals have led to criticisms about its failure to make adjustment a significant criterion in its allocating of funds.

Concerns about limited adjustment are further borne out by low participation in the Enterprise Development Program. The program, launched in 1977, was intended to support firms proposing innovative or adjustment projects that would make them more viable or internationally competitive. Between 1977 and 1982 textile and clothing firms accounted for less than 0.2 per cent of investments supported by the progam.

In 1981 primary responsibility for encouraging adjustment was vested in the new Canadian Industrial Renewal Board (CIRB). The board's approach had three dimensions. First, the Sector Firms Program (SFP) offered strategic and financial assistance to companies modernizing and improving their international competitiveness. The program was to focus on firms with "proven performance, satisfactory financial resources and strong restructuring plans" (CIRB 1986). Second, the Business and Industrial Development Program (BIDP) sought to strenghten and diversify the economic base of seven designated regions that were heavily dependent on textile, clothing, and footwear industries. Finally, the Labour Adjustment Program (LAP) used existing and special Employment and Immigration Department programs to provide textile and clothing workers with training, mobility assistance, and early retirement benefits.

Senior business executives dominated CIRB. This reflected the policy objective of reducing the role of government in the economy and providing more scope for market and private-sector decision making. The performance of the sector-firms program suggests that the board's approach was selective. CIRB received 1,114 applications for all types of business assistance, but 954 companies accounted for 1,069 applications that requested capital assistance with restructuring and modernization. While 500 firms received assistance, 85 per cent of the $267 million disbursed supported capital spending on modernization for 335 firms. Total investment per firm was substantial, averaging $5.3 million for the 112 textile firms and $1.8 million for the 186 clothing firms (CIRB 1986). Clearly the program targeted a select group from the approximately 3,500 firms in the textile and clothing sector. With the end of the program in 1986, bilateral restraints again became the principal element of policy.

In summary, policy has played an ambiguous role in facilitating restructuring in the clothing sector. Over the past quarter-century, special protection has evolved, ostensibly to provide manufacturers with an opportunity to develop business strategies for competing with emerging international competitors. For many Canadian firms the combination of risks and uncertainties, along with technological and financial constraints, has slowed adjustment. However, a small subset of firms, such as those using CIRB modernization assistance, offers a glimpse of the possibilities for the clothing sector's restructuring in a high-wage economy.

Outward Processing and Domestic Restructuring

A second restructuring strategy pursued by clothing producers in high-wage economies has involved off-shore processing. Innovations in transport and communication technology have permitted firms to fragment production processes geographically, creating a new international division of labour (Froebel, Heinrichs, and Kreye 1980). This strategy has been less significant in Canada because of the absence of large firms having the resources needed to develop international production systems. Nevertheless the relationship between domestic production and importing has become more complex as both retailers and manufacturers have become more directly involved in importing.

The retail-clothing sector in Canada is characterized by a high level of economic concentration. Specialized clothing chains and department stores have continued to displace independent clothing stores and account for over 75 per cent of sales. The purchasing power of these concentrated groups can be used effectively against the fragmented clothing producers and, more important, lets the large buyers engage directly in off-shore sourcing of clothing.

Manufacturers are well aware of their inability to compete with low-cost producers in some lines of production. Thus in order to fill out their product lines for domestic retail buyers, some manufacturers contract directly with low-cost suppliers, enabling them to present a package of goods sourced both at home and abroad.

The effects of these activities by retailers and manufacturers can be seen in statistics on imports. The share of imports handled by traditional importers/wholesalers has declined, while the portion accounted for by retailers and clothing manufacturers has increased. Manufacturers alone account for about 25 per cent of clothing imports, and the unit value of these imports is lower than the unit value of products imported by retailers and traditional importers. This situation is consistent with a strategy in which clothing manufacturers fill in the bottom end of their product price range with inexpensive imports.

Domestic Restructuring

Finally, clothing manufacturers have reorganized labour processes and introduced innovative technologies in an effort to restructure domestic production. The diversity of products and production processes encountered in the sector allows for considerable variation in restructuring strategies. To illustrate the range of possibilities, I contrast evidence on use of strategies aimed at minimizing labour cost with those directed at increasing productivity by upgrading technology.

Traditional methods of garment making are labour intensive, and efforts to minimize labour costs have produced some of the worst labour exploitation and abuse in the history of industrialization (Johnson 1982). Sewing, which constitutes the heart of the garment-production process, can be undertaken in virtually any physical surroundings. While factory production is one possibility, sewing can be parcelled out along chains of production leading through contractors and subcontractors to homeworkers. Manufacturers can reduce overhead and direct labour costs by shifting production out of the factory into more "flexible" contracting and homeworking channels. Pursuit of this strategy has led to emergence of "hollow corporations" that perform largely organizational and managerial functions from the top of these pyramidal structures (Lipsig-Mummé 1987; Dagg and Fudge 1992). As clothing operatives become more remote from organizational centres, their economic status tends to become weaker and more ephemeral. Though provincial legislation addresses labour exploitation, the "minimum employment standards" that have emerged have been criticized for their limited scope and weak application (Johnson 1982; Grant 1992). The incentive to minimize labour costs is even more intense in times of economic crisis, and evidence suggests that contracting and homeworking have grown.

First, contracting has increased steadily during the recent economic crisis. Men's and womens's clothing contractors currently account for one of every five clothing-sector workers. Expansion of contracting is consistent with accounts of restructuring that relies on a contingent labour force as a basis for increasing capital accumulation. Second, 60 per cent of contracting occurs in women's clothing, where it has displaced factory production. From 1983 to 1987, while the number of women's clothing factories fell by 38 per cent, the number of contractors declined by only 8 per cent (Grant 1992). Contracting facilitates flexible production in fashion segments based on small-batch production. Since these features characterize women's clothing, contracting would be expected to be more apparent there. Finally, contracting is no longer as concentrated in Quebec as it was through the 1970s, when the province accounted for fully 90 per cent of the total. Historically, contracting, which involved dispersion of jobs from urban settings, was viewed with favour by the Catholic clergy, since it slowed the drift of a redundant rural population to the cities. However, in recent years contracting has grown rapidly in Ontario, which now accounts for 20 per cent of the total, while other clothing nodes account for the residual 5 to 10 per cent.

Homeworking is presented by some as an opportunity for workers (women) to earn income while attending to child care. However, it can also allow manufacturers to minimize overhead and direct labour costs. Though it is subject to regulation under provincial labour law, activity levels are greatly under-reported, exposing those workers to the abuses associated with an underground economy.

While comprehensive documentation of the nature and extent of homeworking is lacking, its use seems to have increased of late. In 1983, an investigation estimated that 30,000 homeworkers were employed in the Quebec clothing industry (Grant and Rose 1985). Unionized clothing workers were well aware of the devastating effect that homeworking was having on their ranks. The International Ladies' Garment Workers' (ILGWU) responded by striking Montreal-area plants for the first time in 43 years (Lipsig-Mummé 1987). However, the long-run decline in union density in Quebec's clothing industries continued; 56 per cent of workers was members in 1968, and only 33 per cent in 1990 (Grant 1992).

Homeworking also appears to be increasing in Ontario. A 1992 ILGUW study estimated that though only 70 homeworkers were registered in Ontario as many as 4,000 were working in the clothing sector (Galt 1992). The exploitive nature of homeworking was also revealed. Instances of sewers receiving $10 to produce designer-label clothes that retail for as much as $400 were reported. In another study, interviews with 30 homeworkers revealed an average wage of $4.50 per hour, well below the provincial minimum wage (Dagg and Fudge 1992). Moreover, because homeworkers are treated as self-employed workers rather than industrial employees, they are not eligible for many benefits that would normally increase employers' costs under the Employment Standards Act.

In summary, some clothing manufacturers appear to have used contracting and homeworking to reduce costs and increase competitiveness. While the strategy may help some producers in particular production and product circumstances, it holds little promise as a general approach to revitalizing the industry. In the high-wage Canadian economy, even limited solutions are likely to involve a commitment to technological upgrading.

Compared with manufacturing in general, in the clothing sector mechanization and automation have developed slowly, and the effects have been unevenly felt among industries as well as at different stages of production within industries. Several factors have contributed to the technological lag, including the fragmented industrial structure, technical problems associated with automating the handling of fabric, and continued availability of cheap labour.

When international competition ultimately began to attract production to lower-cost labour sites in developing countries, producers in many segments of clothing production discovered that competitive advantage involved far more than labour costs. On the demand side, markets were changing. As the economy stagnated, consumers allocated a declining share of disposable income to clothing. Moreover, markets have become highly segmented, placing greater emphasis on design, managerial skills, and quick response, all factors that work against sourcing in most low-cost, offshore locations.

On the supply side, machinery manufacturers began to recognize the potential for extending application of technologies developed in other manufacturing sectors to clothing production. Thus technology came to be viewed as a means for restoring the competitive advantage of some types of clothing production in the industrialized countries (Hoffman and Rush 1988; Mytelka 1991).

Technologically based restructuring requires substantial capital investment; for example, an integrated pre-assembly system involving CAD and automated grading, marking, and cutting can cost more than $1 million (Mather 1993). Few Canadian firms have the financial capacity to undertake such initiatives. However, equipment costs have fallen over the past decade, and more modest investment programs can aim at partial automation. Automated marking and grading equipment can be acquired for $100,000. This equipment offers substantial savings in labour costs through "deskilling," savings in material costs resulting from efficient pattern layout, and, perhaps most important, dramatically shortened production cycles. Automating the transfer of materials in the sewing room can require an outlay of $2,000 to $4,000 per workstation, with some products requiring as many as 150 operations (Mytelka 1991). Finally, electronic data systems linking manufacturers with retailers and suppliers enable rapid response to market demand. Manufacturers can replenish retail stocks quickly while minimizing their inventory of supplies and product (Harris 1991).

While the potential for productivity gains is significant, a recent study of technology use in apparel making concluded that the Canadian industry lags as much as a decade behind European and five years behind U.S. companies (Kurt Salmon Associates 1991). Capital investment in clothing has consistently been much lower than that recorded in manufacturing generally. Between 1977 and 1990 the net capital stock in constant 1986 dollars increased by only 27 per cent in the clothing sector, compared with 169 per cent for all manufacturing. The ratio of capital investment to value added in clothing was about 1 to 2 per cent per year, compared with more than 10 per cent for all manufacturing.

However, as the discussion of the CIRB program suggested, an emerging subset of firms appears to be intent on incorporating the new technologies. These include Peerless Clothing, the Montreal suit maker. Peerless adopted machine-oriented manufacturing developed in Europe after the Second World War in response to the shortage of tailors (McKenna 1992). Investments in computerized equipment in excess of $1 million per annum for the past 14 years have produced one of the most advanced plants in the world.

While men's clothing, with its longer production runs, accounts for 55 per cent of net capital stock in all clothing industries, the more fashion-oriented women's sector also has its technologically advanced firms. Mel Alper of Katescorp, a manufacturer of women's sportswear, has commented on his firm's use of automation to coordinate activities throughout the corporate system: "Our particular strength is organizational in that we have very close and intensive communications between a very talented merchandising and design team who create innovative fabrications and style concepts and a very technologically competent manufacturing team who can decipher these concepts and produce them in a relatively efficient manner, on time, at the appropriate cost and with a superior quality of construction" (Alper 1992).

The new technologies are becoming more widely adopted, according to Hector Laplante of CAM GGT Canada: "Automation is accepted and is succeeding because all manufacturers can benefit without sacrificing economy. Modular CAD and CAM systems each fully capable of helping a manufacturer plan or control key operations can be linked as the manufacturer grows. They cost far less than ever and do more to help you, the manager, be more effective" (Laplante 1992).

As the foregoing discussion suggests, restructuring in the clothing sector has been uneven. While I have contrasted efforts to minimize labour costs through increased use of contracting and homeworking with efforts to increase productivity through technological upgrading, these strategies are not necessarily mutually exclusive. As Mytelka (1991) observes, there is not one competitive way to produce a product; rather several production systems may coexist, and individual firms may combine different production strategies within their overall operation. However, in Canada's high-wage economy, a strategy relying solely on minimizing labour costs appears limited. A technologically based restructuring strategy, however, offers no guarantee of success and needs to be developed within a more comprehensive business plan. Thus the future of Canadian clothing production within the current North American and global context remains uncertain.

THE CHALLENGE OF ECONOMIC INTEGRATION

Implementation of the FTA in January 1989 dramatically altered the context of clothing production. The first three years of the agreement witnessed significant shifts in industry performance. Trade with the United States increased sharply, rising from $297 million to $540.7 million. However, the Canadian clothing surplus with the United States declined from $78 million to $64 million as American producers doubled their share of Canadian imports from 5 to 10 per cent. The domestic industry declined precipitously, with the number of establishments falling by 40 per cent, employment by 24 per cent and industry GDP by 20 per cent in constant 1986 dollars. At the same time investment reached historic highs even compared with the years when the CIRB program was in force.

Clearly the industry was undergoing significant restructuring. However, determination of the effect of freer trade on restructuring is made more difficult by the appreciation of the Canadian dollar and introduction of the GST during this period. Moreover, while freer trade was supposed to produce a more market-driven economy, the effect of public policies remained crucial in the clothing sector.

Clothing spokespersons have argued that freer trade's primary objective of enhancing access to the American market was not achieved in the clothing sector. To be eligible for "free trade" entry to U.S. markets, Canadian clothing had to meet a stringent, two-stage transformation test. Garments must not only be cut and sewn in Canada but must be made from fabric sourced in North America. Because of the limited range of fabrics produced by Canadian textile manufacturers, 60 per cent of clothing made from woven fabric uses imported fabric. International sourcing of fabric is an important element of competitive strategy for Canadian apparel producers. Clothing manufacturers, already encumbered with high tariffs on imported fabric, complained that "free" trade was having the perverse effect of limiting their access to American markets. Negotiators attempted to recognize the historic pattern of fabric imports by establishing quotas for garments that failed to meet the test, but the volume of permitted exports is extremely limited and will be exhausted in the very near future, thus limiting the growth of Canadian clothing exports.

The Canadian clothing sector hoped that these problems would be redressed in the negotiations for the North American Free Trade Agreement (NAFTA). Instead the situation was exacerbated by the introduction of a three-stage transformation rule which requires that not only the fabric but the yarn from which it is produced be sourced in North America. NAFTA negotiators have argued that the limitations of three-stage transformation will be offset by more liberal quotas. Whatever the reality, the allegedly market-driven phenomenon of trade liberalization is in fact subject to administrative and bureaucratic fiat. Moreover, the inability of the clothing sector to have its position adopted reflects, in Mahon's view, its relatively weak voice in the policy-making structure (Mahon 1985a, b).

Implementation of NAFTA has also raised concerns about the potential effects of outward processing on Canadian clothing production. American textile firms produce and cut fabric which is shipped to low-cost–labour locations in Mexico and the Caribbean Basin for sewing. The finished garments are then reimported to the United States under special tariff regulations. The U.S. experience with outward processing reveals its devastating impact on domestic apparel employment (Dagg 1990). Moreover, problems associated with verifying labels and tracking the origin of goods have been voiced. For example, Canadian officials have noted the 263 per cent increase in underwear imports from the United States since 1988. Significantly, U.S. imports of underwear from Mexico have increased by 461 per cent over the same period (Apparel Manufacturers-Marketing Association 1992).

In February 1993 the federal government unveiled its Fashion Apparel Sector Campaign (FASC), which it had developed in conjunction with industry representatives. FASC has two major elements. First, to address problems stemming from the fragmented structure of the industry, a Canadian Apparel Federation (CAF) has been created as a central coordinating body. Second, FASC will incorporate an inter-firm linkage program that will financially support cooperative projects among designers, retailers, and textile and apparel manufacturers.

The focus on the fashion segment of the industry is intended to set a course for clothing production in Canada. Background studies that led to FASC identified an emerging fashion-oriented group whose profile is distinct from the traditional image of Canadian apparel manufacturers (Canada, IST, 1989). Compared with industry norms, fashion-oriented firms are strong in design capability, produce high value–added products, are more export oriented, and are not as affected by import penetration.

The need to restructure continues to be stimulated by international competition both continentally and globally. Some firms appear to be adapting, but, as the data reveal, substantial dislocation is associated with restructuring. Marginalized entrepreneurs and workers have always had a presence in garment making (Waldinger 1986; Phizacklea 1990). The challenge will be to facilitate adjustment of those displaced in a manner that is economically, socially, and politically acceptable.

ACKNOWLEDGMENTS

I would like to acknowledge the financial assistance of SSHRC, which supported this research.

REFERENCES

Aggarwal, V.K. 1985. *Liberal Protectionism: The International Politics of Organized Textile Trade*. Los Angeles: University of California Press.

Alper, M. 1992. Address to the Canadian Apparel Federation Inaugural Conference on Technology, Ottawa, May.

Apparel Manufacturers-Marketing Association of Ontario (AM-MAO). 1992. *An Analysis of the North American Free Trade Agreement and Its Implications for the Canadian Apparel Industry*. Ottawa: AM-MAO.

Canada. Industry, Science and Technology Clothing Division (IST). 1989. *Fashion Apparel Sector Strategy*. Ottawa.

Canada Textile and Clothing Board. 1980. *Textile and Clothing Inquiry*. Ottawa: Supply and Services.

– 1985. *Textile and Clothing Inquiry*. Report to the Minister of Regional Economic Expansion. Ottawa: Supplies and Services.

Canadian Industrial Renewal Board (CIRB). 1986. *Fourth and Final Annual Report*. Montreal.

Canadian International Trade Tribunal. 1990. *An Inquiry into Textile Tariffs*. Ottawa: Supply and Services.

Dagg, A. 1990. "Keeping the Jobs at Home." *Globe and Mail*, 7 July.

Dagg, Alexandra, and Fudge, Judy. 1992. "Sewing Pains: Homeworkers in the Garment Trade." *Our Times*, 11 no. 3, 22–5.

Froebel, F., Heinrichs, J., and Kreye, O. 1980. *The New International Division of Labour: Structural Unemployment in Industrialized Countries and Industrialization in Developing Countries*. Cambridge: Cambridge University Press.

Galt, V. 1992. "Protection for Home Based Workers Promised." *Globe and Mail*, 2 Oct.

Gannagé, Charlene. 1986. *Double Day, Double Bind: Women Garment Workers*. Toronto: Women's Press.

GATT. 1984. *Textiles and Clothing in the World Economy*. Geneva: GATT.

Grant, Michel. 1992. "Industrial Relations in the Clothing Industry: Struggle for Survival." In Richard Chaykowski and Anil Verma, eds., *Industrial Relations in Canadian Industry*, 220–43. Toronto: Dryden.

Grant, Michel, and Rose, Ruth. 1985. "L'encadrement du travail à domicile dans l'industrie du vêtement au Québec." *Relations Industrielles*, 40 no. 3, 473–94.

Harris, C. 1991. *How Will You Survive?: Cost Reduction and Time Based Strategies for Canadian Manufacturers*. Montreal: Kurt Salmon Associates.

Hoffman, Kurt, and Rush, H. 1988. *Micro-electronics and Clothing: The Impact of Technical Change on a Global Industry*. New York: Praeger.

Johnson, Laura C. 1982. *The Seam Allowance: Industrial Home Sewing in Canada*. Toronto: Women's Educational Press.

Kurt Salmon Associates. 1991. *Level of Technology Utilization by Apparel Companies in Canada, the United States and Europe*. Ottawa: Department of Industry, Science and Technology.

Laplante H. 1992. "Apparel Automation: How Technology Enhances the Industry's Competitiveness." Paper presented at Canadian Apparel Federation Inaugural Conference on Technology, Ottawa, May.

Lipsig-Mummé, Carla. 1987. "Organizing Women in the Clothing Trades: Homework and the 1983 Garment Strike in Canada." *Studies in Political Economy*, 22, Spring, 41–71.

McKenna, B. 1992. "Canadian Suit Firm Threatens to Unravel NAFTA Talks." *Globe and Mail*, 5 Aug.

Mahon, Rianne. 1985a. *The Politics of Industrial Restructuring: Canadian Textiles*. Toronto: University of Toronto Press.

– 1985b. "Unravelling Canada's Textile Policy." In Allan M. Maslove, ed., *How Ottawa Spends Your Tax Dollars*, 90–113. Ottawa: Carleton University Press.

Mather, Charles. 1993. "Flexible Technology in the Clothing Industry: Some Evidence for Vancouver." *Canadian Geographer*, 37, 40–7.

Mytelka, Lynn Krieger. 1991. "Technological Change and the Global Relocation of Production in Textiles and Clothing." *Studies in Political Economy*, 36, 109–44.

Phizacklea, Annie. 1990. *Unpacking the Fashion Industry*. London: Routledge.

Rush, H., and Soete, L. 1984. "Clothing." In Ken Guy, ed., *Technological Trends and Employment: Basic Consumer Goods*, 174–222. Brookfield, Vt.: Gower Publishing.

Seward, Shirley B. 1990. "Immigrant Women in the Clothing Industry." In Shiva S. Halli, Frank Trovato, and Leo Driedger, eds., *Ethnic Demography: Canadian Immigrant, Racial and Cultural Variations*, 343–62. Ottawa: Carleton University Press.

Trela, Irene, and Whalley, John. 1990. "Unravelling the Threads of the MFA." In Carl B. Hamilton, ed., *Textiles Trade and the Developing Countries: Eliminating the Multi-fibre Arrangements in the 1990's*, 11–45. Washington, DC: World Bank.

Waldinger, Roger D. 1986. *Through the Eye of the Needle: Immigrants and Enterprise in New York's Garment Trades*. New York: New York University Press.

Williams, D. 1987. *Canadian Adjustment Policy: Beyond the Canadian Industrial Renewal Board*. Ottawa: North-South Institute.

CHAPTER THIRTEEN

Restructuring in a Continental Production System

JOHN HOLMES

The automobile industry is of prime importance within the Canadian economy because of its size and because it is the backbone for many other Canadian manufacturing sectors, including steel, rubber, plastics, aluminum, and glass. Nearly one in every seven Canadian manufacturing jobs depends directly on it (Taskforce 1983, 1), including more than 160,000 in the industry itself.[1] The major contribution of motor vehicles and parts to Canada's international trade – nearly one-quarter of all merchandise exports and imports – is explained by one specific aspect of the historical development of the industry. After the signing of the Canada–United States Automotive Products Trade Agreement (the Auto Pact) in 1965, automobile production in the two countries became fully rationalized and integrated into one production and marketing system.

The continental form of this production system has several analytical consequences. First, we can understand the present-day structure and geography of the industry in Canada only by placing it into the broader context of the whole North American industry. Second, we cannot subsume the industry under the popular image of Canadian manufacturing as little more than branch-plant operations. Third, during the 1980s the auto industry in Canada was restructured as an integral part of the North American auto industry in response to formidable competitive pressures from Japanese manufacturers. This restructuring has involved technological regeneration of both product and process technologies, challenging the common perception of the auto industry as a mature consumption-good industry.

Thus the twin themes of this chapter are the role of technological renewal in the restructuring of this key sector and the consequences for Canada of restructuring in an industry that is an integral part of a geographically extensive, continental production system.

THE AUTOMOBILE INDUSTRY IN CANADA

Size and Structure

In 1990, Canada, with a total output of almost 2 million vehicles, was the sixth-largest producer of motor vehicles in the world, accounting for 4.5 per cent of world produc-

Table 1
Profile of the Canadian automotive industry, 1989

	Total	Assembly*	Parts[†]	Trucks and bus bodies[‡]
Number of establishments	1,082	30	673	373
Value of shipments ($million)	44,791	27,519	15,445	1,827
% in Ontario	93.2	94.0	96.8	55.2
Number of employees				
Total	168,034	55,392	96,454	16,188
Production	139,327	43,890	82,162	13,275
Non-production	28,707	11,502	14,292	2,913
Per establishment	155	1,846	143	43
Value added (manufacturing)				
Total ($million)	12,177	4,855	6,592	730
Per production employee ($)	87,399	110,617	80,232	54,991
Average annual earnings per production employee ($)	33,248	42,449	29,767	24,369
Union density of hourly paid employees[§]	79	95	53	–

Sources: Statistics Canada, *Transportation Equipment Industries*, Cat. No. 42-251, annual.
* SIC 3231.
[†] SIC 3251–9.
[‡] SIC 3241–4.
[§] Estimates for 1990 from Herzenberg (1992).

tion and 16 per cent of North American (United States, Canada, and Mexico) output. There is a sharp contrast in industrial organization between assembling motor vehicles and manufacturing parts and accessories: for example, in 1989 assembly accounted for 61 per cent of the industry's shipments by value but only 34 per cent of total employment in the Canadian industry, while the pattern for parts was the mirror image of this (Table 1).

Both sectors are dominated by the "Big Three" (General Motors, Ford, and Chrysler), which account for over 90 per cent of vehicle production, 41 per cent of the value of parts produced, and 50 per cent of the industry's employment. During the second half of the 1980s, however, Asian-owned assembly plants – "transplants" – became a significant feature of the industry (Table 2). In 1991 the four transplants in Canada assembled just over 364,000 vehicles (almost 20 per cent of total Canadian production). Over 80 per cent of the cars and trucks assembled in Canada by both the Big Three and the transplants are exported to the United States.

Ontario accounts for over 90 per cent of employment in vehicle assembly and over 95 per cent in parts manufacturing, which makes the industry crucial to the economy of that province. Except for two assembly plants in Quebec and a small plant in Halifax, Nova Scotia, that assembles less than 8,000 Volvos a year from kits, all cars and light trucks are assembled in southern Ontario between Windsor and Oshawa (Figure 1).

Table 2
Asian assembly transplants in Canada

Company	Location	Date opened	Announced capacity	Production 1991	Employment at full capacity	Investment ($million)
Honda of Canada Mfg. Inc.	Alliston, Ontario	November 1986	100,000	99,150	1,400	400
Toyota Motor Mfg. Canada Inc.	Cambridge, Ontario	November 1988	70,000	87,834	1,000	400
Hyundai Auto Canada Inc.	Bromont, Quebec	January 1989	100,000	28,201	1,200	450
CAMI Automotive Ltd. (GM/Suzuki)	Ingersoll, Ontario	April 1989	200,000	168,862	2,000	620

Sources: Canada, IST (1992); *Automotive News* (28 Dec. 1992).

Evolution before 1980

From the 1930s until the mid-1960s production throughout the international motor-vehicle industry was organized on a national basis, with domestic markets in major auto-producing countries protected from large-scale imports and supplied by locally based (though often foreign-owned) producers (Altshuler et al 1984; Dunn 1987). In Canada, significant tariffs on imported automotive parts entering the country and a 60 per cent Commonwealth-content rule on automobiles made and sold in Canada led to development of a domestic auto industry, which was a relatively inefficient smaller replica of the U.S. industry (Holmes 1983). In an industry emblematic of mass production and very high minimum efficient scales of operation, Canadian auto-makers, severely limited by the scale of the domestic market, experienced declining international competitiveness.

The "competitive crisis" of the industry, which came to a head in the early 1960s, was resolved through the Auto Pact of 1965. While this arrangement created an integrated continental market for automotive products, it also contained safeguards designed to ensure that a certain portion of combined Canadian and American vehicle and parts manufacturing took place in Canada.[2] The result was duty-free, two-way movement between the two nations of new vehicles and original equipment (OEM) parts, though the basis for granting duty-free treatment has been completely different on each side of the border.[3]

The restructured Canadian auto industry of the 1970s was competitive and efficient by North American standards and enjoyed significant growth and expansion (Table 3) but was limited in scope. The Auto Pact gave Canada a disproportionately large share of the more labour-intensive stages of the manufacturing process – vehicle assembly and low-value parts production – while research and development (R&D), product engineering, and production of high-valued parts was concentrated in the United States (Holmes 1983).[4] As a result, since 1965 Canada has continuously registered large trade surpluses in finished vehicles and large trade deficits in parts and components (Table 4

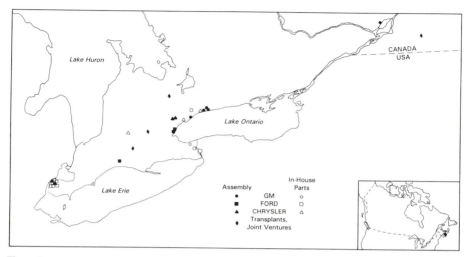

Figure 1
Automobile assembly and Big Three in-house parts plants in Canada, 1993. *Sources*: Canada, IST (199
Automotive News (1993).

Table 3
Production and employment figures for assembly and component sectors in Canada, 1961–91

		Vehicle production		Employment	
Year	Value of shipments ($billion)	Total (000s)	% of North American total	Vehicle assembly	Components
1961	1.3	387	5.8	25,100	20,800
1966	3.2	897	9.5	42,500	34,800
1971	5.9	1,371	11.4	42,300	43,800
1976	11.4	1,644	11.8	49,100	47,300
1981	17.3	1,241	15.6	44,300	55,700
1982	19.0	1,251	13.8	42,200	53,600
1983	24.7	1,516	14.7	47,600	64,700
1984	33.3	1,874	14.2	54,300	77,400
1985	37.8	1,953	14.1	56,900	84,400
1986	38.5	1,866	12.5	53,500	85,200
1987	36.9	1,635	13.9	52,000	88,800
1988	44.4	1,975	15.2	52,700	94,300
1989	44.5	1,916	15.2	55,400	96,300
1990	42.8	1,905	16.4	52,700	88,000
1991	41.9	1,867	17.5	48,900	76,200

Source: Canada, IST (1992).

Table 4
Trade in automotive products with United States and overseas, 1965–91

| | Trade | | | | Trade surplus (deficit) ($million) | | | | | | | | |
| | Exports | | Imports | | United States | | | Overseas | | | Total | | |
Year	Total ($million)	% to U.S.	Total ($million)	% from U.S.	Vehicles	Parts	Total*	Vehicles	Parts	Total*	Vehicles	Parts	Total*
1965	383	65.3	1,099	87.4	(59)	(646)	(711)	(23)	11	(5)	(82)	(635)	(716)
1970	3,521	92.8	3,454	88.7	1,193	(980)	204	(99)	(31)	(137)	1,094	(1,011)	67
1975	6,519	90.6	8,422	91.7	665	(2,380)	(1,821)	11	(26)	(82)	676	(2,406)	(1,903)
1980	11,480	89.8	14,073	87.8	2,065	(4,195)	(2,045)	(525)	65	(548)	1,540	(4,130)	(2,593)
1981	14,417	88.3	16,580	87.2	3,230	(5,079)	(1,728)	(943)	214	(435)	2,287	(4,865)	(2,163)
1982	17,684	92.9	15,478	87.7	7,368	(4,774)	2,853	(973)	25	(647)	6,395	(4,749)	2,206
1983	21,637	96.5	19,964	88.2	7,395	(4,303)	3,286	(1,345)	(359)	(1,620)	6,052	(4,657)	1,673
1984	30,674	97.3	27,626	86.6	10,841	(5,159)	5,935	(1,830)	(1,048)	(2,887)	9,011	(6,207)	3,048
1985	34,515	98.0	33,611	85.8	10,563	(5,926)	4,965	(2,891)	(1,152)	(4,071)	7,682	(7,078)	904
1986	35,363	97.5	35,249	83.2	10,780	(6,058)	5,170	(3,732)	(1,343)	(5,056)	7,048	(7,401)	114
1987	33,379	97.6	34,852	82.3	8,370	(4,818)	3,903	(3,893)	(1,449)	(5,376)	4,477	(6,267)	(1,473)
1988	36,881	97.5	37,797	84.5	12,008	(8,204)	3,973	(3,633)	(1,284)	(4,889)	8,375	(9,488)	(916)
1989	36,309	97.5	36,061	81.7	12,060	(6,354)	5,927	(3,641)	(2,080)	(5,679)	8,419	(8,434)	248
1990	35,726	97.8	33,605	78.1	14,595	(6,156)	8,706	(3,869)	(2,495)	(6,585)	10,726	(8,651)	2,121
1991	34,238	97.7	31,925	73.4	16,372	(6,621)	10,036	(4,736)	(2,750)	(7,723)	11,636	(9,371)	2,313

Sources: Canada, 1ST (1992); Statistics Canada, Cat. Nos. 65-202 and 65-203, annual.

* Includes tires, tubes, and re-exports.

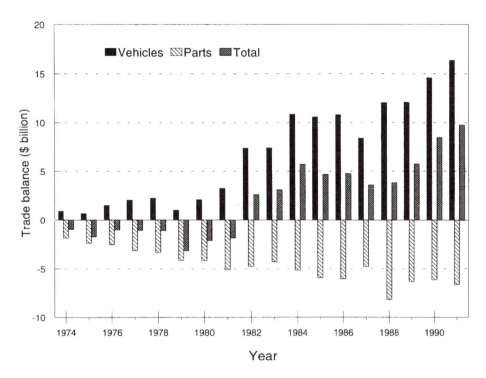

Figure 2
Canadian-U.S. automotive trade balances ($billion) under the Auto Pact, 1974–91. *Sources*: Statistics Canada, Cat. Nos. 65–202 and 65–203, annual.

and Figure 2). Furthermore, given the enormous difference in the size of the U.S. and Canadian markets, the economic vitality of the industry in Canada is now highly dependent on the decisions of the Big Three as to which specific vehicle platforms are to be built in Canada and the strength and composition of vehicle demand in the U.S. market.

RENEWAL AND RESTRUCTURING
DURING THE 1980S

The long-standing expectation that growth and prosperity in the North American auto industry would continue indefinitely was shattered during the 1980s as the industry entered a period of serious competitive decline. In 1978 North American automakers accounted for over 85 per cent of vehicle sales in the North American market, but by 1990 this figure had declined to about 67 per cent (representing a loss of well over 2 million vehicles a year). The 1979 and 1990 markets in North America were almost identical in size – at approximately 15.5 million units – yet in 1990 the Big Three produced only 77 per cent as many units as in 1979. Even in the most profitable year of the decade, 1984, profits were lower than in most of the 1970s, and after 1984 there was a steady decline. In the United States over 240,000 jobs in the automotive industry were lost between 1978 and 1990.

The 1980s was a decade of success in North America for Japanese auto companies, which not only captured a large and growing share of the North American vehicle market but also created, through direct investment in the United States and Canada, a fully developed and competitive automobile production system. Consequently, the less-competitive, traditional North American industry was subjected to extensive re-structuring aimed at creating a "new manufacturing culture."

The Roots of the Competitive Crisis

During the 1970s, a combination of poor productivity and the continuation of the contractually entrenched practice of granting auto workers annual real wage increases sharply increased unit production costs, which further weakened the competitive position of the Big Three (Holmes 1987). Meanwhile, Japan's exports of motor vehicles almost tripled, from 2.1 million units in 1973 to nearly six million in 1981, and by 1982 over a third of all Japanese vehicle exports were directed to North America. For the first time, the Big Three experienced the full force of international competition for their own North American market. Japan's share of this market rose quickly. In Canada it increased from 9.7 per cent in 1975 to 25 per cent in 1982 (Table 5), and while Canada began to enjoy a significant positive balance in automotive trade with the United States after 1981, the deficit on its annual overseas balance of automotive trade rose from a mere $82 million in 1975 to $2.9 billion by 1984, mainly as a result of a flood of imported Japanese cars (Table 5 and Figure 3).

The rapid growth of the Japanese market share and the deep recession of 1981–82 led the Canadian government to take steps to keep auto production and investment in Canada (Holmes 1991). In 1980, in exchange for helping Chrysler Canada avert bankruptcy, the Canadian government exacted two promises from the company: that Chrysler would not close any of its Canadian plants before 1985 and that the company would invest a further $1 billion in Canada. In 1981, both the u.s. and Canadian governments moved to stem the tide of Japanese imports by negotiating voluntary restraint agreements (VRAs) with the Japanese automakers covering their exports into the North American market. But Canada went much further. To encourage Asian and European firms to source parts in Canada, it introduced a program of individual "export-based duty remission orders," which allowed vehicle importers duty-free entry of one dollar's worth of automotive product (essentially finished vehicles) for each dollar of Canadian value added (CVA) that was exported. In the mid-1980s, it lobbied Asian producers to build assembly plants in Canada by offering them individual "production-based duty remission orders" for imported parts based on the company's progress toward attaining the CVA and assembly/sales ratios demanded of Auto Pact producers. These agreements were seen as facilitating eventual incorporation of Asian transplants into the Auto Pact.[5]

Table 5
New-vehicle sales in Canada, by origin, 1965–91

	All vehicles			Passenger cars			
		Place of manufacture			Place of manufacture		
Year	Number (000s)	North America (%)	Off-shore (%)	Number (000s)	North America (%)	Japan (%)	Other (%)
1965	831	90.7	9.3	709	89.4	0.4	10.2
1970	774	80.3	19.7	640	77.6	10.2	12.1
1975	1,317	87.1	12.9	989	84.5	9.7	5.9
1979	1,396	89.1	10.9	1,003	86.1	8.0	5.9
1980	1,266	85.9	14.1	932	79.5	14.8	1.5
1981	1,191	75.4	24.6	904	71.6	23.0	5.5
1982	920	71.3	28.7	713	68.6	25.0	6.4
1983	1,081	75.6	24.4	843	74.1	21.0	5.0
1984	1,284	77.8	22.2	971	74.7	17.6	7.7
1985	1,530	74.5	25.5	1,137	69.9	17.5	12.6
1986	1,510	74.5	25.5	1,091	69.9	18.1	12.0
1987	1,528	72.9	27.1	1,065	65.8	22.8	11.4
1988	1,566	75.7	24.3	1,057	68.6	23.1	8.3
1989	1,483	74.0	26.0	993	67.9	24.7	7.4
1990	1,318	71.4	28.6	884	65.6	27.1	7.2
1991	1,288	71.5	28.5	873	65.6	27.3	7.1

Sources: Canada, IST (1992); Statistics Canada, *New Motor Vehicle Sales*, Cat. No. 63-208, annual.
Note: "North America" = built in North America (including Mexico); "off-shore" or "other" includes captive imports.

The VRAs sent a strong message to Japanese producers either to conform to the "access via investment" rule of the traditional postwar international auto-trade regime (Dunn 1987) or to face more overt protectionist sanctions. As a result, Japanese companies began to invest in new assembly and component plants in North America. By 1991 there were 13 Asian-managed assembly plants in the United States and Canada employing 30,000 workers and producing close to 2 million vehicles, or 18 per cent of the total number of vehicles built in Canada and the United States annually. Though Canada secured about 20 per cent of assembly transplant investment (Table 2) it attracted only a handful of the over four hundred transplant and joint-venture parts plants built to supply the transplant assemblers.

The transplants, using American and Canadian workers in Japanese-managed plants, introduced new standards of production and competition to the North American auto industry. Thus the VRAs had the effect of shifting the competitive threat from imported vehicles built in Japan to vehicles produced by the Japanese transplants on the Big Three's own "turf".

Of the range of competitive strategies developed by the Big Three during the 1980s to respond to the Japanese challenge, by far the most important was a restructuring program aimed at technological renewal through plant modernization. Out of 46 U.S.

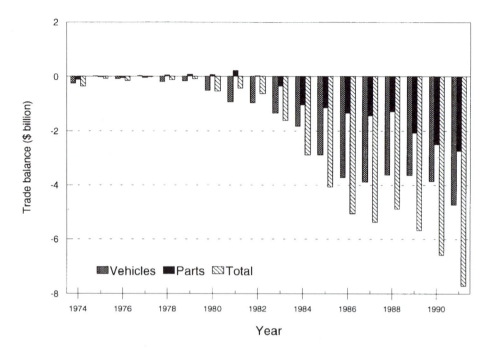

Figure 3
Canadian-overseas trade balances ($billion), automotive products, 1973–91. *Sources*: As Figure 2.

plants owned by the Big Three in 1990, 32 had been completely retooled or con-structed since 1980 (Automotive Select Panel 1991, 17). In Canada, new capital investment in the auto industry during the 1980s exceeded $14 billion (Canada, IST 1990).

Technological Change and the New Manufacturing Culture

The Fordist system of production organization and marketing perfected in the inter-war period and pursued with brilliant success for close to half a century by the Big Three was overwhelmed at the end of the 1970s by Japanese competitors. During the 1950s and 1960s in Japan, domestic motor-vehicle manufacturers developed and then perfected, through trial and error, a new system of flexible mass production. This Japanese Production System (JPS) wreaked havoc with Detroit's previous notions of "best practice" and transformed management's thinking and the nature of technologi-cal renewal within the North American auto industry during the 1980s. Increasingly, this renewal came to focus on the task of creating a "new manufacturing culture," modelled on the JPS, to replace the traditional, Fordist approach.

Conventionally, the logic of each approach is defined in terms of five significant elements – market strategy, production machinery, work organization and industrial relations, organization of the production chain, and general principles of production management – which together constitute a production system, since each element makes sense only in terms of the others. Despite a fair degree of consensus regarding the nature of the Fordist production system,[6] vigorous debate continues around aspects of the JPS and, in particular, introduction and use of JPS-style methods in North American plants.

The Japanese Production System

The market constraints in postwar Japan, which called for a diversity of products manufactured in relatively small quantities, and the defeat of industrial unionism in 1954, which allowed unrestrained exercise of managerial authority, were the basis for the production system developed by Toyota in the 1950s and 1960s. Demands for stringent economizing in use of materials and workers formed the point of departure for development of this new way of organizing production. As with the earlier development of Fordism in North America, the individual features of the new JPS were often not unique, it was their combination into an integrated and mutually reinforcing system that was novel.

Manufacturing in the smallest possible batches replaced the American philosophy of "optimal batch size,"[7] and new materials-control methods, the kanban system within the plant and just-in-time (JIT) deliveries from suppliers, were developed to manage this. The resulting "lean" manufacturing system, with virtually no inventories, streamlined the flow of materials through the manufacturing process and had a number of consequences. First, there was a dramatic reduction in the changeover times required for such things as die changes and the resetting of press lines, which were reduced from hours to a matter of minutes. Second, because of the ever-decreasing inventory buffers between steps in the manufacturing process, quality control became critical, and "zero defects" at every step of the process became the goal. Third, production of small batches led to the need for a highly flexible workforce within the plant, with rapid and frequent redeployment of personnel according to the production needs of the moment. To ensure such flexibility, Toyota invested heavily in further job simplification. Fourth, to achieve minimum use of labour, cycle times were rationalized, multi-machine tending was introduced into the machine shops, and routine inspection and maintenance became the responsibility of the production worker. In the absence of union-imposed ceilings on line speeds and staffing, these changes greatly heightened work intensity. Control on the shop-floor was achieved through a variety of methods, including promises of long-term job security, strict discipline, and self-directed, team-based endeavours to improve morale, foster commitment, and coincidentally raise output.

Toyota's production system assigned as much as possible of parts manufacture to suppliers that were independent but closely associated with Toyota. Over time these suppliers became organized hierarchically into a production pyramid, in which a few large firms coordinated production of complete subassemblies, which they delivered directly to the assembly plant. These "first-tier" suppliers in turn had connected to them a number of parts suppliers, which in turn contracted jobs out to even smaller firms. Many of the first-tier suppliers became advanced component firms with extensive research and development capabilities, while at lower levels in the pyramid low wages and high labour-market flexibility remained essential competitive advantages.

Thus Japanese auto-makers achieved striking increases in productivity in a "mature" industry through extraordinary and incessant attention to shop-floor production management practices – work simplification, machine set-up procedures, quality control, and ordering and handling of materials.[8]

Many analysts claim that this type of lean production represents a break from Taylorism and a fundamental transformation of classical mass-production work.[9] As Berggren (1992) points out, however, the empirical evidence provided by case studies of Japanese plants, transplants, and the "lean production" plants of the Big Three paints a radically different picture. Factory layout, production control, and quality standards have certainly changed, and personnel policies such as employee involvement, pay-for-knowledge, quality circles, job security, and team organization, which emphasize the personal commitment of all production associates, have been introduced. Yet the character of the work itself has not changed, and "if anything, the rhythm and pace of the work on the assembly line [are] more inexorable under the Japanese management system than [they] ever [were] before. Off line jobs ... are geared strictly to the main line by means of just-in-time (JIT) control. Idle time is squeezed out of each work station through the application of kaizen techniques, while work pressure has been intensified and staffing drastically reduced in the name of eliminating all 'waste' " (Berggren 1992, 5).[10] In summary, while the lean production system undoubtedly represented a major advance in productivity, it was, if one considers working conditions as well, a double-edged sword to at least the degree that the classical Ford system was.

"Lean Production" in the North American Automobile Industry

By the late 1980s there was a widely held belief in the industry that the Big Three and their suppliers had to adopt Japanese management techniques and production methods in order to survive. Initially the success of companies such as Honda and Toyota had been ascribed to unique features of Japanese culture, but the growing number of Japanese auto plants in the United States and Britain during the 1980s demonstrated that key elements in Japanese management and production methods could indeed be transplanted. In particular, NUMMI, which opened in 1986, clearly showed that North American plants, using a combination of Japanese management methods and American workers, could obtain levels of productivity and quality similar to those achieved in Japan (Krafcik 1988). The performance of the transplants was even the more striking, since it occurred when the advanced, automated-technology strategy, launched by General Motors in the early 1980s in the hope of catching up with the Japanese, was beginning to flounder as new, high-tech plants, such as Hamtramck-Poletown in Detroit, continued to experience inefficiency and low quality.

The transplants in Canada and the United States are similar to each other in several ways. All of them manufacture Japanese products, and many critical components are still produced in Japan and imported.[11] Local suppliers satisfy much stricter demands for quality, frequency of delivery, and overall commitment than ever before, and there have been efforts to develop an embryonic, pyramidal supplier structure, including first-tier suppliers. All the assembly plants use mechanically controlled, high-paced assembly line systems, and except for Diamond Star, which uses robots, assembly is still almost completely manual.

All the transplants use variants of the Toyota system. They have lean staffing and high intensity of work, driven by use of Japanese techniques such as the kaizen method and andon system. New forms of teamwork help achieve the same goal. The productivity and cost advantage of the transplants have been improved by the stringent system of personnel selection, based on social-psychological profiles and intelligence testing as well as measures of manual dexterity. With the enormous number of applicants (in some plants, up to a hundred for each available opening) the transplants could choose elite workers: young, strong, intelligent, highly motivated, and inclined to cooperate. The youthful nature of the workforce has resulted lower costs for pensions and health benefits (Howes 1991).

Except for NUMMI (located in the San Francisco Bay area of California), Japanese auto-makers built assembly transplants in small rural towns in the U.S. midwest and upper south and in Ontario (Mair, Florida, and Kenney 1989; Reid 1990; Woodward 1992) – mostly rural regions with high unemployment – and offer higher wages than other manufacturing jobs in the area. NUMMI and Mazda converted and reopened plants that had been closed by the Big Three and rehired their workers. Though this fact is not mentioned by Womack, Jones, and Roos (1990), these conditions surely affected the transplants' ability to develop a new factory regime based on strict selection of workers and severe pressure for performance and discipline.

Through the 1980s, as they came to understand the workings of the new Japanese production-management system, the Big Three, led by Ford, began to introduce its organizational practices into their own plants. In fact, the joint-venture transplants – NUMMI, Mazda, and Diamond Star in the United States and CAMI (a GM-Suzuki venture) in Canada – became learning laboratories for the North American partners. Combinations of these practices constitute the new, lean manufacturing culture being developed by the Big Three and their suppliers.

The transformation of the Big Three plants does not represent a simple, straightforward cloning of the JPS. Just as when Fordism spread to Europe and Japan and became hybridized, so introduction of elements of the JPS into North America involves their adaptation to and integration with existing institutional structures, particularly the supplier system and industrial relations; see Wood (1992) on this point.

Pay systems and teamwork in both the transplants and Big Three plants, while different from traditional Big Three practices, differ also from those in the home plants of the Japanese auto-makers. There is a tremendous diversity in production management practices within the North American industry – among companies, among plants within the same companies, and among countries (Holmes and Kumar 1991), and mounting evidence suggests that high quality and plant efficiency can be achieved in different ways. In short, there is no single, dominant "best practice" in the industry.

The Restructured Industry

The impact of this restructuring is difficult to assess, since exact or even meaningful comparisons of productivity and efficiency between any two factories are notoriously difficult to make. Conventional wisdom throughout the 1980s, reflected most recently in the influential MIT auto study, (Womack, Jones, and Roos 1990), was that the Japanese transplants were the lowest-cost producers in North America. However, recent reports have pointed to a significant closing of the gap in both productivity rates and production costs between North American and Japanese auto-makers during the 1980s.

There is no doubt that the transplants still enjoy a clear advantage in unit labour cost. The hourly wage rates paid by the non-union transplants are on a par with union rates paid in Big Three assembly plants, but the former experience substantial savings in benefit costs because of a much younger labour force and use only two-thirds as many person-hours as the Big Three to assemble a vehicle. However, the transplants still import many of the components they assemble, and, with the value of the yen appreciating against the U.S. dollar, the cost of components purchased by the transplants was recently estimated to be about 50 per cent higher than that for the Big Three (Economic Strategy Institute 1992). The same report suggested that, on balance, in 1992 Ford and Chrysler's production costs were lower than those of the transplants.

It is no longer particularly useful to differentiate between Japanese and North American auto-makers. Whereas in 1980 the variance in quality and productivity was relatively small within each group but large between them, by 1992 the distinction had become blurred; Ford, for example, by 1992 was one of the world's lowest-cost producers, while GM was one of the highest (Economic Strategy Institute 1992). Similarly, a 1991 listing of the 25 most productive automobile plants in North America contained all the transplants, but the four most productive were owned by Ford and the highest-placed GM plant was ranked number 20 by industry sources (Harbour and Associates; see Economic Strategy Institute 1992, 12). Canadian assembly plants have consistently ranked high on this list and have also enjoyed high ratings for the quality of their products.

Besides the 13 new Asian-managed assembly transplants that have been built, the Big Three since 1979 have closed 20 assembly plants, opened eight plants in new locations, replaced five old plants, and converted eight car assembly plants to the production of light trucks (Rubenstein 1992, 3). The majority of the remaining assembly plants were extensively refurbished and retooled to allow introduction of new model platforms and increased productivity.

The restructuring reintegrated and reinforced the traditional automotive-production cluster centred on the Great Lakes region, which spans the American midwest and southern Ontario. Even when U.S. assembly activities decentralized to the west coast and south in the 1950s and 1960s, the automotive-component–supply industry, and particularly those manufacturers producing drive-train components and body stampings, remained heavily concentrated in Michigan, Ohio, and Ontario (Glasmeir and McCluskey 1988). With the increasing adoption of JIT manufacturing methods during the 1980s, this geographical concentration of parts factories aligned along I-75 and I-65 in the United States and Highway 401 in Canada exerted a strong locational influence on investment decisions surrounding the closing of existing assembly plants and opening of new assembly capacity. Virtually all the new plants opened in the last decade were built in this region. By contrast, a disproportionate number of the closed plants were outside the region. The core auto-making region has extended southward into Kentucky and Tennessee, through the opening there of assembly plants by Nissan, Toyota, and GM Saturn and a large number of component transplant factories (Rubenstein 1992). The "Canadian auto industry" is simply the northeastern end of this region (Figure 4).

While all the assembly plants closed by the Big Three during the 1980s were located in the United States, Canadian Big Three capacity increased with the opening of the new Chrysler plant in Bramalea and the expansion and modernization of GM's Oshawa complex. As a result of this and Asian investment, Canada's share of total North American vehicle production had risen to 17.5 per cent by 1991, though Canada's share of the total North American vehicle market remained at around 8–9 per cent.

Though helped by a model-mix of large cars and mini-vans that sold well in the 1980s, the high quality ratings enjoyed by the products of Canadian plants, and the more recent vintage of much of the capital stock in Canada (in part, a function of post–Auto Pact investment), the strong performance of the industry in Canada during the 1980s is attributable mainly to a significant labour-cost differential in Canada's favour. This differential is a consequence of both the lower value of the Canadian dollar against the U.S. currency and the lower cost to the employer in Canada of employee benefits such as health care coverage. Given that productivity and efficiency in Canadian plants are judged to be at least on a par with, if not slightly ahead of, those in Big Three U.S. plants, these labour-cost savings (which in the mid-1980s amounted to around $7.50 per person-hour worked) gave the Canadian plants a decided advantage.[12]

The Canadian components and accessories sector also grew substantially during the 1980s, with employment rising from 52,000 in 1980 to 96,300 in 1989 (Table 3), and by 1991 Canada's share of North American auto parts production stood at 13 per

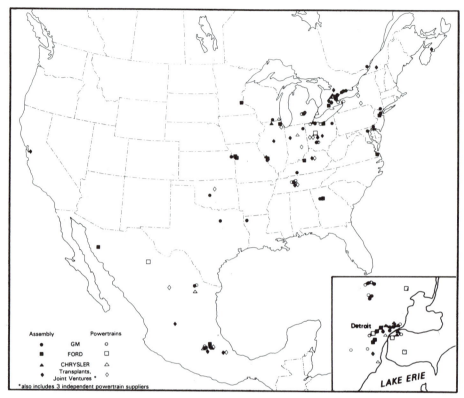

Figure 4

Automotive assembly and powertrain plants in North America, 1993. *Sources*: Canada, IST (1993); Automotive News (1993). "Transplants, joint ventures" includes three independent powertrain suppliers.

cent. This growth was surprising, given that U.S. production costs were being driven down through intense wage competition and the growth of a significant non-union segment in the industry, which offset Canada's currency advantage.[13] At least part of the expansion in Canada was accounted for by the rapid growth of a small number of independent Canadian firms (such as Magna International, A.G. Simpson, Tridon, and Woodbridge), which emerged as international competitors within their particular product areas. These firms, which were atypical of the Canadian parts sector as a whole, established an infrastructure and technological capability that allowed them to take full advantage of the changing relationship between assemblers and suppliers (Anderson and Holmes 1995).[14] These component firms were aided by both the Canadian Federal government's export-based duty-remission orders and other federal and Ontario government industrial assistance programs. During the 1980s there was a decided shift in government policy from an emphasis solely on the assembly sector to both assembly and parts production. The rise of these firms and the disinvestment of a number of large, multinational parts manufacturers resulted during the 1980s in an increase to around 20 per cent of the share of Canadian parts industry shipments accounted for by independent Canadian-owned firms.

In summary, the widespread restructuring and technological renewal of the North American auto industry meant that by 1990 the newly established transplant industry and the restructured operations of both Ford and Chrysler were competitive by international standards. The extensive reorganization and reshaping of the North American auto industry also had geographical consequences. First, they further reinforced a relatively well-defined and integrated auto-production region spanning Ontario and the American midwest and upper south. Second, and for a variety of reasons, in Canada the traditional industry was spared the dislocation that accompanied U.S. plant closings, and both vehicle assembly and parts manufacturing flourished during the 1980s.

THE FUTURE: KEYS TO SURVIVAL

Notwithstanding a decade of wide-ranging restructuring, the traditional North American automobile industry still faces serious problems. The economic downturn at the beginning of the 1990s was even more pronounced than that of a decade earlier. The depressed market for new vehicles, combined with a continuing loss of market share (Table 5), forced GM to embark belatedly on a restructuring that will see elimination of further assembly capacity, rationalization of parts production, and significant downsizing of the GM workforce.[15]

The future of the assembly sector in Canada appears relatively secure, since most plants were extensively modernized during the 1980s, are regarded as efficient, make products with high quality ratings, and have product commitments for the foreseeable future. As long as a significant production-cost differential continues between the United States and Canada (currently around $9 per hour worked), Canada will probably continue to increase its proportion of the Big Three's North American vehicle production. Of course, some of Canada's competitive edge in attracting automotive production could be eroded if there were to be sustained upward movement in the value of the Canadian dollar.[16] In the wake of the Canada–United States Free Trade Agreement (FTA) and the new content rules contained in the North American Free Trade Agreement (NAFTA), further significant Japanese investment for assembly in Canada is unlikely.

The prospects for the components sector in Canada are far less encouraging. Business prospects are shrinking because of imports and transplant vehicles, but North American parts suppliers must develop new products while they rationalize and adapt to increasing competition. Notwithstanding the strong performance of a few innovative Canadian component manufacturers over the last decade, most parts suppliers in Canada are weak competitors. Over the next five to ten years, three developments that potentially raise the most serious threats to the components sector are increasing growth of the Japanese transplants' market share at the expense of the Big Three; continental free trade and the momentum toward integration of Mexico into the North American auto industry; and the more general issue of which wage/skills strategy to pursue.

The Growing Importance of the Transplants

If the trends of the late 1980s continue, whereby vehicles produced by the transplant assemblers displace those built by the Big Three rather than simply substituting for imports built in Japan, jobs will be displaced in the U.S.- and Canadian-owned parts sectors as the transplants continue to expand their share of total production. This is because the transplants will continue for a number of years to source a significant proportion of their components from outside North America or buy them from transplant suppliers. As the Japanese auto-makers complete the building of their North American supplier base, imported parts will be replaced by parts from transplant suppliers, which are located overwhelming in the United States (and increasingly in Mexico).

The rapid growth in the number of U.S. transplant suppliers during the 1980s increased the competitive pressure on Canadian-based parts producers in two ways. First, with over four hundred new transplant supplier plants alone, the U.S.-based parts industry now has a higher proportion of new, modern facilities, typically with lower production costs. Second, not only are almost all the transplant suppliers non-unionized but the 1980s also saw the collapse of unionization and pattern wage setting in the traditional U.S. independent parts sector (Herzenberg 1991). Consequently, U.S. parts manufacturers feel that they can now obtain "low labour costs and a committed and flexible labour force" (Booz, Allen, and Hamilton 1990, I-6) without deserting the traditional Great Lakes manufacturing region.

Freer Trade

Since the rationalized and integrated Canadian and American auto industries already have operated under a virtual "free trade" environment for close to thirty years, the FTA had much less immediate impact on the auto industry than on many other industrial sectors (Holmes 1991). However, it did affect the Auto Pact. By preventing Canada from further granting "Auto Pact Status" to new manufacturers and requiring phasing out of the export- and production-based duty remission orders introduced during the 1980s to entice Asian investment, the FTA eliminated an incentive for non–Auto Pact manufacturers (that is, transplant firms such as Toyota, Hyundai, and Honda) to expand Canadian operations or use Canadian-produced parts.[17] Also, to the extent that the FTA made ineffective the incentives to Auto Pact producers to comply with requirements concerning Canadian value added, it further undermined the position of Canadian-based parts manufacturers.[18]

Of far more potential significance to Canadian parts manufacturers is the further expansion of the already substantial role played by Mexico in the North American auto industry (Holmes 1993). The export-oriented production of low-cost, low-technology parts in the low-wage maquiladoras along the Mexican-U.S. border is well known, and most of the 140 (in 1989) maquiladora plants in the transportation-equipment sector are operated by U.S. companies (Shaiken 1990). Less well known, however, the Mexican auto industry also includes production of engines and other high-value drive-train components for export to the rest of North America, Europe, and Japan and automobiles assembled for export to the United States and Canada. In 1988 Mexico exported 1.43 million engines, including 933,000 to the United States (compared to 992,000 exported by Canada to the United States), and in 1991 it exported over 365,000 assembled vehicles, 85 per cent of them destined for its northern neighbours. In northern Mexico a new, integrated auto-production region is beginning to crystallize as stamping and plastic-injection–moulding plants producing body parts are added to the engine plants and the export-oriented assembly plants of Ford and GM.[19]

Over the last decade, a radical shift in Mexican economic policy has favoured promotion of an export-led, market-oriented strategy, while the multinational auto-makers have re-evaluated the future role of Mexico within the North American auto industry. Mexico not only offers a low-cost production location for the manufacture of competitively priced, entry-level vehicles and labour-intensive parts but also represents the one geographical segment of the continental market with potential for significant market growth over the next few decades.[20] Thus the momentum toward full integration of Mexico into the continental auto industry appeared irreversible even before negotiation of NAFTA.

Given the size of the automotive sectors in Canada and Mexico in comparison with the United States, and the historical development of Canada's role within the North American auto industry, Canada is likely to be much more sensitive than the United States to expanded investment and production in the Mexican auto industry. Mexico has replaced Canada as the continental industry's least-cost production site – a competitive advantage enjoyed by Canada for close to thirty years under the Auto Pact. Since the late 1980s there have been well-publicized instances of plants producing such low-technology, labour-intensive parts as wiring harnesses, electrical equipment, and hand-sewn seating, attracted by lower wage rates, that have relocated from Canada to Mexico and the U.S. southeast.

The Canadian fear is that the relative costs of doing business in Ontario will not decline, tangible value (such as superior quality or superior technical capabilities) will not be offered for its high costs, and the southward drift may soon extend to the core of manufacture of higher-value–added parts (Automotive Select Panel 1991). It is not clear, however, just how vulnerable Ontario's remaining auto-parts industry is to competition from more distant, low-wage regions such as Mexico. One study estimated that 62 per cent of the province's auto-parts manufacturing is in product lines that JIT assemblers prefer to have near their plants (less than two hours away). Only 6 per cent is in lines that can be easily transported long distances (Ontario, Ministry of Treasury and Economics, 1991). The high concentration of assembly capacity in the Great Lakes area means that the parts industry can still exploit southern Ontario's locational advantage.[21] Nevertheless, the parts industry in Canada will continue to come under increasing competitive pressure as a result of both the changing relationship between assemblers and parts manufacturers and the presence of lower-wage producers on the American side of the border.

Strategic Choices

Echoing ongoing debates regarding the future of Ontario's manufacturing economy as a whole, two competing business strategies have emerged for the auto-parts industry that offer quite distinct scenarios and contrasting sets of institutional arrangements with regard to assembler-supplier relations, labour relations, and government policies (Herzenberg 1991).

The "high wage–high skill" approach would see the development of integrated concentrations of assembly and supplier plants in the Great Lakes region and northern Mexico. The underlying competitive strategy would involve making North American plants (whether American-, Japanese- or European-owned) more productive by fostering both the development of worker skills and the technological capacity of firms to innovate. In turn, "the high performance of the auto industry in these regions would be accompanied by a decline in the North American auto trade deficit with Japan; by employment growth, rising real wages and an expanding market in Mexico; and by stable hours of employment (and larger numbers of jobs if average hours decline) and modest wage growth in U.S. and Canadian assembly *and* parts plants" (Herzenberg 1991, 47).

In the alternative, "low wage–low skill" scenario, intense wage competition already evident in the U.S. supplier segment of the industry would become more deeply entrenched, spread to Canada, and lead an increasing number of assemblers to use Mexico as a low-wage, low-skill production base. Hard bargaining on wages and parts prices and movement to low-wage, non-union regions would undermine attempts to foster cooperative relations with workers and suppliers in established plants in the United States and Canada. Modern assembly plants would become high-tech "islands of automation" amid a sea of suppliers competing on the basis of low wages. Low unionization rates and low productivity growth associated with low-wage strategies would limit wage increases throughout the industry and across the continent.

Unfortunately, certain features of the past development of the parts sector in Canada stand in the way of moving toward high wages high skills. Perhaps the two most critical barriers are the low levels of existing investment in research and development (R&D) by most Canadian parts firms and the associated lack of a skills development strategy within the industry. Though Canada has accounted for an increasing share of North American vehicle and parts production, R&D expenditures as a percentage of sales by the automotive sector in Canada remain low compared with those in the United States. The business strategy of most firms in the Canadian auto-parts sector, for at least the past thirty years, has been one of differentiation on the basis of low manufacturing costs. Their approach to technology has been described as " 'adopt and adapt', whereby firms imported state-of-the-art machinery and then capitalized on lower Canadian labour costs" (Science Council of Canada 1992, 16). This strategy was successful in the post–Auto Pact period, but it has been called into serious question by the rapid restructuring of the 1980s and the efforts by assemblers to create the "new manufacturing culture."

Though the Canadian auto-parts firms that had superior growth rates during the 1980s, such as Woodbridge, Magna, Court, and the ABC Group, relied on innovation as a basis for differentiation, they are atypical. Only 2 per cent of Canadian parts firms have strategies of technological innovation based on either proprietary-product technology or manufacturing processes (Science Council of Canada 1992, 16). The branch plants of the large multinational parts makers and the Big Three perform very little R&D in Canada; for example, in 1986 Canadian-owned independent firms, though producing only 17 per cent of total parts shipments, conducted 46 per cent of Canada's automotive R&D (Science Council of Canada 1992, 15).

This situation has its roots in the restructuring that took place under the Auto Pact, when R&D, product development, and key production decisions in Canadian-based plants were taken over by the U.S. parent firms. Furthermore, the recent, relatively weak profitability and cash-flow situation experienced by auto-parts manufacturers as a result of the recession, and compounded by the assemblers' demands for continuous reductions in the price of components, means that, in the short run at least, there is only limited potential for increased R&D in the Canadian auto-parts sector.

Some initiatives are being taken in the Canadian auto-supplier sector to improve skills training, for which there is significant support in Canada from both employers and labour. Employers believe that training is essential for competitiveness because the new manufacturing culture is increasing the proportion of more highly skilled workers in the auto-parts industry.[22] For labour, training may provide workers with greater employment security. The recently created sectoral training board is working for establishment of a general training credential for the industry.

The Canadian labour movement has argued also that the government should support establishment of institutional constraints on low-wage competition in order to move the auto-supplier industry toward a high skill – high wage future. One such constraint would be to extend the economic terms of pattern-setting collective agreements to small employers through regional or sectoral master agreements similar to those that exist in countries such as Germany. Much of the opposition from within the Canadian labour movement to continental freer trade through the FTA and NAFTA arises out of the fear that the business community will press for creation of a "level playing-field" on which to conduct business. A result of this would be pressure on Canadian workers to sustain "competitiveness" and protect jobs by accepting the low wage – low skill strategy and the segmented model of labour relations that emerged in the United States during the 1980s.

Certain features of the auto industry make it unique among Canadian manufacturing industries. It has already made the transition from a branch-plant industry serving a sheltered domestic market to one competing in a North American market subject to intense global competition, but it did so in two stages, over the course of a quarter-century. Since its incorporation into a continental production and marketing system in the 1960s, its fortunes have been inextricably linked to the American economy and shaped by the corporate strategies of a handful of transnational corporations. During the restructuring and technological regeneration that took place in the industry during the 1980s, in response to the challenge posed by Japanese auto-makers, it fared well and increased its share of North American automobile production. Even so, the future of the Canadian auto-parts sector remains extremely uncertain, due in no small measure to the weak technological capacity of many Canadian-based manufacturers. Viewed as a whole, however, the automotive industry will probably for the foreseeable future remain both a significant and technologically sophisticated branch of Canadian manufacturing and a mainstay of Ontario's economy.

NOTES

1 Under the present standard industrial classification (SIC) employed by Statistics Canada, the automotive industry is part of major group 32 (Transportation Equipment Industries). It is made up of three main branches, which correspond to the following four-digit industries: motor vehicle assembly, SIC 3231; manufacture of motor-vehicle parts and accessories, SICs 3251–7 and 5259; and manufacture of trailers, motor homes, and bus and truck bodies, SICs 3241–4. It is common to treat the third branch as a distinct industry. Analysis and discussion in this chapter focus on the first two branches, referred to as the automotive or automobile industry.

2 While each manufacturer was free to supply the Canadian market from either domestic or U.S. production, the volume of sales that the manufacturer could make in Canada was linked to the volume of Canadian production. Furthermore, a Canadian value-added (CVA) requirement was designed to ensure that Canadian production amounted to more than mere assembly (Holmes 1991).

3 To qualify for duty-free treatment under the Auto Pact, new vehicles and OEM parts imported into the United States have to come from Canada and must contain at least 50 per cent North American content. Eligibility for duty-free treatment of vehicles and parts entering Canada is based not on the goods' origin but on the fulfilment by the importing manufacturer of the performance criteria (see note 2, above). As long as these criteria are met, it does not matter where the new vehicles or OEM parts originate or where they were shipped from. This interpretation was used as one of the inducements to attract new Asian investment to the auto industry in Canada during the 1980s. At the same time, General Motors, Ford, and Chrysler were annually importing over $3 billion in parts into Canada from third countries and saving about $300 million in duty because of their "Auto Pact Status" (Wonnacott 1988).

4 The more labour-intensive nature of parts production in Canada is reflected by the $38.28 value added per person-hour in Canada in 1989, compared with $50.83 per person-hour in the United States.

5 Both types of duty-remission order were consistent with the Canadian government's long-standing view of the Auto Pact as a more general policy instrument. It would accord "Auto Pact Status" to any manufacturer, regardless of nationality, that started manufacturing vehicles in Canada and met criteria analogous to the Auto Pact requirements (Johnson 1993).

6 In an earlier paper (Holmes 1987), I set out and compared the key elements of each model.

My characterization of the traditional, Fordist system of production remains essentially unchanged. However, after further research and reflection, I have modified certain aspects of my characterization of the JPS.

7 An enormous amount of confusion exists around the question of whether the JPS represents an abandonment of the basic principles of mass production. Berggren (1992) points out that much of this uncertainty arises because observers fail to distinguish between production in small batches (the hallmark of the JPS) and short total production runs. Toyota made having small inventories a priority and therefore chose to concentrate on small-batch manufacturing. The object, however, was to produce the greatest possible cumulative volume of each product. Thus "it cannot be said the Toyota system is the antithesis of mass production; it is, on the other hand, the antithesis of large-batch manufacturing" (Shingo 1981, 112).

8 Alice Amsden (1990) explains this strategic focus on the shop-floor by arguing that Japan, as a late industrializer, based its industrialization on borrowed technology; even leading firms had to deal with wage competition without the help of any technological advantage by focusing on "learning."

9 For example, in their influential book, *The Machine That Changed the World*, Womack, Jones, and Roos (1990) maintain that Japanese auto companies have blessed the world with a completely new approach to manufacturing – lean production – which "will supplant both mass production and the remaining outposts of craft production in all areas of industrial endeavour to become the standard global production system of the twenty-first century" (278).

10 At the same time, Berggren (1982) notes that certain features of lean-production personnel practices, such as job security, removal of distinctions between blue-collar and white-collar workers, and emphasis on pride in work, have proved highly attractive to American workers.

11 The latter explains the shift in the source of growth in the automotive-trade deficit with Japan during the 1980s from assembled vehicles to parts (see Table 5 and Figure 3).

12 One significant side-effect of the uneven performance of the auto industry in the United States and Canada during the 1980s was the breakup of the United Autoworkers union (UAW) and formation of the independent Canadian Autoworkers (CAW) in 1985 (Holmes and Rusonik 1991).

13 In the 1980s, pattern wage and benefit bargaining in even very large U.S. independent parts suppliers (IPS) firms broke down. As a result hourly wages in unionized auto parts plants in 1989 were only 67 per cent of those in assembly firms, with the equivalent figure for non-union IPS plants being 51 per cent. By contrast, wages at Canadian parts plants remained more than 80 per cent of those in assembly plants and decreased only marginally during the 1980s (Herzenberg 1991, 22).

14 The success of these firms was based on their technological capacity to develop innovative products, but, at least in the case of Magna, it was also combined with a non-union, low-wage industrial relations strategy (Anderson and Holmes 1995).

15 The extent of the restructuring begun in 1992 by GM will inevitably cause plant closures in Canada. In addition to the previously announced closure of its Scarborough van assembly plant, GM has announced planned closing of parts plants in St Catharines, Windsor, and Oshawa as well as the lay-off of 1,400 workers from the Oshawa assembly operations to be achieved by increasing efficiency.

16 The parts sector views a Canadian dollar worth no more than 80¢ U.S. as critical for Canadian competitiveness (Booz, Allen, and Hamilton 1990, III-26).

17 The FTA thus created two classes of manufacturers: those with "Auto Pact Status" (GM, Ford, Chrysler, Volvo, and CAMI, which received such status just before the FTA came into effect in 1989, and a number of small, specialty-vehicle manufacturers) and those with "non–Auto Pact Status." The fact that the Big Three and CAMI import vehicles and parts from third countries duty free puts them at an advantage over other producers in Canada such as Honda, Toyota, and Hyundai.

18 However, as Johnson (1993, 14) notes: "this concern is probably more theoretical than real." There is one significant inducement that remains under the FTA for existing Auto Pact producers to meet the Canadian performance requirements demanded by the Auto Pact: that of duty remission on vehicles and parts entering Canada from third countries (Holmes 1992).

19 The Ford assembly plant opened at Hermosillo in 1986 uses advanced Japanese-style production management and work organization techniques to assemble an entry-level vehicle for sale in the United States and Canada. Ford Hermosillo has attained high levels of productivity and achieved a product-quality rating higher than the best volume-assembly Japanese plants or the best North American transplants (Womack, Jones, and Roos 1990, 87).

20 The potential contradiction between growth of the Mexican internal market and maintenance of a low-wage manufacturing economy is rarely addressed.

21 By the same token, however, southern Ontario is poorly positioned to supply the embryonic auto-production complex that is forming in northern Mexico to assemble entry-level vehicles for the North American market.

22 A recent study by Employment and Immigration Canada (1991) estimated that by 1995 the proportion of unskilled workers in the auto-parts industry would have fallen from 62 per cent in 1985 to 33 per cent.

REFERENCES

Altshuler, A., et al (1984). *The Future of the Automobile*. Cambridge, Mass.: MIT Press.

Amsden, A. 1990. "Third World Industrialization: 'Global Fordism' or a New Model?" *New Left Review*, 182, 5–31.

Anderson M., and Holmes, J. 1995. "High-Skill, Low-Wage Manufacturing in North America: A Case Study from the Automotive Parts Industry." *Regional Studies*, 29 no. 7.

Automotive News. 1993. *Automotive News 1993 Market Data Book*. Detroit: Crain.

Automotive Select Panel. 1991. *Study of the Competitiveness of the North American Automotive Industry*. Toronto: Coopers and Lybrand.

Berggren, C. 1992. *Alternatives to Lean Production: Work Organization in the Swedish Auto Industry*. Ithaca, NY: ILR Press.

Booz, Allen and Hamilton. 1990. *A Comparative Study of the Cost Competitiveness of the Automotive Parts Manufacturing Industry in North America*. Toronto: Automotive Parts Manufacturers Association.

Canada. Industry, Science and Technology (IST). 1990. *Statistical Review of the Canadian Automotive Industry: 1989*. Ottawa.

– 1992. *Statistical Review of the Canadian Automotive Industry: 1992 Edition*. Ottawa.

– 1993. *Statistical Review of the Canadian Automotive Industry: 1993 Edition*. Ottawa.

Dunn, J.A. 1987. "Automobiles in International Trade: Regime Change or Persistence?" *International Organization*, 41 no. 2, 225–52.

Economic Strategy Institute. 1992. *The Future of the Auto Industry: It Can Compete, Can It Survive*? Washington, DC: Economic Strategy Institute.

Employment and Immigration Canada. 1991. *Report of the Automotive Parts Human Resources Study*. Ottawa: Minister of Supply and Services.

Glasmeir, A., and McClusky, R.E. 1988. "US Auto Parts Production: An Analysis of the Organization and Location of a Changing Industry." *Economic Geography*, 64, 142–59.

Hart, M. 1990. *A North American Free Trade Agreement: The Strategic Implications for Canada*. Halifax: Institute for Research on Public Policy.

Herzenberg, S. 1991. *The North American Auto Industry at the Onset of Continental Free Trade Negotiations*. Economic Discussion Paper No. 38, U.S. Department of Labor, Bureau of International Labor Affairs Washington, DC.

Holmes, J. 1983. "Industrial Reorganization, Capital Restructuring and Locational Change: An Analysis of the Canadian Automobile Industry in the 1960s." *Economic Geography*, 59, 251–71.

– 1987. "Technical Change and the Restructuring of the North American Automobile Industry." In K. Chapman and G. Humphrys, eds., *Technical Change and Industrial Policy*, 121–56. Oxford: Basil Blackwell.

– 1991. "The Globalization of Production and Canada's Mature Industries: The Case of the Auto Industry." In D. Drache and M. Gertler, eds., *In the Era of the New Competition*, 153–80. Montreal: McGill-Queen's University Press.

– 1992. "The Continental Integration of the North American Automobile Industry: From the Auto Pact to the FTA and Beyond." *Environment and Planning A*, 24, 95–119.

– 1993. "From Three Industries to One: The Integration of the North American Auto Industry." In M. Molot, ed., *Driving Continentally: National Policies and the North American Auto Industry*, 23–61. Ottawa: Carleton University Press.

Holmes, J., and Kumar, P. 1991. *Divergent Paths: Restructuring Industrial Relations in the North American Automobile Industry*. Queen's Papers in Industrial Relations, 1991–4, Queen's University, Kingston, Ontario.

Holmes, J., and Rusonik, A. 1991. "The Break-up of an International Labour Union: Uneven Development in the North American Auto Industry and the Schism in the UAW." *Environment and Planning A*, 23, 9–35.

Howes, C. 1991. "The Benefits of Youth: The Role of Japanese Fringe Benefit Policies in the Restructuring of the US Motor Vehicle Industry." *International Contributions to Labour Studies*, 1, 113–32.

Johnson, J.R. 1993. "The Effect of the Canada–U.S. Free Trade Agreement on the Auto Pact." In M. Molot, ed., *Driving Continentally: National Policies and the North American Auto Industry*, 255–84. Ottawa: Carleton University Press.

Katz, H.C. 1986. "Recent Developments in U.S. Auto Labour Relations." In S. Tolliday and J. Zeitlin, eds., *The Automobile Industry and Its Workers: Between Fordism and Flexibility*, 282–304. Cambridge: Polity Press/Basil Blackwell.

Krafcik, John F. 1988. "Comparative Analysis of Performance Indicators at World Auto Assembly Plants." MS thesis, Sloan School of Management, Massachusetts Institute of Technology, Cambridge, Mass.

Mair, A., Florida, R., and Kenney, M. 1989. "The New Geography of Automobile Production: Japanese Transplants in North America." *Economic Geography*, 64, 352–73.

Ontario, Ministry of Treasury and Intergovernmental Affairs. 1991. *Ontario Economic Outlook*. Toronto: Queen's Printer.

Reid, N. 1990. "The Spatial Location of Japanese Automobile Production in North America." *Industrial Relations Journal*, 21, 49–59.

Roy, F. 1991. "Recent Trends in the Automotive Industry." *Canadian Economic Observer*, 4.1– 4.14.

Rubenstein, J. 1992. *The Changing US Auto Industry: A Geographical Analysis*. London: Routledge.

Science Council of Canada. 1992. *Sectoral Technology Strategy Studies: No. 2, The Canadian Automotive-Parts Sector*. Ottawa: Minister of Supply and Services.

Shaiken, H. 1990. *Going Global: High Technology in Mexican Export Industry*. Monograph Series No. 33, Center for U.S.-Mexican Studies, University of California at San Diego.

Shingo, S. 1981. *The Toyota Production System*. Tokyo: Japan Management Association.

Taskforce on the Canadian Motor Vehicle and Automotive Parts Industries. 1983. *An Automotive Strategy for Canada*. Ottawa: Minister of Supply and Services.

Womack, J. 1989. *The US Automobile Industry in an Era of International Competition: Performance and Prospects*. Working Papers of the MIT Commission on Industrial Productivity, Vol. 1. Cambridge, Mass.: MIT Press.

– 1990. "North American Integration in the Motor Vehicle Sector: Logic and Consequences." Paper for International Forum, "Mexico's Trade Options in the Changing International Economy," Mexico City.

Womack, J., Jones, D.T., and Roos, D. 1990. *The Machine That Changed the World*. New York: Rawson Associates.

Wood, Stephen. 1992. "Japanization and/or Toyotaism." *Work, Employment and Society*, 5, 567–600.

Wonnacott, P. 1988. "The Canada-U.S. Free Trade Agreement." *Trade Monitor*, No. 2 (March 1988).

Woodward, D. 1992. "Locational Determinants of Japanese Plants." *Southern Economic Journal*, 58, 690–708.

CHAPTER FOURTEEN

High-Tech Canada

JOHN N.H. BRITTON

As markets demand products embodying advanced technologies, so the success of firms, industries, and regions reflects how well they embrace the new.[1] We can see this most clearly in the shifts in the terms of trade in favour of higher-value goods and services based on new technologies. Viewed in this way, competitive advantage, which is necessary for economic growth, requires generation and expansion of technologically innovative activities. The distinctive features of this model in the late twentieth century are its global range of application and the speed with which advances are being made in a variety of technologies. The pervasive impact of one of these fields of industrial knowledge – information technology, or IT – is also a notable feature of this era (Freeman and Perez 1988; Freeman 1991).

It is important to understand how well Canada's new technology-based activities have prospered. In the light of the small scale of this economy, its openness to trade and capital flows (see in this volume Norcliffe, chap. 2, and MacPherson, chap. 4) and the access that competitive producers have to the Canadian market, the country's economic development depends on increasing technology-intensive exports. A necessary condition for this form of development is indigenous technological capability, as demonstrated by generation of new and improved products. Over time transfer of knowledge between firms leads to the accumulation of technological expertise within particular regions and activities, creating positive externalities (Krugman 1986; Malecki 1991). New technological knowledge becomes available for use by other firms in the form of new or improved products, services, corporate experience, and labour skills.

The capacity to innovate generates the ability to export new, technology-based products. Canada's goal has been to secure trading conditions that will allow these technological strengths to emerge or to grow, and the Canada–United States Free Trade Agreement (FTA) and the North American Free Trade Agreement (NAFTA) may be seen as moves in this direction. There will, however, always be need for technology imports, though the means by which these are organized may vary considerably. Technology in the form of industrial knowledge may be imported through market arrangements, through the internal channels of multinational firms, or through business alliances. Importing of product concepts, parts and sub-assemblies, production systems, and capital equipment (all of which embody new technology) is characteristic of the way many Canadian firms have sought to increase their productivity. The long-term weaknesses of this method as a major route to innovation have been recognized for some time and are discussed in other chapters.

The primary tasks of this chapter are to identify the elements of technological strength in the Canadian economy and to place them in a broader international and national economic context. I establish the scale of Canada's high-technology activities initially by referring to industries with intensive research and development (R&D). I use Canada's international ranking in R&D expenditures, and the technological trade balance between Canada's imports and exports of goods and services, to make comparisions of its performance. Then I identify Canada's industrial successes through analysis of the sets of high-tech activities undertaken by individual firms. I also assess the location of Canada's high-tech firms and examine the recent history of this sector, seeking to provide an appraisal of the potential for the existing location pattern to change in favour of non-industrial cities. As well, I re-examine structural factors thought to shape Canada's technological development. Here I note the legacy of a high level of foreign direct investment because of the influence that NAFTA and global economic changes are projected to have on the choices open to multinational corporations and on Canada's industrial structure.

CANADA'S STRENGTHS: AN AGGREGATE VIEW

High-Technology Activities

Canada's high-technology (three-digit) industries are identified using high values of the ratio of R&D to sales (Table 1). As used here, the minimum value of this ratio to delimit research-intensive industries (2.7 per cent) includes the Scientific and professional equipment industry, considered for some time part of the Canadian set of these activities (Science Council 1981); international norms, however, would divide industries with high and medium R&D intensity at about 4 per cent (Malecki 1991).

Canada's Telecommunications equipment and Aircraft and parts industries are of such large scale that together they account for 34 per cent of R&D expenditures in manufacturing, and the high-technology group of manufacturing industries as a whole increased its share of industrial R&D from 49 per cent to 64 per cent (1975–92). While this growth implies increasing specialization and focusing of R&D expenditures, the high-technology group of industries contributes only 14 per cent of manufacturing value added and therefore has limited direct effect on the output structure of the manufacturing sector.

Table 1
R&D expenditures for selected industries, Canada, 1991

Industry*	R&D expenditures ($million)	R&D/sales
Manufacturing		
Telecommunication equipment[†]	693	22.1
Other electronic equipment	361	13.5
Aircraft and parts	445	13.0
Business machines	284	3.1
Machinery	90	3.4
Scientific and professional equipment	57	2.7
Drugs and medicines	234	5.3
All manufacturing	3,205	1.8
Services		
Engineering and scientific	423	18.7
Computer services	234	17.9
All services[‡]	1,304	1.6
Total for all industries	4,853	1.9

Source: Industry Canada (1994a).

* Industries were selected if R&D/sales ≥ 2.5 per cent for 1989.

[†] Includes Telecommunications equipment and electronic parts and components.

[‡] Includes Finance and insurance.

Telecommunications equipment, Aircraft and parts, and Business machines are Canada's leaders among advanced technology-based activities. To put this into international perspective, however, it is worth noting that only Canada's telecommunications equipment industry undertakes R&D at a level above the average R&D/value-added ratio recorded for the G7 countries (Statistics Canada 1993b). Machinery is both the smallest industry of the group (Table 1) and a relatively weak performer, despite investment in modern, reprogrammable (flexible) production tools needed in all sectors of the economy (see also, in this volume, Barnes, chap. 2; Hayter, chap. 6; Wallace, chap. 7; Cannon, chap. 12; Holmes, chap. 13; and Gertler, chap. 15). Among the industries that lie between these extremes of performance, Drugs and pharmaceuticals are of interest, since press reports in recent years suggest that a consequence of federal increases in patent protection for new drugs is the planned expansion of the R&D labs of foreign proprietary drug companies.

The largest recent change in the industrial structure of technology-intensive activities has been expansion of R&D by the service sector between 1980 and 1992, from 14 per cent of the Canadian total to 33 per cent. This growth reflects the increasing importance of software design and the producer services as sources of technological competitiveness. With 42 per cent of service-sector expenditures, the largest performer of R&D is Engineering and scientific services, an activity that reflects

Canada's established expertise in various energy-production technologies and in urban and long-distance transportation systems. There have also been some declines in the share of R&D undertaken especially by Mining and Oil wells (from over 6 per cent of total R&D to less than 2 per cent).

The Technology Gap

Since the 1970s, the technology gap – the deficit in high-technology trade – in Canada's international trade has been well documented in the economic literature (see also Norcliffe, this volume, chap. 2). In merchandise trade the deficit increased by nearly $1 billion (I use 1981 dollars as my reference in this section), or 19 per cent, over the period 1979–87 (Statistics Canada 1989b) and then in half this length of time, 1990–93, it grew by another 19 per cent, a $2-billion increase (Industry Canada 1994b). Canada's trade position varies considerably across product markets. It is strongest in Telecommunications equipment and Aircraft and parts, where the trade balance fluctuates between small surplus and deficit positions, but it is weak in Computers (37 per cent of the high-tech trade deficit), Machinery (30 per cent), and Electronic equipment (16 per cent).

There is also a substantial deficit ($1 billion) in the trade account, which covers payments for consultancy services, royalties, management services, and R&D – that is, for explicit transfers of technology. Among these services it is only in consulting that Canada has a small surplus because of its success in civil-engineering design, especially for resource and transportation projects. In other service categories, Canada's deficit is increasing, but trade data provide only an indication of the direction of technology flows, since available statistics are a partial representation of the transactions involved. In particular, a "significant number of subsidiary companies in Canada are not specifically charged for the services supplied for them by their foreign parent companies": this is an important consideration because flows related to direct investments are about 80 per cent of the total (Statistics Canada 1990).

The most troubling aspect of Canada's export performance is that technology-intensive products have increased to only 14 per cent of manufactured exports, compared with 38 per cent in the United States, 28 per cent in Japan, and 24 per cent in Asian newly industrializing countries (NICs) (Economic Council of Canada 1989). Canada's share of world exports of high-technology products has declined, from 3.5 to 2.6 per cent, while Japan's has more than doubled, to 17 per cent, and Asian NICs are now at about 9 per cent.

Who Does R&D?

The technology gap embedded in the pattern of Canada's international trade has a long-term relationship with its low level of industrial R&D. Though expenditures on R&D increased substantially during the 1970s and 1980s (by 112 per cent 1971–91, in constant dollars), this was not a competitive advance. Canada spent only 1.49 per cent of GDP on R&D in 1991, only 0.17 percentage points more than the proportion in 1971, and in this respect it differed sharply from other industrial nations; the average of the ratios for Germany, Japan, Sweden, and the United States increased from 2.1 per cent of GDP (1974) to 2.8 per cent (1990).

Structural differences in the sectoral origins of Canada's GDP may provide some of the explanation for Canada's apparent under-investment in R&D. Over the long term, for example, Canada has made only comparatively small expenditures on R&D in defence industries. Since Canada has made choices that differ considerably from those of a number of its competitors, business funding of R&D is a more useful indicator of how competitive Canadian industry really is. In the United States, business funds R&D at about 1.4 per cent GDP (1989), in Japan, at 2.2 per cent; and in Sweden, at 1.7 per cent; while Canada operates at the comparatively low level of 0.6 per cent (Industry Canada 1994b). Resource sectors with low ratios of R&D expenditures to sales (for example, 0.5 per cent in Mines and oil wells and 0.4 per cent in Paper and allied products) are another structural difference with some G7 countries, but these sectors are equally important in Sweden, which maintains a higher aggregate rate of expenditure on R&D.

Other than in defence, governments have been alert to the problem of limited industrial R&D, and Canada, in terms of public funds for R&D, is in the bottom half of the OECD league, providing (at 0.6 per cent of GDP) slightly more than the United Kingdom and the United States, but much less than most OECD countries; Sweden funds about 0.85 per cent of GDP, and Germany about 0.75 per cent (Statistics Canada, 1993b). Canada's public funds are spent on R&D in industry, in universities, and in government laboratories, the latter reflecting a long-run policy response to the difficulties of a small economy with many small firms that have limited R&D capacity. It has been known for some time, however, that transfer of technology from federal labs is flawed. With more orthodox economic voices being heard, these facilities are often perceived as absorbing public dollars that could be used to support industrial R&D, more closely linked to market demands.

In practice in Canada, the federal role in performing R&D has been cut from 29 per cent of the total in the early 1970s to 16 per cent, and thus recently there has been no growth in the number of scientists and engineers employed in federal R&D. Nevertheless, the public sector is the main source of funds for university R&D (43 per cent in 1991). Very recently, business has increased its contribution to 7.5 per cent (1991) of the funds spent on university R&D (from only 0.4 per cent in 1981) partly because federal policy has tied some of its assistance to matching funds provided by businesses. The federal and provincial governments also provide grants and tax credits to assist industry in undertaking R&D (see Britton 1991), but the question remains why Canadian businesses do not perceive the need to make larger expenditures on R&D that are more commensurate with those of the leading industrial countries.

In answering this question the history of foreign subsidiaries in Canadian industrial development is important. In manufacturing, for example, 52 per cent of expenditures on R&D are made by foreign firms, and this share is even higher in the economy as a whole (61 per cent) because of the impact of foreign investments in the service sector (Statistics Canada 1993a). Nevertheless, in every technology-intensive industry for which data are available, foreign firms in Canada allocate a substantially smaller fraction of their sales to R&D compared with Canadian firms. This proportion ranges between 35 per cent and 77 per cent of the level for indigenous firms in Electronic parts, Other electronic equipment, Business machines, and Drugs and drops to 5 per cent for Machinery (Statistics Canada 1993a). This pattern has a substantial influence on the aggregate performance of technology-intensive industry, since foreign control is on average about 70 per cent, based on sales in these industries.

Canadian firms undertaking R&D are much smaller on average than foreign subsidiaries and operate under completely different conditions of capital availability and experience in the innovation process. For that reason it is not surprising that the elasticities of domestic firms' R&D expenditures with respect to sales are less than those of foreign firms (Industry Canada 1994b). These data suggest that Canadian research-intensive firms have problems in expanding their R&D as sales increase and that R&D policies might be revised specifically to accommodate the particular difficulties of small Canadian firms (see also Britton 1991, for transaction-cost arguments for the same recommendation). The data, however, do not throw much light on the problem of why, given international norms, foreign firms have relatively low R&D/sales ratios.

In order to establish whether large foreign subsidiaries have similar or lower levels of R&D/sales than domestic firms of similar size, foreign and domestic firms are compared in information technology, drawing on firms in telecommunications equipment, electronics, and computer hardware/software. Restricting the comparison to firms with revenues of $200 million or more in 1993, there are 11 foreign firms and 5 domestic companies. The unweighted average domestic ratio of R&D to revenue for this group (11.2 per cent) is nearly twice as large as that for the foreign group (5.7 per cent), showing that ownership differences in R&D intensity found at the aggregate level occur even among the largest firms. Comparisons using those firms undertaking at least $20 million of R&D yield a Canadian unweighted average (17 per cent) that is considerably higher than that for foreign firms (10.8 per cent).[2] Evidently, the collective inequalities in research intensity associated with ownership are real, not a matter of scale, and are an outcome of corporate policy.

Surveys show that the largest 485 performers of R&D undertake 88 per cent of Canada's industrial R&D expenditures (Statistics Canada 1989a), demonstrating a clear positive relationship among R&D, competitiveness, and company size, though the main performers of R&D represent a very broad technological spectrum, which suggests that Canada may well be spreading its resources thinly. It is still surprising to realize that 1993 expenditures as small as $20 million earned a place among the top 50 performers of R&D and that the expenditure of the 100th-ranked company was as small as $7.2 million. The prevailing pattern of R&D is thus one in which there are many small and medium-sized firms (predominantly Canadian-owned) and a small number of larger Canadian and foreign firms.

SPECIALIZATIONS

Macro-economic data are very useful in establishing the basic pattern of Canada's technological development, but the discussion above points to several ways in which industry and trade data provide limited grounds for locating Canada's technological capabilities. At the outset I argued that there are strong spillover and spread effects associated with development and adoption of new technology. These do not respect industrial definitions; for example, technologically advanced or progressive firms exist in industries otherwise not distinguished by high R&D/sales ratios. Moreover, the industrial classification is insensitive to new combinations of technologies that influence the output of firms, moving them in directions that are novel for their industries. This is an important issue, particularly as more than one-third (36 per cent) of Canadian R&D is performed by firms classified in industries excluded from Table 1.

To delve beneath standard industrial definitions requires survey data for individual enterprises making new, technology-based products or providing related services, regardless of the industry in which they are normally classified. The data used here are taken from the Cantech database (Hutchinson Research 1991) on the high-technology products of Canadian firms. These products are recorded for over 2,300 firms (under foreign and domestic control) according to 31 high-technology service, software, or hardware product groups, following a classification developed by Corptech, a u.s. innovator in this area of data assembly. This information permits a more extensive view of Canadian output than has been possible previously.

Cantech, constructed from a national coverage of large firms and from geographic cluster samples of firms of all sizes, provides representative coverage of Canadian high-technology firms. These data have allowed each firm to be classified in terms of the combination of its main products.

Technology Clusters

The product data (Table 2) have been generalized by identifying the way the 31 product technologies are associated in the activities of the sample firms, and similarities between firms in their joint output of products (including services) have been used to define clusters of technologies (Table 3).[3] Eight technology clusters have been found, including a generic cluster – firms supplying Subassemblies and Consultant services, which are inputs to many other economic activities. The search for these groups of products was pursued without any requirement that each firm be assigned to only one cluster, though this proved to be the way two-thirds of the firms in the sample were assigned. The other firms belong to pairs of clusters, and analysis of these connections shows all clusters to be connected in one system, with Energy-Environmental (E-E) firms being central.

The clusters record the response of Canadian industry to the rise of new technology paradigms (Freeman and Perez 1988) such as Automation, Biotech-Pharmaceuticals (B-P), E-E, and Information Technology (IT). To a large extent these clusters

Table 2
High-tech products

Product technologies	Number of firms producing	Product technologies	Number of firms producing
Goods		*Services*	
Defense/military equipment	26	Artificial intelligence	7
Manufacturing equipment	26	Energy	14
Pharmaceuticals	34	Material	29
Medical equipment	59	Test/measurement/control	30
Photonic equipment	59	Automation	41
Biotechnology systems	68	Chemical	43
Chemicals	89	Education/training	50
Transportation equipment	89	Telecommunications	55
Factory automation	127	Electronics	71
Advanced materials	132	Biotech	72
Telecommunications	161	Transportation	89
Computer hardware	162	Environmental	116
High technology nec	166	Manufacturing	177
Energy/environmental equipment	198	Engineering/design	182
Test/measurement/control equipment	235		
Subassemblies/subsystems	241	Total number of firms	2,344
Computer software	724		

Source: Hutchinson Research, 1991 data.

(part A – Table 4) reflect emergence of new technology-based firms, which are predominantly small enterprises (Rothwell and Zegveld 1982; 1985). This seems particularly true for B-P and IT. The relatively large number of small high-tech firms is valuable as a base for growth and as a source of product ideas and components, though the management of these firms often has limited experience and access to venture capital, and so debt finance may be a problem.

The concentration of R&D expenditures among firms shows up strongly in the survey, with only 5 per cent of firms employing 100 or more R&D staff and most (57 per cent), fewer than five. There is, however, a direct relationship, across the technology clusters, between the size of firms and the size of R&D labs. Those clusters in which there is a higher incidence of R&D labs employing 100 or more (Table 4, part B) are the activities that have higher proportions of large firms (more than 500 employees). These are also the clusters with higher proportions of businesses established before 1970, and the two groups of technology clusters generally are associated with different age profiles of establishments. In those that have high proportions of small firms, most enterprises have been founded within the last 20 years.

The growth of the number of Canadian high-technology enterprises slowed dramatically after 1981–85, a pattern represented in all clusters. It is also true that the clusters with higher proportions of large firms contain more companies dating from before 1971. These are the clusters (Table 4, part B) that owe their origin to dominant

Table 3
Clusters of high-tech products

	Total number of firms	
	This cluster	This cluster exclusively
Biotech-pharmaceutical	111	187
Information technology (IT)	804	1,155
Automation technology	107	378
Chemicals and services	41	121
Advanced materials and services	58	157
Transportation	26	116
Energy-environmental technologies (E-E)	247	709
Generic technologies	174	590

Source: As Table 2.
Note: The clusters contain product technologies as follows:
Biotech-pharmaceutical: Biotechnology products and equipment, Pharmaceuticals, Biotech services, Medical Equipment, Medical Services.
IT: Computer hardware, Computer software, Telecommunications, Computer services, Telecommunication services, Services for artificial intelligence.
Automation technology: Factory Automation equipment, Manufacturing equipment, Automation services, Manufacturing services, Electronics services.
Chemicals and services: Chemicals, Chemical services.
Advanced materials and services: Advanced Materials, Material Services.
Transportation: Transportation equipment, Transportation services.
E-E: Energy/environmental equipment, Energy services, Design and engineering services, Test/Measurement/Control Equipment, Environmental services, Test/Measurement services.
Generic technologies: Subassemblies/subsystems, Other high tech, General Consulting services, Educational/training services, Financial services, Membership organization, General R&D services, Other services.

Table 4
Size distribution of firms in high-tech clusters

	% firms in each cluster							
	Number of employees				Year founded		R&D staff	
	<20	<50	>499	>999	After 1975	Before 1969	>99	<5
PART A								
Automation	46	68	5	2	50	36	4	63
Biotechnology	52	64	4	1	62	24	3	47
Information technology	62	79	4	2	76	10	3	58
Energy-environmental	46	68	6	3	53	32	5	58
PART B								
Chemicals	26	44	13	10	38	53	12	47
Advanced materials	27	41	13	9	38	53	10	52
Transportation equipment	26	42	19	10	34	45	14	37
Generic	38	55	11	7	54	33	9	50
Total	51	69	7	4	63	24	5	57

Source: As Table 2.

technologies of earlier waves of innovation, and in these cases firms tend to be both older and larger. These clusters record firms that adopted new-technology – based products or that evolved technologically within their product mix. Some Canadian resource firms, for example, have moved with the times (but Hayter, this volume, chap. 6, points out that in the forest industries many have not), while firms manufacturing advanced materials (see also Wallace, this volume, chap. 7) are found in minerals production.

Some technology clusters have strong connections with recognized "industries," but the implication is also clear that the conventional industries provide an inadequate identification of products based on new technologies. Equally important, the traditional separation among hardware, software, and services artificially disintegrates the ensemble of activities undertaken in many IT firms.

Information Technology

In terms of number of firms, the information technology (IT) cluster is the largest. Its size reflects the many small service, software, and hardware suppliers and the substantial degree of co-production of computer services, software, and hardware. Firms making telecommunications equipment and systems and providing related services are the core of this technology cluster, and an earlier survey of software producers (Statistics Canada 1990b) found that 60 per cent of expenditures on software R&D are made in the telecommunications and business machines industries. IT is Canada's prime field of technological innovation, and a substantial proportion of its firms have a sustained international presence. This is a significant capability if the conventional wisdom is correct that there will be major extensions of this field of innovation.

In terms of the industries listed in Table 1, above, activities in the IT cluster include Other electronic equipment, Business machines, and Computer services. Together these account for about 35 per cent of Canadian R&D expenditures, but Northern Telecom (NT) alone performs about 21 per cent of all Canadian industrial R&D. Known for its telecommunications systems, NT operates the largest software laboratory in the country, and Bell-Northern Research, its R&D subsidiary, devotes 80 per cent of its R&D to software development (Science Council of Canada 1992, No. 15).

The importance of NT places telecommunications at the core of the IT cluster. NT is also the largest Canadian designer/producer of semi-conductors (Science Council of Canada 1992, No. 14), necessary for attaining an advanced position in telecom technology. This, however, is no longer a sufficient condition, and firms such as NT now integrate their products with services such as network design and planning and operational analysis and support. These services present extensive opportunities for software development.

Software development is now responsible for 25 per cent of Canadian industrial research expenditures, and the largest share of this (44 per cent) is undertaken in telecommunications and electronics. By contrast, smaller expenditures on software R&D are made by the computer hardware industry (16 per cent), while the computer services industry accounts for 15 per cent, and other service activities, such as engineering and scientific services, are responsible for 12 per cent (Statistics Canada 1990b). IBM, which is the second-largest software firm in Canada, adds considerably to the depth and complexity of the IT cluster. Its software lab in Toronto is the corporation's third largest, and based on its specialization in systems integration it is being recognized as a growth centre.

Firms in telecommunications are differentiated in terms of the market segments that they serve.[4] Large system houses such as NT supply the backbone technologies (including software) to a market of public (communications) carriers. Telephone companies have been the traditional customers, but major new markets provided by alternate (such as cable TV) and cellular networks are expanding because of the joint effects of innovation and progressive deregulation. By comparison, product development companies are usually specialized, small and medium-sized firms (such as Gandalf and Newbridge). When viewed on their own, these software innovators may have "disjointed capabilities" (ISTC 1991b), but other firms have emerged to integrate systems built in part from the output of these companies. Familiar computer companies such as IBM and telecom equipment suppliers function as systems integrators, but strong competition comes from a group of "non-aligned" firms (such as SHL Systemhouse), which can supply independent advice and design communications systems for individual institutional customers.

Energy-Environmental (E-E) Technologies

This cluster, the second largest, is a composite of strengths concerned with environmental and industrial testing and monitoring services and equipment supply, including engineering and design services related to energy projects. There is considerable technological diversity among firms in this cluster, with emergence of Canada's environmental expertise being closely associated with engineering and related consulting activities, which have been significant in energy projects, mining, forestry, transportation, and municipal services. Though production of capital equipment appears less developed, Canadian firms design plants for paper recycling and electricity co-generation (a significant capability because of the rapid shift in environmental standards and markets); extraction technology for heavy oil, oil sands, and sour gas; large-scale hydro-electric projects; and large (CANDU) nuclear reactors (Science Council of Canada 1992, No. 12). Many small enterprises in the cluster probably reflect the recent industrial diversification of western Canada (see also Gertler, this volume, chap. 15), but the many fewer established firms in electricity-generation equipment have not lost their importance.

LOCATION OF HIGH-TECH ACTIVITY

Locational analysis of Canadian technology-intensive industry is subject to the constraint of limited and shrinking published data from official sources. It is possible, however, to establish the broad locational foundations of Canada's technological development using aggregate data, while CANTECH (see Hutchinson Research 1988) permits inquiries for a few urban centres where the full size range of firms has been surveyed.

Canadian urbanization is the key to understanding the location of technology-intensive activity. It is highly localized, reflecting the geography of secondary manufacturing, with Ontario (LQ = 1.5) contributing 56 per cent of Canadian R&D and Quebec 27 per cent, the balance being in British Columbia and Alberta. Ontario is dominant in every activity of industrial R&D except Aircraft and parts, where it slipped from 41 to 29 per cent (1975–89) in favour of Quebec, which is in second rank in all other industries. The pattern for the future also seems to be set, since the number of high-tech establishments generally has grown faster in the central provinces, though the increase in the number of Alberta firms manufacturing non-business machinery and emergence of pockets of expertise in IT in British Columbia represent interesting exceptions (ISTC 1991b).

At the urban scale, Toronto (26 per cent of 1990 industrial R&D), Montreal (23 per cent), and Ottawa (17 per cent – 1986 data) are the major locations of high-tech manufacturing production. Following the provincial pattern, Toronto is the largest and strongest in most high-technology industries, employing twice as many as plants in Montreal, the exception being the Aircraft and parts industry, in which Montreal is half as big again. In Drugs and medicine the two centres are at about the same scale. Localization of foreign subsidiaries in Toronto in most industries helps determine its larger average size of establishments. These and large indigenous firms have responded to the size and depth of the Toronto labour market and to its other agglomeration advantages, which derive from its greater industrial scale and diversity.

Over the long term, emerging industrial locations have been the outcome of a variety of influences, including new technology-based activities (Freeman and Perez 1988; Storper and Walker 1989), and internationally in recent decades this has proved particularly true in electronics and the IT cluster. In Canada, however, it is difficult to perceive locational forces overturning existing locations. From time to time it is suggested in the business press that Ottawa is a contender as a new technology complex, and it is a significant centre of public-sector R&D – the substantial concentration of government R&D labs there claims 27 per cent of the national total. This large share of public R&D and the federal government market have stimulated a variety of private-sector investments. The key to the success of the Ottawa region, according to Steed (Steed and De Genova 1983; Steed 1987), has been its generation of new technology-based enterprises from a handful of original Canadian core firms in telecommunications, computers, software and computer services, and electronics. Nevertheless, the amount of employment generated by Ottawa's firms is only about one-tenth that of Toronto's employment in technology-intensive activities.

The frequent references in the press to Ottawa's high-technology development are frustrating because of the difficulty of undertaking comparative analysis across the urban system. Fortunately, CANTECH data (see Hutchinson 1988) provide adequate coverage of the technological structure of Toronto, Ottawa, Calgary, and Edmonton, though interpretations of the survey data are heavily influenced by the incidence of small firms. In order to develop an aggregate context for these data, it is useful to note that Toronto has over 800 R&D units, about four times as many as the Ottawa region and 2.5 times as many as Alberta.

If product cycle theory applied at the regional level we would see production in industrial activities that began in the major Canadian urban centres such as Toronto trickle down to other places as lower-cost production bases and markets are identified (Thompson 1989). This does not, however, appear to be a useful aggregate model in Canada. Over the long term, the absolute industrial scale of the Toronto region has grown with that of the national economy.[5] New technology-based activities have found Toronto attractive – the educational and producer service infrastructure on the input side and the large governmental and private-sector markets on the other being major stimuli for new firms. Moreover, Toronto, rather than being a source of "spin-off" firms for other centres, appears to attract new technology-based firms that have started in other centres such as Kitchener-Waterloo and Ottawa (Steed 1982). Toronto's industrial structure is an evolving amalgam of producer services and manufacturing activities; as the manufacturing share of employment has fallen (and productivity has risen) there have been notable employment increases in software and the producer services.

Toronto's industrial history, in terms of technology-intensive activities, is reflected in its higher proportion of pre-1960s firms, but growth spurts during the 1960s, 1970s and 1980s led to emergence of new technology-based firms, which became a common locational experience in major centres across the urban system. Toronto did absolutely and relatively better up to the recession of the 1990s than other locations, suggesting the clear advantages of industrial agglomeration as recognized through birth/survival of new firms.

There are differences in technological specializations among urban centres, in addition to variations in their scale of industrial development and in their recent experience in generation of new technology-based firms (incubation). In Toronto and Ottawa the largest number of firms in the sample are found in the IT, Generic, and E-E clusters, while for Calgary and Edmonton the order favours E-E, IT, and Generic technologies, reflecting the local petroleum industry. The time at which the IT cluster experienced its major growth may be used to establish distinct differences in incidence of development. The primary phase of innovation in microelectronics and computer systems occurred in the Boston and San Francisco regions in the mid-1960s, but it did not replace E-E as the pre-eminent generator of new firms in Ottawa and Toronto until 1971–75. By contrast, in Calgary and Edmonton E-E continued to be the main source of growth in the number of firms until the late 1970s. Only then did the wave of microelectronics applications finally affect production there.

None of the three smaller centres has had the benefit of an industrial past, though Ottawa has seen considerably more technological development because of several quite distinct advantages, and thus its history is scarcely a model of development for other urban regions. As noted, federal government ministries, especially Communications, have been important sources of contracts, but reductions in federal expenditures have taken their toll since the mid-1980s when the rate of new-firm generation in Ottawa dropped to two-thirds, that of Toronto. This should not be taken as a prediction of the technological future of Calgary and to a lesser extent Edmonton, as these centres may find a more reliable stimulus for environmental industries in the raising of environmental standards applied to the resource industries of the western provinces.

CONCLUSION

Though Canada's traditional exports are of reduced relative importance in a world trading system that places high value on technology-intensive products, they remain of great significance to the economy, particularly because Canadian industry underperforms in terms of growth of new technology-based activity. The internationally low level and stability of the R&D/GDP ratio are congruent with this view. In terms of output, there is a continuing deficit in the trade balance in technology-intensive products, and this provides one external (trade-based) evaluation of changes in the effectiveness of the Canadian industrial system over time. The simple interpretation is that Canadian high-tech industry continues to be insufficiently developed to meet Canadian industrial and consumer demand, and this situation appears related to the generally low level of business expenditure on R&D and industry's limited ability to innovate.

A more complex interpretation involves the wide distribution of technological capability in the Canadian economy. Some of this has been developed in firms whose main industrial activity is in one of the resource or primary processing industries or in one of the older secondary manufacturing activities. Canada has some strong technology-intensive companies (both indigenous and foreign), and identification of technology clusters has broadened systematic knowledge of Canada's development in a way not possible from data that describe industries defined in the standard way. The results confirm that technological development of firms occurs by means of their skill in combining related product technologies. This means that the opportunities for spillover effects – made operational through labour mobility, joint ventures, and production and service subcontracting – are likely to be increased.

Telecommunications equipment and Aircraft and parts stand out in R&D expenditures, trade performance, and growth, and these are the industrial cores of two technology clusters. In the services, too, there has been clear growth in R&D. It is worrying, however, that the Machinery industry (along with the Automation cluster) is a weak performer and surprisingly small; these characteristics impose difficulties on domestic user-industries. There is a larger problem, too; without a strong machinery industry the adoption rate for new techniques is slower and the spread effects weaker. In time, however, the Environmental cluster of firms may prove to be a new source of stimulation for these activities, since it involves strong producer-service and hardware-production components. Most parts of the economy are expected to make more substantial responses to a broad pattern of rising, legislated standards applied to air and water pollution, energy conservation, ease of product recycling, and reduced packaging.

Canadian governments have sought to devise policies that will stimulate industrial R&D expenditure. The continuing low value of R&D/GDP, however, reflects the policies' generally limited impact. One reason is that Canada is distinctive among the G7 in the very large, even dominating, presence of foreign subsidiaries within technology-intensive industries. This represents success in attracting scarce investment capital and human capital from other countries, but it is a form of investment that has problematic aspects for Canadian technology development. Imports of intermediate inputs and technology in other forms are proportionally large. While the resulting trade burden is obvious, the opportunity costs in Canadian technology development associated with unrealized domestic input linkages are more subtle and difficult to quantify, though probably no less significant.

Part of the conventional rhetoric has been that foreign subsidiaries represent a successful means of effecting technology transfers into Canada, but there are flaws in this stance that have been explored (Britton and Gilmour 1978). Though such firms undertake more than half of Canada's industrial R&D, when this fact is related to their even greater proportion of sales it becomes evident that they spend less in comparison with Canadian firms.

Existing policy instruments such as tax credits for R&D have assisted development of innovation-related activities in domestic firms and subsidiaries, which are required to bid within corporations for world product mandates based on their local expertise and innovative potential. Despite these and other incentives, it has been difficult for some subsidiaries (and parents) to shed the long-term relationship that gave subsidiaries limited discretion to invest in innovation. Consequently, there is evidence that fully functional units undertaking research, development, design, and engineering remain either a low priority or unattractive for some U.S. corporations.

Other policies, such as extended patent protection, hold out some promise of increasing R&D and production. In the Drug industry, there is a clear attempt to acquire new investments in expanding technologies and to see partnerships developed between university and corporate research labs. Above all, NAFTA may stimulate expenditures on innovation by both foreign and Canadian firms. More foreign corporations may now decide that it is logical for their Canadian subsidiaries to undertake more R&D and development of new products so as to gain further access to the skills and knowledge of Canada's technology centres. But from every rational perspective the effects of NAFTA on high-technology industry are very unpredictable because the agreement represents a significant change in the political economy of Canadian development and trade.

NAFTA makes substantial demands on the majority of Canadian firms because it involves significantly new skills for much of Canadian industry, though for some large firms that have been competitive internationally for some time Canada has become only one of many R&D and production loci. This locational diversity of R&D has much to do with the conditions of global competition and the maturity of Canada's corporations. Most small and medium-sized indigenous firms, however, serve only domestic markets and do not yet have the experience to make any rapid adjustments. Nevertheless, the election of a Liberal federal government seems to have brought forth a more pragmatic understanding of the difficulties encountered by small firms and of the need to overhaul the policy regime for them (Government of Canada 1994). Small firms are particularly disadvantaged by their scale if they have not developed sound networks to get access to producer services and to provide responses to their markets (Britton 1991; 1993; MacPherson 1987). The large number of small firms in each technology cluster indicates that there is considerable entrepreneurial activity, though there has been a sharp decline in recent years in the number of firms that have been generated, and the venture capital industry reports a shortage of firms attractive to investment.

The structure of all of Canada's high-technology centres includes large firms and small enterprises. Toronto and Montreal, however, have the advantage of substantial agglomeration economies and the accumulation of technological infrastructure. Given the scale of development there, surprisingly little is known about differences in the generation of new technology-based firms or the local geographical circumstances of high-technology growth, and the limitations of public statistical data make that research a difficult task. The record suggests, however, that the majority of new firms will be born in the major centres or move there in order to capture positive externalities. By implication, other smaller and aspiring high-tech locations will probably remain among the myriad North American cities discovering that it is impossible to develop the critical mass that will facilitate high-tech development.

ACKNOWLEDGMENTS

I am pleased to recognize the financial support for the project reported here provided by a grant from the Social Sciences and Humanities Research Council of Canada, which funded the research assistance of Michael Skelly and Sandra Young. I am grateful also for the comments on an earlier draft by Meric Gertler.

NOTES

1 This chapter draws on some of the data analysis used in a paper, "Specialization versus Diversity in Canadian Technological Development," which I wrote for an issue of *Small Business Economics* edited by Maryann Feldman.

2 Source: *Report on Business Magazine*, Sept. 1994 (data reported from Evert Communications Ltd, Ottawa). Two problems occur with these data: First, R&D funded by customers performed under contract is not included. This may be significant (up to 19 times the indicated amount) for Spar Aerospace, which is listed as 5.7 per cent. Second, R&D for Ericsson was reported (1990) as $9.1 million (*Financial Post*, 22 June, 1992, S23), the rate of

increase to $62.3 million (1993) was abnormal for the sector, and there may be a reporting error – Ericsson enters the second comparison. Subtracting Spar Aerospace and/or Ericsson would tilt the results in the reported direction.

3 The clustering process was started with a preliminary assignment of firms to products associated with core Canadian technologies identified in earlier work by the Science Council of Canada and Statistics Canada (1989a). Additional products were added as required or new clusters were formed, the heuristic rule being that products were added to the definition of a cluster when a minimum proportion (40 per cent) of the firms already included made the additional product. Only a small number of product groups could not be assigned because of their limited development.

4 This paragraph relies on Science Council of Canada (1992), No. 1.

5 This is not to deny the decline of traditional manufacturing industries in Toronto and the closure of many assembly plants since the region's peak manufacturing employment in the early 1980s. Recent recessions have taken a substantial toll of inefficient plants. As noted in this volume (Britton, chap. 1), the recession of the early 1990s was particularly tough on Toronto, since its effects were compounded by the impact of the FTA and the expectation and reality of NAFTA.

REFERENCES

Britton, John N.H. 1989. "Innovation Policies for Small Firms." *Regional Studies*, 23, 167–72.

– 1991. "Reconsidering Innovation Policy for Small and Medium Sized Enterprises: The Canadian Case." *Environment and Planning C*, 9, 189–206.

– 1993. "Canada under Free Trade: Defining the Metropolitan Agenda for Innovation Policy." *International Journal of Urban and Regional Research*, 17, 559–77.

Britton, John N.H., and Gilmour, James M. 1978. *The Weakest Link – a Technological Perspective on Canadian Industrial Underdevelopment*. Background Study No. 43, Science Council of Canada.

Economic Council of Canada. 1989. "High-tech Trade – Serious Weaknesses to Solve." *Au Courant*, 10 no. 2, 10–11.

Freeman, C. 1991. "Networks of Innovators: A Synthesis of Research Issues." *Research Policy*, 20, 499–514.

Freeman, Christopher, and Perez, Carlota. 1988. "Structural Crises of Adjustment, Business Cycles and Investment Behaviour." In Giovanni Dosi et al., eds., *Technical Change and Economic Theory*, 38–66. London: Pinter Publishers.

Government of Canada. 1994. *Breaking through Barriers: Forging Our Future*. Ottawa.

Hutchinson Research. 1988. *CANTECH DATABASE High Technology Information System Manual*. North York, Ont.

Industry Canada. 1994a. *Resource Book for Science and Technology Consultations, Vol. I*. Ottawa.

– 1994b. *Resource Book for Science and Technology Consultations, Vol. II*. Ottawa.

Industry, Science and Technology, Canada (ISTC). 1991a. *Selected Science and Technology Statistics*.

— 1991b. *Telecommunications Equipment*.

Krugman, Paul R. 1986. *Strategic Trade Policy and the New International Economics*. Cambridge, Mass.: MIT Press.

MacPherson, Alan. 1987. "Industrial Innovation in the Small Business Sector: Empirical Evidence from Metropolitan Toronto." *Environment and Planning A*, 20, 953–71.

Malecki, Edward J. 1991. *Technology and Economic Development*. London: Longman.

Rothwell, Roy, and Zegveld, W. 1982. *Innovation and the Small and Medium Sized Firm*. London: Pinter.

— 1985. *Reindustrialization and Technology*. Harlow: Longman.

Science Council of Canada. 1981. *Hard Times, Hard Choices*. Ottawa.

— 1992. Sectoral Technology Strategy Series. Ottawa: Minister of Supply and Services:

No. 1. *The Canadian Telecommunications Sector*.

No. 7. *The Canadian Banking Sector*.

No. 12. *The Canadian Electric-Power Sector*

No. 13. *The Canadian Consulting Engineering Sector*.

No. 14. *The Canadian Electronics Sector*.

No. 15. *The Canadian Computer Software and Services Sector*.

Statistics Canada. 1989a. "Industrial R&D and Key Technologies." *Science Statistics*, 13 no. 4, Cat. No. 88–001.

— 1989b. *Science and Technology Indicators 1988*. Cat. No. 88-201.

— 1990a. *Indicators of Science and Technology*, 2 no. 4, Cat. No. 88-002.

— 1990b. "Software Research and Development (R&D) in Canadian Industry, 1988." *Science Statistics*, 14 no. 5, Cat. No. 88-001.

— 1992. "Industrial Research and Development Expenditures, 1983 to 1992." *Science Statistics*, Cat. No. 88-001.

— 1993a. "Factors Affecting Spending on Research and Development (R&D) Performance by Firms in Canada, 1990." *Science Statistics*, 17 no. 2, Cat. No. 88-001.

— 1993b, *Selected Science and Technology Statistics, 1992*. C1-04/1992.

Steed, G.P.F. 1982. *Threshold Firms: Backing Canada's Winners*. Background Study No. 48, Science Council of Canada.

Steed, G.P.F., and De Genova, D. 1983. "Ottawa's Technology-oriented Complex." *Canadian Geographer*, 27, 263–78.

— 1987. "Policy and High Technology Complexes: Ottawa's 'Silicon Valley North.'" In F.E.I. Hamilton, ed., *Industrial Change in Advanced Economies*, 261–9. Wolfboro, NH: Croom Helm.

Storper, Michael, and Walker, Richard. 1989. *The Capitalist Imperative: Territory, Technology and Industrial Growth*. Oxford: Basil Blackwell.

Thompson, C. 1989. "High-technology Theories and Public Policy." *Environment and Planning C*, 7, 121–52.

CHAPTER FIFTEEN

Capital, Technological Change, and Regional Growth

MERIC S. GERTLER

In economic doctrines of all stripes, one of the seemingly eternal verities is that capital plays a crucial role in determining the path of growth and change in industrial econo-mies.[1] Contemporary work in geography, regional science, and regional economic planning has reinforced this view, attesting to the key influence of investment – and investors – on the economic vitality of localities and regions. However, until compara-tively recently, both our theoretical apparatus for understanding the role of capital in regional development and our collected body of empirical evidence to illuminate this process were rather limited (Gertler 1984b; 1987a). The Canadian geographical litera-ture, like the American, European, and Asian, was notably underdeveloped, at least until the mid-1980s.

Since then, our grasp of empirical phenomena such as regional rates of investment, size of accumulated capital stocks, and (implied) interregional investment flows has improved significantly (Gertler 1986; 1987b; Anderson and Rigby 1989; Rigby 1991). At the same time, our understanding of the investment process has developed along new lines of theoretical inquiry. At the macroeconomic scale, economists have developed new approaches to the study of growth theory, in an attempt to incorporate some of the most important attributes of the capitalist system – such as technological change and scale economies – which had hitherto defied easy incorporation into accepted theories of macroeconomic growth (Romer 1986; 1990a; 1990b). At the microeconomic scale, the results of a set of detailed, firm-level studies in several countries (including Canada) have forced reappraisal of the accepted wisdom con-cerning the relationship among investment, productivity, and firm performance (Meurer, Sobel, and Wolfe 1987; Gertler 1995a). Within this second avenue of progress, economic geography has made an important contribution to our understand-ing of the process by which fluid, financial capital becomes "fixed" in place as pro-ductive apparatus. Furthermore, this work has suggested important implications for the objectives and shape of industrial policy intervention.

In this chapter, I wish to present the broad outlines, insights, and implications of the work described above. In the first section, I review the recently accumulated evidence on regional patterns of investment and capital formation in Canada, noting also some of the limitations stemming from its largely aggregate, empirical nature. In the second section, I review some of the more significant insights produced by the "new growth theory" and explore their implications for our study of investment in Canada from a spatial perspective. The intent here is to penetrate beyond the more traditional, aggregate measures of capital and to explore the Canadian geography of investment based on more realistic renderings of this central concept. In the third section, I push this line of thinking even further, using the recent firm-level research of geographers and other social scientists. The objective here is to indicate the strong links between capital per se and the socio-spatial context in which investment and technical change take place. I argue, making reference to recent Canadian empirical evidence, that a geographical perspective on the process of investment in manufacturing establishments sheds useful light on the obstacles to more effective use of new technologies, yielding major implications for the sorts of policies to promote technology transfer and training that have received much attention in recent times (see, for example, Ontario 1992).

SPATIAL AND SECTORAL AGGREGATES

The geography of capital has for many years remained one of the least well-studied aspects of economic geography (Bolton 1980; Dicken and Lloyd 1990). A pervasive scarcity of comprehensive data on this phenomenon at the sub-national scale, plus widespread willingness to accept certain theoretical assertions as fact, are mainly to blame for this state of affairs. Only within the last decade has geographical research begun to address the major lacunae in our knowledge of patterns and processes of capital accumulation at the regional scale (for reviews of this literature, see Gertler 1984b; 1987a; Erickson 1989; Schoenberger 1989; Pudup 1992). North American interest in this phenomenon rose dramatically in the late 1970s with the sudden awareness of major shifts in investment flows away from the traditional destinations (such as the U.S. northeast and mid-Atlantic manufacturing belt) to regions such as the U.S. south and west. Theoretical work by geographers such as Massey (1978; 1984), Walker and Storper (1981), and Harvey (1982) offered explanatory insights into the trends documented by Birch (1979), Bluestone and Harrison (1982), and others. However, the smattering of empirical analysis that had been performed was either based on case studies of plant shutdowns and relocations or used employment change as an indicator of inter-regional shifts in economic activity.

The first empirical studies of this "new" geography of capital accumulation in the United States began to emerge in the early and mid-1980s, for both metropolitan areas (Varaiya and Wiseman 1981) and states (Browne, Mieszkowski, and Syron 1980; Gertler 1984a; McHugh and Widdows 1984). These studies were motivated by the desire to see if the processes of inter-regional capital shifts identified by case-study analyses were general and sufficiently widespread to influence aggregate spatial trends in capital formation. Based on their findings, it became apparent that the inter-regional flows of capital described above had in fact been under-way for considerably longer than had been made evident by the case-study approach so popular during the late 1970s. Indeed, major shifts from the U.S. northeast and midwest to the sunbelt could be identified as having begun at least as early as the 1950s. Similarly, these aggregate studies revealed a distinctive trend toward markedly greater capital intensities in manufacturing operations, particularly in regions experiencing significant declines in manufacturing employment.

Taken together, these findings produced insights that would qualify the "deindustrialization" thesis promoted by Bluestone and Harrison. First, while the decline in manufacturing employment in the U.S. "rustbelt" could be linked to redirection of capital to other locations, it was compounded by the tendency of remaining producers to introduce new, labour-saving investments. Hence, despite the popular image of firms shutting down ageing plants in high-wage, union-dominated locations, producers had not altogether stopped investing in such regions. (Indeed, Gertler [1984a] documented that in no instances had the regional stock of accumulated capital actually experienced an absolute decline; rather, regional decline was best measured in relative terms, as investment rates in the rustbelt states – while still positive – failed to match those in the south and west). Second, because inter-regional capital mobility did not emerge suddenly in the mid- to late 1970s, but instead has constituted a force of change in U.S. economic geography over many decades, these empirical findings are more consistent with the approach of Storper and Walker (1989), who make much of capitalism's "inconstant geography."

These vivid empirical trends and the accompanying theoretical debates soon generated considerable interest "north of the border." This interest was piqued further by the economic devastation being wreaked upon many Canadian localities during the 1981–82 recession. Some scholars sought to apply intact Bluestone and Harrison's theoretical and empirical framework to Canadian case studies (Norcliffe, Goldrick, and Muszynski 1986), while others were somewhat more cautious in using this approach to understand spatial economic change in some of Canada's largest metropolitan economies (Gertler 1985; see also MacLachlan, this volume, chap. 11).

The release of provincial estimates of accumulated capital stocks for major sectors of the Canadian economy by Statistics Canada (Garston 1983) made possible a more systematic, aggregate analysis of inter-regional differences in rates of capital accumulation for the period 1955–83 (Gertler 1986). Furthermore, because these data covered not only manufacturing but also primary and service activity, they permitted some interesting intersectoral comparisons. A number of noteworthy empirical trends have emerged from this work, some of which stand in sharp contrast to the American experience over a similar period. First, the kind of dramatic shifts in inter-regional distribution of manufacturing capital stock that characterized the American case were not as strongly evident in Canada, at least through 1983, the end of the capital-stock time series. Manufacturing constituted the most stable sphere of economic activity in this period. Despite Alberta's rising share, the interprovincial ranking changed very little. However, primary activity (mining, petroleum, agriculture, and fisheries) exhibited considerably more volatility. In addition to Alberta's meteoric rise, there were especially notable fluctuations in the fortunes of Quebec and British Columbia. The experience in the service industries showed somewhat less volatility, with Ontario's share of the national capital stock rising, apparently at the expense of Quebec. Once again, Alberta's share of the accumulated national capital stock in these activities showed a substantial increase.

The net result of these trends in the three broad sectors – as reflected in regional shares of the total national stock of accumulated capital – was, with two exceptions, one of general stability in inter-regional distribution. Quebec and Alberta, however, traded positions in the interprovincial ranking (with Alberta in 1980 usurping Quebec's number-two position behind Ontario), largely on the basis of changes in their primary and tertiary sectors. Meanwhile, Ontario experienced a slow, gradual decline in its national share, as Alberta's rose.

These trends are entirely consistent with major, well-documented regional economic events taking place during this period. Alberta's rise through the ranks, beginning in the early 1970s, coincides with the first OPEC oil shock of 1973 and is further propelled by major government investments in energy-related megaprojects such as the Athabaska Tar Sands (on the latter, see Weaver and Gunton 1986). The tremendous stimulus to petroleum exploration and extraction activity resulting from the subsequent dramatic rise in world oil prices is apparent not only in the primary sector but also especially in services. The response in Alberta's manufacturing sector is considerably more muted, reflecting structural impediments to manufacturing growth that consistently frustrated provincial attempts at economic diversification (for an excellent discussion of these barriers, see Mansell 1987).

British Columbia's relatively volatile primary sector also reflects a similar vulnerability to fluctuations in global prices for its chief export commodities such as lumber, paper products, coal, and other minerals (see Barnes, this volume, chap. 3, for more extensive elaboration of this theme). The apparent upswing in Atlantic Canada's share of primary capital stock in the early 1980s can also probably be linked to government-initiated development of the Hibernia off-shore oil project. The volatility experienced in Quebec (particularly in its primary and service sectors) first surfaces in the latter half of the 1960s and appears to coincide with the rise of the separatist political movement in that province.

A more recent study (Bougrine 1992), using a similar framework with an updated version of the database, has brought this analysis forward to 1990. It confirms continuation of similar trends, with one notable exception. The gradual decline in Ontario's share of total capital stock begins to reverse itself in the second half of the 1980s, reflecting the provincial economy's sustained boom from 1983 to 1989. Alberta's rate of accumulation slows dramatically during the same period, while British Columbia experiences a significant rise in its share of the nation's capital stock, driven in part by a surge in manufacturing investment.

The predominant conclusion to be drawn from this analysis is that the most important shifts in the geography of regional investment in the Canadian economy are strongly influenced by exogenous global fluctuations in the prices for Canada's major commodity exports and also to some extent by public-sector investment. Also of considerable importance are major political shifts in Quebec in the 1960s and 1970s. Indeed, a causal time-series analysis of regional manufacturing investment rates (Gertler 1987b) produced results that suggest further qualifications to the deindustrialization thesis. Labour costs alone appear to exert little influence on regional investment rates. Instead, their impact is most evident in inducing changes to regional capital intensity over time. When average labour productivity is held constant, upward movements in regional wage rates produce consistently strong subsequent upswings in capital-labour ratios (for similar results from a more recent study covering the same period, see Rigby 1991). In short, as regional efficiency wages[2] increase, manufacturing firms appear to respond not by modulating the rate of investment but by changing qualitatively the nature of subsequent regional investment.

Furthermore, regional rates of unionization were found to exert no consistent direction of influence on investment rates: in some jurisdictions, such as British Columbia, where relations between management (with the frequently active support of government) and workers could be accurately characterized as adversarial and conflict-ridden through much of the study period, unionization rates did reduce subsequent rates of investment flow. In other provinces, such a relationship either failed to materialize or produced a positive association between unionization and investment.

These results, along with those concerning the effect of regional wage rates, support a more recent view of the influence of capital-labour relations on regional economic trajectories that has been shaped by European studies of employment relations and economic change. Scholars such as Streeck (1985) (see also Mahon 1991 and Drache and Gertler 1991), writing on the German experience, have observed that – contrary to conventional wisdom – high wage rates secured by effective union participation in bargaining can set in train operation of a "virtuous circle," through which manufacturers are induced to seek routes to competitive success that are not based on cheap wages. This high-wage option is based on effective use of advanced process technologies by highly trained workers. This strategy of capital-intensive, technologically advanced, high-wage production leads to greater productivity, which then supports still-higher wage rates. Seen in this light, the bargaining between workers and plant owners will actually produce outcomes that are mutually beneficial.

The overriding stability evident in regional shares of national capital stock is also consistent with additional results produced by Gertler (1987b). This analysis found evidence that a neoclassical model of inter-regional convergence of economic welfare, driven by the inter-regional mobility of investment in search of ever-higher rates of profit, is not well substantiated by the Canadian data. Rather, there remain surprisingly strong impediments to the free flow of capital between regions, and these result mainly from strong differences in the sectoral composition of each region's economy. In short, within the Canadian context, inter-regional mobility normally translates into intersectoral mobility, and there appear to be enduring entry costs that discourage investment in different industries (and places).

Further recent analysis of the inter-regional dynamics of manufacturing investment in Canada (Rigby 1991) supports this insight by showing a marked tendency for profit rates in certain provinces (especially Ontario and Quebec) to remain consistently higher than those in other parts of the country. Rigby's conclusion (363) is clear and unequivocal: "These results do not support the argument that regional profit rates tend toward equilibrium ... In 1955 the average profit rate of manufacturing firms in Atlantic Canada was approximately 65 per cent of that enjoyed by producers in Quebec and Ontario. In 1984 the manufacturing profit rate in Atlantic Canada was only 37 per cent of the average in Quebec and Ontario." Rigby also suspects that a significant determinant of this lack of inter-regional equilibration is the enduring difference in sectoral mix across the country.

With hindsight, it is evident that these recent aggregate analyses of the geography of capital accumulation have made some major contributions to our understanding of the processes of regional growth and stagnation across Canada, as well as offering some useful refinements to prevailing theories concerning capital mobility and regional economic change.

Unfortunately, however, the econometric mode of analysis of aggregate data used in this work introduces certain limitations. Chief among these is the inability of this kind of analysis to tell us much about the qualitative characteristics of regional investment and how these are changing over time. Hepworth (1989) points out, for example, that a growing proportion of new investment expenditure by firms in virtually all sectors of the economy is being directed toward what he calls information capital – that is, information technologies related to computers and telecommunications. One of the important implications of this trend is that it becomes progressively less appropriate to argue that a region's economic fortunes remain strongly tied to the volume and type of investment within its own borders. The advent of cheap inter-regional telecommunications networks has created a situation in which the services associated with major computer investments in one region can be readily transmitted "over the wire" to users in other regions. While the more recent trend toward smaller-scale, PC and local-area, network-based computing in business (in step with the precipitous decline in prices and the rapid expansion in capabilities) suggests that major, centralized investments in mainframe computers are becoming increasingly less important, inter-regional flows of capital services probably remain significant. Further shortcomings related to the excessively aggregate nature of these analyses – and some of the insights stemming from a more disaggregated and qualitatively varied conception of capital – are explored in the next two sections of this chapter.

"NEW" GROWTH THEORY AND OTHER DISAGGREGATE APPROACHES

Within the past decade, across a number of social sciences, scholars have begun to discredit the traditional interpretation of capital outlined in the previous section, offering to replace it with alternative conceptions. These developments have been most evident within the "new" economic theory of growth, within industrial sociology (and what has come to be called "socioeconomics"), and, most recently, within industrial geography. Taken together, this work poses an important set of challenges to existing views of capital and its role in the development of capitalist economies. In this section, I review some of the basic insights of this recent work. I then explore its implications for our understanding of the geography of capital accumulation in Canada.

As I have indicated in an earlier review (Gertler 1984b), capital remains one of the most slippery and contentious concepts within economic theory, for a number of reasons. Traditionally, both economic theory and empirical explorations have defined capital stock as an undifferentiated mass of accumulated capital – typically, the sum of past investment expenditures on machinery/equipment and plant, minus depreciation of these investments over time, reflecting their declining productivity. In empirical analyses of the "sources of growth" in capitalist economies, production functions that model an economy's output as a function of labour and undifferentiated capital inputs have consistently found large "residuals," which have normally been attributed to "technological change." A large proportion of the variance in output thus appears to be explained by an unspecified, mysterious concept that remains exogenous to the system – as if technology cascaded down onto firms like manna from heaven.

In recent years, economists have begun to address these deficiencies by proposing (and verifying) new models in which technological change is treated as endogenous (see especially Romer 1986; 1990a; 1990b). This has been achieved in a variety of ways (see also the following section). Romer (1986) began by proposing that technology be viewed as being embodied within recently innovated machinery and equipment. Hence technology and physical capital are viewed as strongly complementary, such that "an increase in the rate of growth of physical capital necessarily leads to an increase in the rate of technological change" (Romer 1990a, 338). Based on this conception, one would expect to find a strong positive correlation between a country's investment expenditures as a share of gross domestic product (GDP) and its rate of growth in productivity. Moreover, the most relevant component of investment is that which is devoted to machinery and equipment acquisition, excluding additional sums spent on plant construction, as it is machinery investment that most directly reflects technical progress. This relationship has been confirmed in cross-national comparative and historical studies by Romer and others who have adopted his theoretical framework (see De Long and Summers 1991; De Long 1992). It is also consistent with Scott's (1989) recent reformulation of growth theory, which sees modern industrial technology as being inescapably bound up with (or "embodied in") industrial machinery itself.

This recent work strongly implies that empirical analysis of capital accumulation (at any spatial scale) ought properly to focus on the machinery and equipment component of firms' capital expenditures, rather than employing an aggregate, undifferentiated measure of capital stock. This idea appears to be borne out in recent data for manufacturing investment in Ontario (see Figure 1), which show a notable upward trend in machinery and equipment expenditures as a share of total annual investment

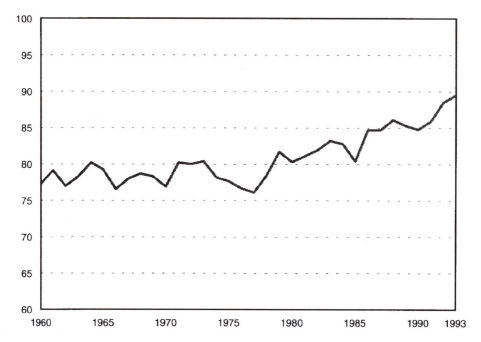

Figure 1

Expenditures on manufacturing machinery and equipment as a percentage of total investment expenditures, Ontario, 1960–93. *Source*: Statistics Canada, *Annual Investment Expenditures*, Cat. No. 61-205, annual. Data for 1992 and 1993 are preliminary actual expenditures and expenditure intentions, respectively.

outlays, especially since 1977. Furthermore, there appear to be straightforward explanations for this phenomenon, based on a sound understanding of the changing nature of the investment goods themselves.

Along these lines, one of the most penetrating recent analyses has emerged from industrial sociology, best exemplified in the work of Block (1990). As part of a comprehensive "critique of economic discourse," Block engages in what amounts to an almost complete deconstruction of the traditional idea of capital. The constant theme underlying his analysis is that there have been extremely important qualitative changes to the nature of prevailing process technologies, yet accepted methods of measurement of the volume of accumulated productive investment fail to capture the full effects of these changes on the economy's productivity. At the core of this change is increasingly widespread introduction of microprocessors and electronic controls in industrial machinery. This change has, in Block's view, brought about a quantum leap in the capabilities of many types of machinery, and these improvements have become available at steadily decreasing prices. The net result is that more effective means of production have become available without a corresponding increase in price – producing what he refers to as "capital savings" – i.e., costless improvements in quality and productivity.

Block, by way of example, reports on a number of research studies in the metalworking industries, in which productivity gains from switching to computerized machinery have been found to be in the range of 200 to 300 per cent (Block 1990, 142–3).

These changes have a number of interesting implications for empirical analysis of investment and its consequences for economic activity. First, because these new, computerized technologies (such as computer numerically controlled, or CNC machine tools) are both more versatile (capable of making a much broader array of products) and more productive, they may reduce significantly the number of machines (hence the volume of accumulated investment) required to generate a given quantity of output. Furthermore, because of decreasing prices, the net effect is likely to be higher potential output at lower total investment cost. It is not difficult to see how, if this is the case, neither the physical quantity of investment (number of machines) nor the dollar value will be a very accurate indicator of the true productive capacity of a region's (or nation's) economy.

A second implication, following partly from the previous point, is that manufacturing firms now require considerably less space within their plants than before to produce a given level of output – in essence, a second form of capital savings. This situation arises not only from replacement of a string of relatively specialized, dedicated machines with a much smaller number of more versatile, programmable machines, but also from another set of changes ongoing within the internal and interfirm organization of production. I refer here to the variety of increasingly common measures to reduce inventories of both inputs and finished products, as manufacturers move toward just-in-time systems of production. This latter change has drastically reduced firms' space needs for warehousing, as well as reducing unrealized investment tied up in production. Taken together, these reductions in required space imply that each new dollar of investment in machinery and equipment will be accompanied by declining volumes of investment in the plant that houses this production apparatus. This hypothesis provides a compelling explanation for why the ratio of annual outlays on machinery and equipment to total investment expenditures should be expected to rise over time, particularly since the second half of the 1970s, when CNC and related technologies began to diffuse more rapidly throughout Europe, Asia, and North America.

These insights suggest that it is increasingly difficult to get an accurate view of ongoing changes within Canada's (or any other advanced economy's) manufacturing regions by analysing aggregate data on capital stock. Even annual regional investment expenditures for machinery and equipment are difficult to interpret, for all the reasons presented above, if they are not also supplemented by a more detailed examination of the actual types of process technologies being installed.

Not surprisingly, social scientists have of late begun to focus more explicitly on such questions, by examining rates of adoption of those new, computerized process technologies that are becoming more important for competitiveness in manufacturing. Geographers have participated in this exercise by conducting surveys to determine adoption rates of advanced manufacturing technologies (AMTs) within and across various regions (see, for example, Oakey, Twaites, and Nash, 1982; Gibbs and Edwards 1985;

Table 1
Adoption rates (%) for selected AMTs, by province, 1989

Province	CAD/CAE		NC/CNC		Pick and place robots		FMC/FMS		Computer networks*	
	W[†]	U[‡]	W	U	W	U	W	U	W	U
Nfld	14	15	12	6	0	2	1	4	12	15
PEI	2	6	23	4	–	–	–	–	14	1
NS	27	16	23	5	–	–	6	3	24	2
NB	26	17	10	14	6	0	10	4	25	14
Que.	50	17	33	17	11	4	19	10	33	14
Ont.	53	18	33	15	21	5	23	8	41	12
Man.	42	17	25	10	4	1	8	6	17	5
Sask.	47	18	11	7	6	1	18	2	10	2
Alta	48	19	9	10	1	0	16	3	34	3
BC	34	13	21	10	4	2	19	4	22	6
Total	49	17	30	14	15	3	21	7	35	10

Source: Statistics Canada (1989).

* Computer network links with suppliers, subcontractors, or customers.

† Weighted by value of shipments represented by adopting establishments.

‡ Unweighted (simple number of establishments adopting as a percentage of total number of establishments in each province).

Rees, Briggs, and Hicks 1985; Rees, Briggs, and Oakey 1986; Oakey and O'Farrell 1992). In Canada, this analysis has been facilitated by Statistics Canada's launching in 1987 and 1989 of national surveys of AMT adoption (Statistics Canada 1989) in parallel with similar surveys conducted in the United States by the Department of Commerce.

The kinds of insights produced from such surveys are reflected in Table 1, which illuminates a number of patterns. First, the speed of AMT diffusion appears to vary notably from one type of technology to another: technologies such as computer-aided design or engineering (CAD or CAE) and CNC seem to have penetrated more widely than have robotics or flexible manufacturing cells or systems (FMC or FMS). Generally, the fastest-diffusing technologies seem to be those that are least complex and most amenable to "stand-alone" application (i.e., not integrated with other machines to form a larger system). Second, the consistently higher figures produced when weighting adoption rates by the value of shipments for which the adopting establishments are responsible indicate that, as might be expected, larger plants are more likely to be early adopters of AMTs than are smaller plants. Third, for each technology reported here, adoption rates (at least when weighted by shipments) are highest in Ontario or Quebec. Consistent with the first trend noted above, interprovincial variation in adoption rates seems to be smaller for the simpler and more "discrete" technologies (CAD/CAE, CNC) and greater for the more complex ones (robots, FMC/FMS). Overall, manufacturers in Ontario and Quebec seem to be the most common users of AMTs, particularly if due weight is given to the share of total provincial shipments represented by those adopters.

McFetridge (1992) has used these data to compare adoption rates of AMTs in Can-

Table 2
Canadian and u.s. establishments (%) using at least one AMT

	Employment size class					
	20–99		100–499		500+	
Industrial sector	Canada	U.S.A.	Canada	U.S.A.	Canada	U.S.A.
Metal fabricating	43	57	79	87	99	97
Machinery	67	76	92	94	99	99
Electrical and electronic equipment	60	72	87	89	94	98
Transportation equipment	41	52	68	84	97	98
Instruments and related products	43	69	93	89	99	99

Source: McFetridge (1992), based on data from Statistics Canada and u.s. Department of Commerce.

ada to those in the United States and other countries.[3] He finds consistent and significantly lower rates of adoption across most technology and industry groups in Canada as compared to the United States, thereby supporting the earlier contentions of Britton and Gilmour (1978) and Steed (1989) about the lagging technological sophistication of much of Canadian industry. Furthermore, and consistent with the data in Table 1, adoption rates vary strongly by size of establishment, with larger plants (in both countries) most likely to be adopters of AMTs (see Table 2). Unfortunately, this analysis provides no geographical comparisons for sub-national units within Canada.

In review, it is evident that, because of important qualitative changes in the nature of industrial machinery and production systems, traditional aggregate measures of capital accumulation have become increasingly imperfect indicators of the scale and quality of new investment. One approach, inspired by the "new" growth theory, is to focus instead only on the machinery and equipment component of annual investment, since this provides a more direct link to current process technologies. However, this approach still uses a numéraire of measurement that is insufficiently sensitive to the kinds of fundamental qualitative change discussed above. Others have pushed this idea considerably further, rejecting altogether the aggregate approach in favour of direct examination of the uptake of each new process technology. In the final section of this chapter, I explore in greater depth the contributions and limitations of this latter approach and conclude by emphasizing the insights that can be brought to bear on these phenomena by an explicitly geographical outlook on use of AMTs.

BEYOND ADOPTION: SPATIAL CONTEXT AND THE SOCIAL CONSTRUCTION OF TECHNOLOGY

One of the central themes to emerge from the "new" growth theory, as noted above, is the idea that technology and investment are intimately linked – indeed, that the latter embodies the former, as new machines and production systems incorporate within them the latest technological developments. This conception of technology and capital, which I refer to here as the "embodied approach," represents a major advance over the earlier (and still widely accepted) conception, which remains oblivious to qualitative changes in the nature of fixed capital. The mainstream view has always assumed that qualitative improvements and the potential for productivity enhancement would be fully reflected in the prices of capital goods traded in the marketplace – in other words, that they would be capitalized into the asset's price at the time of sale. The reality – steadily declining prices alongside rapid increases in capabilities – is difficult to reconcile with this view.

Another implication arising from the "old" conception of capital is that, if one accepts that it can be satisfactorily measured using a single, universal numéraire – i.e., in terms of dollar value – then, from a policy standpoint, it is a straightforward matter to declare the economy "better off" when the volume of investment rises, and vice versa when it falls.

This view, which Block (1990) has dubbed the "intravenous model of capital investment," has dominated diagnoses of economic trends as well as policy prescriptions (152): "Capital is seen as the life-sustaining substance that is injected into the veins of the economy. Too slow a flow of capital makes the economy lethargic, producing slow growth and the buildup of inflationary pressures. With a faster flow, the economy becomes healthier as growth and productivity accelerate." Furthermore, as Block notes, this simple idea underlies some of the most fundamental tenets of contemporary Western capitalism – particularly that the continuous and undiminished flow of profits back to investors is necessary to compensate them for forgoing immediate consumption in favour of investment for future returns. Principles such as these have become enshrined both in economic doctrine and in economic policy, serving as the justification for reduced corporate tax rates and sustained (even growing) inequalities of income and wealth.

Though Block does not comment on the subject, the implications of the embodied view arising out of the "new" growth theory are not all that different from those of the intravenous model. While Romer (1986), Scott (1989), and others in the same school would undoubtedly point to the particular necessity of encouraging investment in machinery and equipment, because of the beneficial effects flowing from the technologies embodied within machinery and equipment, the underlying presumption is still "the more the better." This view, however, presumes that these new technologies will simply implement themselves – that is, that their adoption and effective use are unproblematic. Indeed, some scholars, such as De Long and Summers (1991) and De Long (1992), have taken this idea even further, to argue that acquisition of advanced capital goods creates indirect social benefits for the wider economy by forcing skill increases on workers who will be using the new machinery. Furthermore, they merely cite the seminal theoretical work of Arrow (1962), on learning by doing, as sufficient justification for this view.[4]

Unfortunately for the proponents of the embodied approach, much of the recently accumulated empirical evidence at the level of individual firms in mature industrial economies contradicts this argument. Elsewhere (Gertler 1993) I have reviewed a large international literature which documents widespread difficulties encountered by manufacturers as they struggle to implement advanced process technologies (for Canadian evidence, see Beatty 1987; Meurer, Sobel, and Wolfe 1987). Increases in productivity are anything but automatic, and the costs associated with implementation are considerably greater than first imagined. Indeed, studies (such as those reviewed in the previous section) that simply monitor rates of adoption of AMTs may produce quite misleading results concerning affairs inside the adopting firms. As well, production of advanced machinery and the resulting experiences in implementation exhibit distinctive spatial patterns, which result has led to the suspicion that there are important geographical dimensions to this phenomenon whose illumination is essential to our understanding of the relationship among investment, technological change, productivity enhancement, and firms' performance.

First, difficulties in implementation are especially evident within mature industrial regions of countries such as Canada, the United Kingdom, and the United States, while firms in the industrial regions of Germany, Italy, Japan, and Sweden seem to have enjoyed considerably greater success. Second, the leading sites of innovation and production of many of the most important advanced manufacturing technologies happen also to be found within the latter group of countries. Recent work in industrial economics (see especially Lundvall 1988) suggests that this spatial pattern is anything but coincidental. Because advanced capital goods are typically complex, expensive, and long-lasting, major benefits flow to adopting firms located close enough to the producers of such goods to facilitate effective and frequent interaction. This interaction not only helps to solve operational and break-in problems after-installation but also increases the chances that the original design (or customization) of the machinery will closely fit the user firm's needs and circumstances. Consequently, the resulting hypothesis is that firms (such as those in Canada) aspiring to use AMTs can be expected to encounter significant difficulties when this machinery comes from distant production sites. This argument is also consistent with the earlier analysis of Britton and Gilmour (1978, 142–3), who note the difficulties that arise for independent Canadian firms because of their isolation from sources of innovation in process technology.[5]

In recent empirical work (Gertler 1995a) I have sought to analyse the experience of manufacturers in Ontario that have been trying to implement AMTs. Based on a postal survey covering 170 plants and over 400 instances in which advanced technology was implemented,[6] as well as personal interviews and site visits at 30 of these plants, I have found substantial evidence of significant and continuing difficulties. These are especially evident for smaller and single-plant, Canadian-owned operations, and these difficulties seem to be exacerbated when the source of the technology is foreign (especially overseas). Furthermore, simple physical distance is often a proxy for other effects. Organizational distance, for example, was found to influence implementation: plants belonging to larger, multi-locational firms suffered less from locational isolation from machinery production sites because they had ready access to abundant intra-firm engineering expertise to help them overcome the usual symptoms of "remote" technology implementation. Some may also have benefited from machinery production conducted by the firm at another branch location.[7]

Cultural dimensions, broadly defined, also appear to shape firms' implementation experience and the relationship between AMT users and producers (Gertler 1995a). Language difficulties, for example, become magnified when interacting parties are spatially distant from each other and the subject of conversation is technically detailed and complex. As well, distinctly different workplace cultures may prevail in each region or country in which the transacting parties are located. Here one sees the overwhelming influence of the social context surrounding a user firm and how this may differ from that of the (distant) producer firm.

Of paramount importance is the set of national (and, to some extent, sub-national) institutions and regulatory systems governing labour relations. This framework creates a system of incentives to which firms and workers respond in generally rational ways. Hence, in Germany (a leading site of machinery production), national and regional regulatory frameworks governing labour markets reinforce maintenance of long-term employment relations, thereby encouraging management to depend on internal promotion and to invest heavily in worker training (Streeck 1985; Sorge and Streeck 1988). This system, paired with wider industrial training that provides all workers with an excellent foundation in both the theoretical and practical aspects of advanced process technologies, results in a labour force possessing a high level of skill on entry, which is further enhanced by long years of progressively responsible experience and employer-sponsored training.

In comparison, the Canadian regulatory system provides little encouragement for manufacturing firms to maintain long-term relations with their shop-floor workers, instead favouring extensive use of the external labour market (frequent firings, lay-offs, and rehirings). This system offers few incentives for employers to provide meaningful training, as many of the skills imparted to workers would probably be made available to some other employer in the future (Muszynski and Wolfe 1989). Accordingly, the skills and abilities possessed by workers (and, frequently, managers) do not equal those in German plants in the same industry. Thus serious difficulties may arise when Canadian firms adopt German equipment.

These differing regulatory approaches can themselves be linked to different conceptions of technology (Gordon 1989). The typically Canadian approach (representative of an Anglo-American view) is closely consistent with the embodiment idea discussed above. It assumes that the full technological capabilities of a piece of industrial machinery are contained (or embodied) within the capital good itself. This attitude stands in contrast to the typically European approach, itself encouraged by a labour-market regulatory framework that presumes that "the most sophisticated performance levels can only be achieved through a combination of advanced technology and human labor, not through automation alone" (Gordon 1989, 21) – in other words, that "technology" is socially constructed. This view is also strongly shared by Block (1990), who asserts (145): "Most of the quality improvements in capital depend on the organizational context in which the technology is used ... [O]rganizational variables have become the key factors in shaping the economy's capacity to produce."

These ideas, and the Canadian study of AMT implementation, underscore the short-comings of the embodied approach to capital and technology and demonstrate the rather limited advance that it represents over earlier generations of growth theory in economics. They also suggest that the presumption that investment in AMTs will raise workers' skills is based on an excessively narrow interpretation of learning by doing. An alternative understanding of Arrow's (1962) arguments is far more consistent with the insights just discussed. In this alternative view (Gertler 1984b), Arrow can be seen to be making the case that the ultimate productive capabilities of a piece of fixed capital are not somehow "locked inside" the equipment itself but instead develop over time, through association among the machine, worker-operators, and management – an interpretation entirely consistent with the social construction model of capital and technology presented above.

It is thus not surprising that in his more recent work Romer (1990a; 1990b) has recognized to role of a country's investment in human capital. Explicitly rejecting embodiment and learning by doing (narrowly defined), he concludes (1990a, 339): "Thus contrary to the suggestion of those who place exclusive reliance on capital accumulation to generate long-run growth (and contrary to my previous claims) something else – something that does not vary one-for-one with the rate of investment in physical capital – must be decisive for long-term productivity growth." That "something else" is the quantity of educated human capital dedicated to applied research and development – particularly to production of new designs for capital goods! The second-generation model of endogenous technological change is therefore a distinct improvement over the original.

Despite evidence of some convergence between economists and geographers on this point, Romer (1990b) still expresses unswerving faith in the idea that when new capital goods are imported from abroad (for example, by developing countries with unprotected domestic capital-goods sectors) they can readily be put to effective use. In other words, Romer argues implicitly that the externalities arising from capital-goods production can spill across international borders and between countries of radically differing regulatory contexts and workplace cultures. Romer's macro-level (and aspatial) analysis renders him oblivious to the idea that the most important externalities arise not from capital-goods production alone but from the close interaction between producers and users of capital goods. In this sense, his approach is clearly distinct from the one recently advanced by Tyson (1992), who bases her manifesto for U.S. trade policy on the position that such positive externalities, particularly in production and use of high-technology commodities, are spatially bound. According to Tyson, this insight justifies protective measures to nurture indigenous development of technology-intensive sectors.

CONCLUSIONS

In this chapter, it has been my objective both to document the changing Canadian geography of capital accumulation and to demonstrate the utility of monitoring regional economic change through the lens of capital. I have examined several conceptions of capital, each of which (I have argued) offers a progressively more insightful perspective. They ranged from a simple, aggregate capital stock, denominated in a standard unit of measurement (dollars), through the embodied view, which emphasizes the volume of investment in machinery and equipment, to a focus on the actual types of new capital goods being implemented by firms. As I have moved from aggregate to increasingly disaggregated approaches (and from a macro to a micro scale of inquiry), I have also moved toward a fundamentally social model of technological change. It is important at this point to consider a few implications of this latter view.

First, as one moves closer to a socially constructed model of technology, the contribution made by a geographical perspective becomes clearer. I hope that I have demonstrated how and why "geography matters" and how the spatial perspective can enrich and extend the insights arising from recent work in the economics of growth.

Second, if one accepts that technology is socially constructed and that inanimate machines do not have, locked deep within them, all the capabilities that could possibly arise in their application, then a considerable amount of indeterminacy arises concerning the range of possible local outcomes arising from what appear to be similar volumes of investment, or even similar types of machinery in two different plants (or regions). Outcomes will presumably depend on regulatory and institutional environments and local employment relations.

Third, as I have noted elsewhere (Gertler 1993), the conception of technology as socially constructed poses a serious challenge to the ascendant policy doctrine of free trade and comparative advantage. If, as suggested here, the effectiveness of advanced industrial machinery declines significantly as the distance from the original production site increases, and if the positive externalities arising from machinery production are geographically bound (i.e., maximized when producers and users are able to interact closely), then one can hardly expect unrestricted importation of advanced capital goods to redress the inadequacies of technologically unsophisticated firms in Canada or other mature industrial economies.

Instead, the policy focus must shift to considering a range of action on at least three levels. First, in individual user firms, programs are required to facilitate not only adoption of AMTs but also effective implementation. Second, for interfirm industrial organization, there is an apparent need for policies that both recognize the strategic significance of domestic machinery production and address the relatively underdeveloped state of Canada's machinery industries. Following the lead of newly industrializing countries such as Taiwan and South Korea, as well as the policy regimes in established economic powers such as Germany, Japan, and the United States, it may make some sense to target for development an indigenous machinery-building capability in certain key sectors.[8] Third, at the macro level, there is a need to recast the regulatory framework surrounding workplace practices, to give firms stronger incentives to invest in long-term upgrading of their workers' skills. Under such a new regime, employers would come to view labour not as a cost but as a major (indeed crucial) partner in the drive to create prosperity in Canada's cities and regions.

NOTES

1 I wish to acknowledge the Social Sciences and Humanities Research Council of Canada, which has provided two grants in support of the work reflected in this chapter. I would also like to acknowledge the helpful comments of John Britton and Gordon Clark and the able research assistance of Dan Tassie.

2 The efficiency wage is simply the money wage rate divided by a measure of labour productivity (Kaldor 1970). Hence it is a measure of the "true" cost of labour, taking into account the changing productivity of the workforce. If wages and productivity rise at the same rate, the efficiency wage remains constant. If productivity should increase more slowly than the money wage rate, then the efficiency wage would rise.

3 In his first analysis (Ontario 1989), based on a comparison of Statistics Canada's 1987 survey results with those from the U.S. Department of Commerce's 1988 survey, McFetridge reached the surprising conclusion that adoption rates (broken down by sector and technology) did not differ significantly between the two countries. Adoption rates in Canada as a whole, and Ontario in particular, appeared to be higher as often as they were lower than the corresponding U.S. rates. However, in a more recent analysis based on Statistics Canada's 1989 survey, McFetridge (1992) reverses his conclusions, suggesting that Statistics Canada's survey of 1987 may have suffered from difficulties of data collection in what was then very much a pilot study.

4 In this classic article, Arrow had pointed out that workers frequently learn how to use new techniques through actual on-the-job experience, meaning that the same combination of labour and capital (workers and machines) will normally become more productive with the passage of time. This simple observation held devastating implications for the theory of capital and use of production function analysis, since it asserted the difficulties of accurately measuring the inputs traditionally entered into the production function. For further discussion, see Gertler (1984b).

5 Britton (this volume, chap. 14) also documents Canada's large and chronic trade deficit in machinery.

6 Respondents at each of the 170 plants were given the opportunity to document up to three different types of advanced-technology implementation.

7 Despite this point, two countervailing findings should be noted. First, even when technology users were large, multinational firms, significant difficulties in implementation were not infrequent, suggesting that intraorganizational transfers of expertise and technology were a distinctly second-best substitute for "being there." Second, further analysis revealed that foreign-owned users of AMTs in Ontario were considerably less likely to source their technology locally or to engage in close, collaborative relations with local suppliers, even when it could be demonstrated that sophisticated Ontario suppliers were available (Gertler 1995b). These findings challenge the commonly held view that foreign-owned branch plants exert a positive influence on their host economies by introducing more advanced process technologies and practices.

8 For a more extended discussion of these themes, including a comparative review of policy approaches in a range of industrial countries, see Gertler, DiGiovanna, and Tassie (1995).

REFERENCES

Anderson, W., and Rigby, D. 1989. "Estimating Capital Stocks and Capital Ages in Canada's Regions: 1961–1981." *Regional Studies*, 23, 117–26.

Arrow, K.J. 1962. "The Economic Implications of Learning by Doing." *Review of Economic Studies*, 29, 155–73.

Beatty, C.A. 1987. *The Implementation of Technological Change*. Research and Current Issues Series Report No. 49. Kingston: Industrial Relations Centre, Queen's University.

Birch, D. 1979. *The Job Generation Process*. Cambridge, Mass.: MIT Program in Neighborhood and Regional Change.

Block, F. 1990. *Postindustrial Possibilities: A Critique of Economic Discourse*. Berkeley: University of California Press.

Bluestone, B., and Harrison, B. 1982. *The Deindustrialization of America*. New York: Basic Books.

Bolton, R. 1980. "Multiregional Models: Introduction to a Symposium." *Journal of Regional Science*, 20, 131–42.

Bougrine, H. 1992. "The Role of Capital Formation in Economic Disparities among Canadian Regions: 1961–1990." *Canadian Journal of Regional Science*, 15, 21–33.

Britton, J.N., and Gilmour, J.M. 1978. *The Weakest Link: A Technological Perspective on Canadian Industrial Underdevelopment*. Background Study 43. Ottawa: Science Council of Canada.

Browne, L.E., Mieszkowski, P. and Syron, R.F. 1980. "Regional Investment Patterns." *New England Economic Review*, July–Aug., 5–13.

De Long, J.B. 1992. "Research Summary: Growth Industrialization and Finance." *NBER Reporter*, Summer, 5–11.

De Long, J.B., and Summers, L.H. 1991. "Equipment Investment and Economic Growth." NBER Working Paper No. 3515, *Quarterly Journal of Economics*, 106, 445–502.

Dicken, P., and Lloyd, P.E. 1990. *Location in Space: Theoretical Perspectives in Economic Geography*. 3rd edn. New York: Harper and Row.

Drache, D., and Gertler, M.S. 1991. "The World Economy and the Nation-State." In D. Drache and M.S. Gertler, eds., *The New Era of Global Competition – State Policy and Market Power*, 3–25. Montreal: McGill-Queen's University Press.

Erickson, R. 1989. "The Influence of Economics on Geographic Inquiry." *Progress in Human Geography*, 13, 223–49.

Garston, G.J. 1983. *Canada's Capital Stock*. Economic Council of Canada Discussion Paper 226. Ottawa.

Gertler, M.S. 1984a. "The Dynamics of Regional Capital Accumulation." *Economic Geography*, 60, 150–74.

– 1984b. "Regional Capital Theory." *Progress in Human Geography*, 8, 50–81.

– 1985. "Industrialism, Deindustrialism and Regional Development in Central Canada." *Canadian Journal of Regional Science*, 353–75.

– 1986. "Regional Dynamics of Manufacturing and Non-manufacturing Investment in Canada." *Regional Studies*, 20, 523–34.

- 1987a. "Capital, Technology and Industry Dynamics in Regional Development." *Urban Geography*, 8, 251–63.
- 1987b. "Economic and Political Determinants of Regional Investment and Technical Change in Canada." *Papers of the Regional Science Association*, 62, 27–43.
- 1993. "Implementing Advanced Manufacturing Technologies in Mature Industrial Regions: Towards a Social Model of Technology Production." *Regional Studies*, 27, 665–80.
- 1995a. " 'Being There': Proximity, Organization and Culture in the Development and Adoption of Advanced Manufacturing Technologies." *Economic Geography*, 71, 1–26.
- 1995b. "In Search of the New Social Economy: Collaborative Relations between Users and Producers of Advanced Manufacturing Technologies." *Environment and Planning A*, 27 (forthcoming).

Gertler, M.S., DiGiovanna, S., and Tassie, D. 1995. *The Structure and Strategic Importance of the Machinery, Tool, Die and Mould Sector in Ontario*. Report prepared for the Ontario Machinery, Tool, Die and Mould Sectoral Partnership Project, Ministry of Economic Development and Trade, Toronto.

Gibbs, D.C., and Edwards, A. 1985. "The Diffusion of New Production Innovations in British Industry." In A.T. Thwaites and R.P. Oakey, eds., *The Regional Impact of Technological Change*, 132–63. New York: St Martin's Press.

Gordon, R. 1989. "Beyond Entrepreneurialism and Hierarchy: The Changing Social and Spatial Organization of Innovation." Paper presented at the Third International Workshop on Innovation, Technological Change and Spatial Impacts, Selwyn College, Cambridge, England, 3–5 Sept.

Harvey, D. 1982. *The Limits to Capital*. Chicago: University of Chicago Press.

Hepworth, M.E. 1989. *The Geography of the Information Economy*. London: Frances Pinter.

Kaldor, N. 1970. "The Case for Regional Policies." *Scottish Journal of Political Economy*, 17, 337–48.

Lundvall, B.-A. 1988. "Innovation as an Interactive Process: From User-producer Interaction to the National System of Innovation." In G. Dosi, C. Freeman, R. Nelson, G. Silverberg, and L. Soete, eds., *Technical Change and Economic Theory*, 349–69. London: Frances Pinter.

McFetridge, D.G. 1992. *Advanced Technologies in Canada: An Analysis of Recent Evidence on Their Use*. Ottawa: Canada Communications Group.

McHugh, R., and Widdows, R. 1984. "The Age of Capital and State Unemployment Rates." *Journal of Regional Science*, 24, 85–92.

Mahon, R. 1991. "Post-Fordism: Some Issues for Labour." in D. Drache and M.S. Gertler, eds., *The New Era of Global Competition – State Policy and Market Power*, 316–32. Montreal: McGill-Queen's University Press.

Mansell, R. 1987. "Energy Policy, Prices and Rents: Implications for Regional Growth and Development." In W.T. Coffey and M. Polèse, eds., *Still Living Together: Recent Trends and Future Directions in Canadian Regional Development*, 277–312. Montreal: Institute for Research on Public Policy.

Massey, D.B. 1978. "Regionalism: Some Current Issues." *Capital and Class*, 6, 106–25.

- 1984. *Spatial Divisions of Labour*. London: Macmillan.

Meurer, S., Sobel, D., and Wolfe, D. 1987. *Challenging Technology's Myths: A Report on the Impact of Technological Change on Secondary Manufacturing in Metropolitan Toronto*. Prepared for the Labour Council of Metropolitan Toronto. Toronto.

Muszynski, L., and Wolfe, D.A. 1989. "New Technology and Training: Lessons from Abroad." *Canadian Public Policy*, 15, 245–64.

Norcliffe, G.B., Goldrick, M.D. and Muszynski, L.. 1986. "Cyclical Factors, Technological Change, Capital Mobility and Deindustrialization in Metropolitan Toronto." *Urban Geography*, 7, 413–36.

Oakey, R.P., and O'Farrell, P.N. 1992. "The Regional Extent of Computer Numerically Controlled (CNC) Machine Tool Adoption and Post Adoption Success in Small British Mechanical Engineering Firms." *Regional Studies*, 26, 13–75.

Oakey, R.P., Thwaites, A.T., and Nash, P.A. 1982. "Technological Change and Regional Development: Some Evidence on Regional Variations in Product and Process Innovation." *Environment and Planning A*, 14, 1073–86.

Ontario. 1989. *A Comparison of Canadian and U.S. Technology Adoption Rates*. Toronto: Ministry of Industry, Trade and Technology.

– 1992. *An Industrial Policy Framework for Ontario*. Toronto: Ministry of Industry, Trade and Technology, Queen's Printer for Ontario.

Pudup, M.B. 1992. "Industrialization after (De)industrialization: A Review Essay." *Urban Geography*, 13, 187–200.

Rees, J., Briggs, R., and Hicks, D. 1985. "New Technology in the United States' Machinery Industry: Trends and Implications." In A.T. Thwaites and R.P. Oakey, eds., *The Regional Economic Impact of Technological Change*, 164–94. New York: St Martin's Press.

Rees, J., Briggs, R., and Oakey, R.P. 1986. "The Adoption of New Technology in the American Machinery Industry." In J. Rees, ed., *Technology, Regions and Policy*, 187–217. Totowa, NJ: Rowman and Littlefield.

Rigby, D.L. 1991. "Technical Change and Profits in Canadian Manufacturing: A Regional Analysis." *Canadian Geographer*, 35, 353–6.

Romer, P.M. 1986. "Increasing Returns and Long-run Growth." *Journal of Political Economy*, 94, 1002–38.

– 1990a. "Capital, Labour and Productivity." In M.N. Bailey and C. Winston, eds., *Brookings Papers on Economic Activity: Microeconomics*, 337–420. Washington, DC: Brookings Institution.

– 1990b. "Endogenous Technological Change." *Journal of Political Economy*. 98 no 5, S71-S102.

Schoenberger, E. 1989. "New Models of Regional Change." In N.J. Thrift and R. Peet, eds., *New Models in Geography*, vol. 1, 115–41. Boston: Unwin and Hyman.

Scott, M. 1989. *A New View of Economic Growth*. Oxford: Oxford University Press.

Sorge, A., and Streeck, W. 1988. "Industrial Relations and Technological Change: the Case for an Extended Perspective." In R. Hyman and W. Streeck, eds., *New Technology and Industrial Relations*, 19–47. Oxford: Basil Blackwell.

Statistics Canada. 1989. *Survey of Manufacturing Technologies – 1989 Statistical Tables*. Science, Technology and Capital Stock Division, Cat. No. ST-89-10. Ottawa.

Steed, G.P.F. 1989. *Not a Long Shot: Canadian Industrial Science and Technology.* Background Study No. 55. Ottawa: Science Council of Canada.

Storper, M., and Walker, R. 1989. *The Capitalist Imperative: Territory, Technology, and Industrial Growth.* Oxford: Basil Blackwell.

Streeck, W., ed. 1985. *Industrial Relations and Technical Change in the British, Italian and German Automobile Industry: Three Case Studies.* Berlin: International Institute of Management.

Tyson, L.D. 1992. *Who's Bashing Whom? Trade Conflict in High Technology Industries.* Washington, DC: Institute for International Economics.

Varaiya, P., and Wiseman, M. 1981. "Investment and Employment in Manufacturing in the U.S. Metropolitan Areas 1960–1976." *Regional Science and Urban Economics*, 11, 431–69.

Walker, R.A., and Storper, M. 1981. "Capital and Industrial Location." *Progress in Human Geography*, 5, 473–509.

Weaver, C., and Gunton, T. 1986. "From Drought Assistance to Megaprojects: Fifty Years of Regional Theory and Policy in Canada." In D.J. Savoie, ed., *The Canadian Economy: A Regional Perspective*, 190–224. Toronto: Methuen.

PART FOUR

Services and the Spatial Organization of the Economy

Service industries now employ 72 per cent of the Canadian workforce, having been the sector of greatest employment growth for the past century. There has, however, been a growth spurt since about 1960, when services employed 55 per cent, and in this respect Canada has followed a pattern quite similar to that of all industrial countries. While services were responsible for 77 per cent of net new jobs during the 1970s and 1980s, according to Coffey (this volume, chap. 18), service output was only 60 per cent of GDP in 1984, compared with 56 per cent in 1961.

Services are essentially an urban activity with quite clear, though dynamic spatial regularities that associate bundles of activities with urban centres of different sizes and spacing (this volume, Simmons, chap. 17; Coffey, chap. 18). For this reason, the geography of services is an important determinant of the spatial and economic structure of the urban system, which is hierarchically organized. The proportion of services consumed and provided locally varies with the size of urban centre and responds to service specializations, which have been developed in response to each region's productive sectors. For large parts of the space economy, particularly for peripheral regions and centres at the lowest urban tier, many purchases are made directly or indirectly at higher levels in the urban system. Thus there are personal income flows from the resource regions – for example, from rural service centres in agricultural regions and one-industry towns associated with mining – to regional and metropolitan centres in payment for goods and higher-order services (Simmons 1976).

The geography of demand for services is a combination of purchases by consumers and by businesses. In the latter case of "intermediate" demand for the information-intensive services – finance, insurance, and real estate (FIRE) and "producer" services such as legal, accounting, commercial, and technical consultant services – inputs are sought by retail, wholesale, manufacturing, or resource firms. Semple (this volume, chap. 19) argues that the headquarters of corporations in all sectors, and the locational requirements of the financial services industry (which provides capital for all sectors in all locations), have strongly influenced the development of national – Toronto and Montreal – and regional urban centres. By way of demonstration he singles out the location of the 3,110 largest corporate headquarters to define Canada's quaternary centres, as a subset of places within the urban system.

Nearly 44 per cent of the assets of non-financial corporations are associated with head offices located in Toronto, with Montreal's 22 per cent a distant second, but Semple shows that over 20 other centres scattered across the country control the remaining 34 per cent of assets. Cities outside the central regions generally achieve second rank in specific sectors of economic activity; in services, Winnipeg, Vancouver, and Calgary follow Toronto and Montreal; in manufacturing, Vancouver ranks third on the basis of its forest products industry, and Calgary ranks second in resource assets controlled from its headquarters because it is the centre for the petroleum industry.

The revenues of financial corporations are even more strongly controlled from Toronto and Montreal (48 per cent and 28 per cent, respectively), and the economic focus on Toronto is at its maximum for foreign financial and non-financial corporations (59 per cent). As a result of recent deregulation, foreign-owned financial corporations are rapidly becoming more numerous, and as noted above, Toronto is recognized by international financial firms as Canada's primary centre. Furthermore, interconnection is increasing between foreign companies and the "four pillars" of the Canadian financial establishment, which are merging rapidly: thus Canada's firms are being integrated into the broad spectrum of international financial activity.

Finance and head office functions together with secondary manufacturing production, all localized in the cities of central Canada, have generated substantial "intermediate" demand for inputs from service firms. The consequence, as Coffey explains, is that while 51 per cent of service jobs are located in the eight largest cities, 52 per cent of employment in the fast-growing, high-order services (FIRE and producer services) is found in four large centres – Toronto, Montreal, Vancouver, and Ottawa-Hull. The growth of these activities is a response to increased demand for increasingly sophisticated services. As the range of service products has expanded, large corporations have both expanded in-house provision of high-order services and purchased infrequently used or highly specialized services from firms in the service sector. Small and medium-sized firms have taken greater advantage of links with external sources of inputs, too.

While the largest metropolitan centres are the major financial nodes of the economy and the location of most of the head offices of corporations in all activities, there is a modest countervailing tendency for headquarters in resource or resource-processing industries to leave the centres of capital for centres with better-developed flows of information on particular product markets. Like the headquarters in metropolitan centres these generate demand for high-order business and technical services, and there is evidence that some of this demand is being met in regional centres through development of local specialized firms (Davis and Hutton 1993). There is in addition an increasingly wider distribution of head offices of other service activities, such as insurance, which are marketed across the country.

There has been interest in the feasibility of new service products such as computer software and database services being growth industries in "new" locations. This idea relies on the view that distance has been removed as a prime influence on the location of many information-intensive activities by telecommunications systems, which make peripheral locations less disadvantaged than formerly. Despite the impact of telecommunications in allowing locational separation of information and non-information activities (Hepworth 1989), which can allow placing of mail and phone-ordering facilities and the processing of credit-card receipts in peripheral centres, the spatial concentration of high-order functions is expected to prevail.

As described here, intermediate services – in particular, producer and financial services – are important to the Canadian economy in the same way as auto plants and pulp mills: all are potential or actual export earners on the basis of their product quality, price, and innovation; they are not tied to local markets; and their importance is to be gauged as part of the productive economic base. But in the present structure of the space economy, more jobs in more places are associated with consumer-related economic activities, particularly retail (1.5 million jobs) and wholesale trade (0.5 million), commercial consumer services (2.0 million), and financial services for consumers (0.3 million) – 4.8 million in total, or 40 per cent of the workforce. Simmons demonstrates their strong spatial association with population and per-capita income and suggests that a high degree of stability is to be expected, especially if federal transfer payments continue to support consumer services in regions such as Atlantic Canada. He also argues that shifts in the spatial pattern of total employment, or per-capita income, are likely to be less significant than the impact of inter-regional migration in changing the map of the Canadian market. The influence of migration, however, is likely to be dampened by the smaller number in the highly mobile age bracket that results from the slowing of population growth.

In contrast to the very high locational concentration of producer services, only 25 per cent of consumer-related jobs are found in Toronto and Montreal, though 50 per cent are located in the Windsor–Quebec City corridor. But taking retail activity as an example of the set of consumer-oriented distribution activities, Simmons shows that the economic concentration of control by large, multi-locational firms (retail chains) is associated with the regional basis of their operations and the spatial concentration of their head offices, in Toronto. The proportion of workers in the retail sector responsible for managerial or inventory-related jobs with no direct contact with consumers (now 55 per cent) is increasing, and these non-sales jobs located at head office or warehouse are connected by computer systems with the sales branches, thus allowing centralized control but dispersed retail operations and warehouse locations.

Our account of the structure of the Canadian urban system has focused on the location of different activities and on the influence of telecommunications systems, which seem to be reducing the magnitude of locational costs associated with access to information as they knit Canada's urban centres and those in other countries into an increasingly complex network. However, transportation systems have been central in shaping the Canadian space economy into one with strong east-west links and continue to sustain it with massive internal and external flows of goods. The North American Free Trade Agreement (NAFTA), however, challenges the fit of Canadian transportation systems to new patterns of demand. Lea and Waters (this volume, chap. 16) argue that the effect of liberalization of trade with the United States is part of a larger process through which Canada is accepting the U.S. pattern of deregulation. Starting over a decade ago, U.S. actions in transportation led to Canadian competitive disadvantages in rate setting and rate publication until 1987, when a less-regulated road and rail transport environment was introduced. Regulatory differences between the two countries remain, however – most notably, road transport taxes. Other substantial issues include long-standing rail freight subsidies embedded in the basic economics of specific industries and regions (such as prairie grain), which Lea and Waters explain have been rescinded.

The obvious spatial implication of the NAFTA is to reduce the importance of east-west flows of goods and services in Canada as U.S. suppliers from neighbouring states see opportunity in Canada and as Canadian businesses seek out similar U.S. markets. NAFTA has not introduced sudden major changes to the urban macro-geography of Canada, but over-time the pattern of flows of goods and services will probably become more consistent with a stronger north-south orientation, which means redirection of infrastructural investment toward transborder routes.

REFERENCES

Davis, H. Craig, and Hutton, Thomas A. 1993. "Producer Services Exports from the Vancouver Metropolitan Region." *Canadian Journal of Regional Science*, 14, 371–89.
Hepworth, Mark E. 1989. *Geography of the Information Economy*. London: Belhaven Press.
Simmons, J.W. 1976. "Short-term Income Growth in the Canadian Urban System." *Canadian Geographer*, 20, 419–31.

CHAPTER SIXTEEN

The Role of Transportation in the Canadian Economy

A.C. LEA AND NIGEL M. WATERS

There be three things which make a nation great and prosperous: a fertile soil, busy workshops, and easy conveyance of men and things from one place to another.   Francis Bacon (quoted by Lardner 1855)

Transportation plays a vital role in the economy and society of Canada – more so than in almost any other country in the world. Canada is a geographically large country with a dispersed population and a primarily resource-based economy.[1] Freight transportation is vital for moving unprocessed resources from remote hinterlands to the industrial heartland of the country and to international markets. Inter-city passenger transportation is critical for moving people between major urban centres and agglomerations. This vital importance was recognized before, during, and after Confederation. The opening remarks in the study of Canadian railways by Trout and Trout (1871) emphasize the point: "The material progress of Canada has depended on nothing so much as the means of communication, the facilities for conveying men and goods."

More recently Rostow (1960, 55) has argued that, for most economies, construction of railways is "the most powerful single initiator of takeoffs" into sustained economic growth. Rostow credits Canada with achieving this takeoff during the period 1896 – 1914, when extensive construction of the transcontinental lines had either just taken place or was still occurring. Railway construction stimulates the economy, and, once construction is complete, the railways allow other sectors to achieve greater efficiencies and lower transportation costs. In the present century transportation has continued to have a major influence on the Canadian economy. As such the transportation sector has been the subject of numerous royal commissions, as it has played a central role in the economic policies of various governments.

In recent years the transportation industry has become highly competitive, and transportation has been the subject of relentless efficiency drives. The goal of cost minimization is being pursued with extensive "downsizing" in the rail and airline industries in particular. Even so, transportation remains the third-largest cost in the process of production and distribution (Sampson, Farris, and Shrock 1990). Today it is exceeded, as a factor of production, only by labour and materials. Though the precise percentage of a finished product's cost that is attributable to transportation varies by sector, in Canada's primary industries it usually represents an even greater share.

Passenger transportation in Canada has been highly subsidized in recent years, and in 1989 concern over this subsidization led to appointment of a royal commission to look at intercity passenger transportation in the hopes that it could be made more efficient and less of a burden on the Canadian taxpayer. As is shown below, the Commission's findings and ongoing deregulation of the industry, together with such major economic developments as the North American Free Trade Agreement (NAFTA), have created a new paradigm for Canadian transportation. This paradigm will manifest itself in the form of less extensive networks for many modes and with transportation systems that are oriented north-south, not east-west, so as to take advantage of new economic opportunities south of the border.

CANADIAN TRANSPORTATION
BEFORE 1920

Innovations and Technological Developments

There are five primary modes of transportation: air, road, rail, water, and pipeline. Their relative importance has varied through time and space, depending on their technological development. Various tendencies, patterns, and historical imperatives may be discerned. First, at any given time one mode will usually be dominant over certain distances and for certain products. Passengers are usually willing to pay a premium for fast, efficient transportation. Thus passengers are often the innovators and will be the first to take advantage of the most recent transportation developments and will pay heavily for this privilege. Second, innovations are normally competitive over only a certain range of distances, though their competitive advantage is likely to grow as the technology matures.

Third, radically new technologies are likely to require substantial additional investments in infrastructure, which will tend to restrict spatially the availability of the new forms of transportation. Fourth, new forms of transportation will reinforce existing networks. The original trails and tracks through the bush established by the Native peoples of Canada were the routes first followed by the early colonial travellers – voyageurs and explorers of the seventeenth and eighteenth centuries. These routes connected the first colonial settlements, and as these settlements grew and time passed new modes of transportation tended to link the largest of the existing settlements. New infrastructure, such as the railways of the late nineteenth and the highways of the twentieth centuries, tended to follow the existing routes, which may well have been used in the first instance because of favourable terrain. Thus transportation corridors quickly became entrenched and have remained in some cases largely unchanged to the present day (Whebell 1969).

Canadian Transportation before 1850

Before 1850 transportation in Canada was dominated by three characteristics (Owram 1992): inadequacies of technology and lack of capital, domination by water transport, and the start of the influence of European technology and capital – specifically, use of steamships and construction of transportation infrastructure in the form of canals.

Throughout this period water transportation was paramount for both freight and passengers. Its dominance in the first part of the nineteenth century was underscored by construction of a number of strategically placed canals. These included the Lachine Canal on the St Lawrence above Montreal, three canals on the Ottawa River, and the Rideau Canal. The Ottawa and Rideau canals were both built with public funds for military and strategic purposes following the War of 1812 with the United States, the goal being an alternative water route between Montreal and Kingston. The first Welland Canal, linking Lakes Erie and Ontario, began with private funds and was completed using government monies, as have been all subsequent improvements to inland waterways.

Construction of canals during this era exhibits a number of themes that are paralleled in the growth of other transportation systems in Canada. These include use of public funds to finance construction and to maintain or support operation of transportation systems and infrastructure. The argument that facilities had to be constructed and improved for strategic purposes has also been important,[2] as has the need for some degree of redundancy in Canada's transportation capabilities. If one link, or indeed an entire system, becomes inoperative – for example, because of an accident – then there should be alternative routes that can be used instead. In most industrialized countries there is redundancy, but in Canada, even today, an avalanche on the Trans-Canada Highway or flooding of the Canadian Pacific tracks in the mountainous regions of Alberta and British Columbia can cause serious inconvenience to passenger and freight transportation.

These ideas were incorporated into Kansky's (1963) measure of transportation network complexity – the cyclomatic number, or the number of circuits in a transportation network. The higher the number, the more evolved the transportation system and the greater the redundancy in the network. The implication of this principle involves reinforcement of existing routes.

Canadian Transportation from 1850 to 1920

The first railway in Canada was not built until 1854. At this time road transportation was still poorly developed (Trout and Trout 1871, 26–7). Water transportation had led to development of numerous small ports on the shorelines of lakes Erie and Ontario. But water transportation had two fundamental disadvantages: it was slow, and it was spatially restricted, despite the existence of extensive inland waterways. Thus when the first railways were built from Toronto in the mid-1850s they immediately became the primary means of passenger transportation when speed was essential, and within a decade they were also competitive for freight transportation for all but the bulkiest of commodities (Guillet 1933, 463; Vance 1990, 178–9).

Domination of the railway quickly spread east and west out of Canada West. The first transcontinental, the Canadian Pacific Railway (CPR) line, completed in 1885, required extensive government subsidies in the form of land grants, substantial cash contributions, and a series of special privileges, including remission of duties on construction materials and the right to a 20-year monopoly between the mainline and the U.S. border. This pattern of a wide range of government subsidies and incentives, as opposed to ownership, was to be repeated again and again as new forms of transportation evolved.

The importance of the railways to the economic growth of the nation was formalized in the so-called staples thesis brought to prominence in essays by Mackintosh (1923) and the work of Innis (Innis 1956; this volume Barnes, chap. 3; Hayter, chap. 6). The argument was that a country could achieve prosperity and the beginnings of economic development through exploitation of its natural resources. Demand for transportation and manufactured goods, among other things, would lead to development of a more mature economy.

Mackintosh and Innis, however, were not in total agreement. Mackintosh (1923, 25) perceived that "Canada is a nation created in defiance of geography," maintaining that it was through building of the CPR that "the western barrier was substantially overcome, and a period of phenomenal expansion set in" (24). Innis attributed a major role in the economic development of Canada to the St Lawrence canal system. He did not dismiss the railways but noted: "Canada has become to an increasing extent amphibian" (Innis 1956, 73). As well, "It is difficult to summarize the importance of transportation as a factor in Canadian economic history. We can suggest, however, the overwhelming significance of the waterways and especially of the St. Lawrence" (74).

Innis argued that Canada existed because of geography – that is, because of the St Lawrence waterway. The argument is strangely reminiscent of U.S. debate over the importance of railways in the development of the economy during the mid-nineteenth century (Fogel 1964; Taylor 1965; Davis 1966; Nerlove 1966; David 1969). Most economic historians would now agree that the various transportation systems had a remarkable degree of symbiosis. They existed and survived because of complementary systems with which they interconnected and to which they supplied traffic or from which they received traffic. Innis (1956, 75), a staunch supporter of the continuing importance of water transport, appears to agree with this synopsis when he comments: "The St. Lawrence route, as improved by canals, was further strengthened by the completion of the Grand Trunk Railway and its connections with the seaboard in the following decade [1860s]." This idea of route integration is another perennial theme in Canadian freight and passenger transportation. Thus in a recent symposium on passenger transportation (ITS 1990, 44) a key recommendation was the need for a coordinated system of transportation services. This was seen to be critical for users' satisfaction and for efficiency.

In the late nineteenth century the railways began to dominate transportation over almost all distances. Canadian railway building and opening of the western interior had lagged behind its American counterparts, but as Innis observes Canadians could benefit from U.S. experience and could use tried and tested technology. But they did not learn the lessons that had been taught in the 1880s when line duplication and over-building in Ontario had led to ruinous competition between fledgling railway companies. Thus the railway era closed in ignominy with construction of two additional transcontinental systems at the turn of the century. These two newer systems did not generate sufficient traffic of their own and were merged into the Canadian National Railways (CNR) system during the years from 1917 to 1923. The CNR eventually incorporated five systems: the old Grand Trunk railway and the Grand Trunk Pacific, the Intercolonial, the Canadian Northern, and the National Transcontinental.

Vance (1990, 318–25) has noted that the growth of U.S. railway systems was dominated by the geographical concepts of ubiquity and natural territory. Individual railway companies attempted to dominate an entire area or region, which they came to see as their natural territory. They then tried to provide a reasonably ubiquitous service to all traffic-generating points within the region. In Canada the concept of natural territory was more weakly developed, especially in central Canada, where redundant rail links to the same centres were provided by competing railway companies, and this was true to some degree across the prairies. But by 1920 redundant services were becoming a feature of the past and cities such as Edmonton and Calgary saw themselves as CN or CP centres, respectively. The original attempt of the railway companies to provide almost ubiquitous coverage has caused many of the current problems. With renewed competition from the road system, the railways have engaged in a series of retrenchments. The natural-area principle in a duopoly has also caused difficulties by inhibiting competition and, as is discussed below, may perhaps be resolved only by transfer of ownership of the infrastructure to an independent agency.

CANADIAN TRANSPORTATION FROM 1920 TO 1987

The Rise of the Automobile

For passenger transportation this era saw the rise to prominence of what Flink (1975) has described as "the car culture" and more recently "the automobile age" (Flink 1988). The new technology had been developed at the turn of the century, but it took a while to mature to the point at which it could challenge rail-based transportation, and again it was passenger transportation that succumbed first, just as it had done when railways challenged water-based systems in the 1850s.

The automobile age ushered in important and fundamental differences from railway transportation. Cars were privately owned, they were most competitive (at least to begin with) over short distances, and they soon came to offer the most spatially unrestricted form of transportation. All this was in stark contrast to railways. Motor vehicle transportation became accessible to large sectors of Canadian society and was to transform industrial organization and urban life, eventually leading to the unprecedented growth of suburbia. The infrastructure of the automobile age was to be public property, thus requiring increasing partnership between the federal and provincial governments.[3]

The automobile initiated the concept that users should pay for transportation – not only the cost of the vehicle but also that of providing the extensive new infrastructure.[4] As is seen below, the idea that the user should pay the full cost of transportation has been resurrected in the recent report of the Royal Commission on National Passenger Transportation (1992).

Developments from 1945 to 1987

Two trends characterized developments in road transportation in the years following the Second World War. The first was the increasing provincial expenditures on improvement of the road infrastructure. In 1945 user taxes from motor vehicles were $119.8 million, while expenditures were only $73 million, but by 1960 the situation had been reversed and expenditures at $657 million were greater than taxes at $530 million (Owram 1992, 92). Expenditures on road infrastructure were generally the largest single item for provincial governments. However, this was no longer the case as, by the mid-1970s, health, education, and social services had become more important (see Davis, this volume, chap. 20).

In 1945 the idea of a road stretching across the nation was first proposed at a governmental level, though popular demand for such a road may be traced back as far as 1910. In 1949 the Trans-Canada Highway (TCH) was launched as a cost-sharing agreement between the federal and provincial governments. A second cost-sharing arrangement, Roads to Resources, was initiated in 1958 and the TCH was formally opened in 1962, though it was not fully completed to design specifications until 1970. Then, as the longest national highway in the world (7,821 km) it became the icon for a new era, just as the CPR had been almost 100 years earlier.

Air Transportation

Air transportation in Canada received a major stimulus in 1937 with formation, by Act of Parliament, of Trans-Canada Airlines as a subsidiary of CNR (Goldenberg 1994, 5). By 1948 the airline was serving major cities in Canada and had links with the Caribbean and major European cities. In 1965 it changed its name to Air Canada. The airline's major competition in this period came from Canadian Pacific Airlines which had been organized in 1942 following purchase of Yukon Southern Air Transport and other smaller operations. After 1949 Canadian Pacific Airlines began establishing extensive international connections with centres in the Far East, Australia, and South America. It became CP Air in 1968 while gaining greater corporate autonomy.

In Canada travel by air had become common for business and government workers by 1960, but it was not until arrival of jet aircraft in the early 1960s and the relative drop in the cost of air travel vis-à-vis the consumer price index during that decade that it became universally attractive to the average traveller and the preferred mode of travel for long-distance passengers. Indeed air travel had now combined with the automobile to destroy the railways as the primary provider of passenger transportation. Automobiles had already become the most attractive means of transportation for short journeys, while the airlines now became the most attractive mode for extremely long journeys and were also the mode of choice for shorter journeys for the affluent and for the business traveller.

Vance (1990, chap. 7) has argued that commercial aviation provides "the ultimate ubiquity," in that once the plane leaves the ground it is not constrained by any infrastructure such as the road or rail system. However, the impact of this freedom has not been realized because, as Vance also states, the airport itself represents a significant "geographical anchor" and its effect has been to concentrate this new form of transportation in only a few favoured centres.

CANADIAN TRANSPORTATION SINCE 1987

Though this chapter has argued that Canadian transportation today cannot be understood without a comprehensive understanding of what has happened in the past, it is important to recognize that four developments that have occurred since 1987 are likely to have a profound effect on the way in which transportation in Canada develops in the future. In chronological order, these developments are the National Transportation Act (NTA) of 1987; freer trade, beginning with the Canada–United States Free Trade Agreement (FTA) of 1987; the Goods and Services Tax (GST) of 1989; and the Royal Commission on National Passenger Transportation of 1992. Canadian transportation appears once again to be following or developing hand in hand with changes south of the border.

Deregulation: The NTA

While the Staggers Act of 1980 formally inaugurated an era of deregulation in the U.S. transportation industry, deregulation policies had been introduced in the late 1970s. From that time on there was a marked divergence in transportation policy between Canada and the United States. Initially this new U.S. legislation was expected to have "no significant impact upon Canada's national trade or traffic flow patterns" (Lande 1992, xviii–xx, 3–10). Subsequent studies by the CPR and by Lande showed this to be untrue. Lande in particular was able to show that the U.S. industry was able to introduce new competitive rate structures and that Canadian railways and shippers were losing export markets because of their adherence to a highly regulated status quo.

The Canadian government delayed formal discussion of deregulation until 1985 (Heaver 1985), and it was not until December 1987 that the National Transportation

Act (NTA) was passed. There were three key components of this new legislation (which took effect 1 January, 1988): final-offer arbitration, public-interest equity appeals of rates, and the competitive joint-line rate application (CJLR). According to Lande (1992), policy specialists hoped that these features would help ameliorate the three problems of divergent U.S. and Canadian regulations, the need to encourage competition between the railway duopoly of the CN and CP Rail, and the need to increase efficiency between industries that varied markedly by size, commodity type, and geographical location.

Final-offer arbitration had been proposed by shippers as a method whereby they might appeal railway transportation rates. Whenever shippers appealed the proffered railway rates an arbitrator was to be appointed to choose either the railway's or the shipper's final offer. According to Lande (1992, 118) only six cases went to final-offer arbitration between January 1988 and June 1991, but the mere perception of the likely impact of the new legislation led CN and CP Rail to offer unprecedented rate discounts to hundreds of their larger customers. Moreover, unlike U.S. experience with the Staggers Act, these discounts appear to have been offered with little concern for whether the shippers were in a captive or competitive location (Lande 1992, 119). Nevertheless the confidential nature of the rate setting process and the discounts offered, and shippers' unwillingness to initiate proceedings against the railways, would seem to indicate that final-offer arbitration is not a panacea to carrier-dominated rate setting.

Prior to the NTA of 1987 and subsequent to the Staggers Act, which allowed confidential rates, Canadian carriers had complained that regulations that required that they publish their tariffs allowed American carriers to use their rates as a target that they could beat by providing confidential discounts. The NTA aimed to change this by allowing confidential agreements between carrier and shipper. To protect the shipper the new legislation also allowed for equity appeals. Thus if a shipper merely suspected that a competitor was receiving an unfair rate advantage, it could require the National Transportation Agency to carry out a confidential review of the offered rates. Lande (1992, 135) concludes that despite some merit as a deterrent the equity appeal mechanism has been used only rarely, tends to duplicate the final-offer arbitration provisions of the act, and has additional structural flaws.

The CJLR provisions of the act were designed to protect shippers served at their origin or destination by only one railway. It was strongly opposed by both CN and CP Rail. The legislation allows a shipper, located in such captive markets, to negotiate rates with a second railway company that is in a position to haul the shipper's freight over a portion of the journey. The first railway would then be required to transfer the merchandise at the connection point and to charge the shipper a rate no greater than that offered by the second. The precise details of how such rates are determined are complex, but the primary function of the legislation was to prevent railways from protecting long-haul traffic from captive markets. It was an attempt to circumvent the lack of redundancy in the modern Canadian railway system. Shippers have used this mechanism for stimulating competition between Canada's two railway systems even less frequently than the two previous appeal procedures described above. Lande (1992, 151–2) concludes on a despondent note: "the legislative solution enacted by the Canadian government has so far proven inadequate to offset the traditional passivity of transport users to seek pro-competitive pricing from their carriers or the inherent tendency toward market sharing and cooperative behaviour which a rail duopoly creates." Perhaps the real problem lies with the Staggers Act, which, as long as it allows confidential rate setting, will haunt the Canadian market, creating the need for new regulations and bureaucracies, which, it might have been hoped, would have been a feature of the past.

The 1987 act made provision for four annual reviews of the results of the legislation (National Transportation Agency of Canada 1992), and, making allowances for the recessionary times and the harsh continental competition facing the carriers, these assessments have generally been positive regarding the impact of the new legislation.

Freer Trade

Freer trade, as agreed most recently under NAFTA, affects trade in goods, provision of services, and foreign investments (Johnson and Crowe 1992), and it is anticipated that transportation will be affected in all three areas. Four operating principles are of particular relevance to transportation. First, the most-favoured-nation principle implies that a benefit given to one nation must be extended to other trading partners. Second, the national treatment principle requires that each partner treat the other no less favourably than its own suppliers of goods and services. Third, under the idea of reciprocity, if U.S. carriers, for example, are allowed to operate in Canada then Canadian carriers must be allowed to operate in the United States. Finally, according to the idea of harmonization, both countries attempt to adopt the same rules. This is perhaps the most novel and far-reaching of the four principles in that it calls for adoption of identical transportation legislation in both countries and, in some respects, represents a surrendering of a degree of national autonomy.

While the primary aim of deregulation and trade agreement has been to provide a "level playing-field" for carriers in both the United States and Canada, this has not occurred. The primary grievances of Canadian carriers are the higher provincial and federal fuel taxes in this country and the accelerated capital cost allowances enjoyed by the American carriers. These grievances would appear just, and it would seem to be essential that they be addressed in the immediate future, before the Canadian trucking industry has been permanently crippled.

Some of the most interesting changes will affect the Auto pact (see also Holmes, this volume, chap. 13) since after 1997 Canadian companies will no longer be required to pay any duties on vehicles manufactured by U.S. companies. Restrictions on importing used vehicles were to be removed in 1993, and the threshold for investment reviews for direct acquisitions was raised from $5 million to $150 million in 1992. In the same year the threshold for indirect acquisitions was removed altogether.

North-South trade has increased as a result and is expected to grow at 6 to 7 per cent per annum, but O'Callaghan and Hartgen (1990) argue that the Canadian road network near the border is simply inadequate and in too poor repair to handle this growth. They therefore suggest that if Canadian and U.S. carriers are to take full advantage of NAFTA substantial investment will be required in the Canadian road system.

Impact of the GST

On 1 January, 1991, the federal sales tax (FST) on goods was replaced by the goods and service tax GST. Many carriers are "zero-rated," or, more correctly, provide services that are zero rated, including transportation to or from destinations outside the country. Transportation firms have generally benefited from lower taxes on their supplies and equipment but have suffered from increases in the cost of financial services. They are GST exempt, have not been allowed to claim input credits, and have recouped these costs by charging higher rates for their services. Perhaps most distressing, in an era of deregulation transportation companies are now faced with another level of bureaucracy and form filling. In the past they were not required to charge the FST for their services, but now they have to impose the GST. Those living on the periphery will now pay more for transportation services; the country just in effect became 7 per cent bigger.

Royal Commission

The Royal Commission on National Passenger Transportation, which submitted its final report in 1992, studied a number of issues, including travel patterns in Canada, costs of transportation and who pays them, effects of transportation on the environment, safety concerns for each mode, and accessibility for the physically challenged. Policy objectives and recommendations were to be guided by safety and environmental concerns; by fairness to taxpayers, travellers, and carriers (a euphemism for no subsidies); and by efficiency. The tenor of the commission's final report is succinctly stated in the following extract (Royal Commission on National Passenger Transportation 1992, Summary, 5): "While a passenger transportation system heavily subsidized by the tax payer may have been appropriate for Canada for the past 125 years, it is not the right one for Canada in the 21st century. Now, and in the decades ahead, Canada needs a system supported by the travellers who use it and not by government subsidies, departments and central controls. Passenger transportation should be treated more like a business."

To date, there has been little formal debate on the final report. An exception is the symposium held by the Van Horne Institute (1993), at which Lou Hyndman, chair of the royal commission, defended the report's recommendations. He suggested that media criticism to the effect that the report penalized the young, the old, and the poor had missed the mark. He argued that it was precisely these groups (who normally travelled by bus, the least subsidized of the modes) that had until now been subsidizing users of the other modes. Thus it was the young, the old, and the poor who now stood to gain the most from the report's recommendations.

Hyndman defended the report by noting that the "socialist" governments of France, Italy, and New Zealand, among others, have all implemented similar policies. When, during the first decades of the century, Britain enacted a series of welfare policies, one historian commented that in a sense all Britons were now socialists. It would appear that as the century draws to a close we are all conservative free-enterprisers.

The report eschewed use of transportation for nation building and regional development. This had been the motivation behind the great railway-building enterprises – the Intercolonial in Eastern Canada and the Canadian Pacific and other transcontinental lines to the west. It had even been the rationale for construction of the Trans-Canada Highway. Specifically, recommendation 4.2 states in unequivocal terms that governments should "pursue nation-building and regional development through other programs, rather than using the passenger transportation system." One commentator at the symposium argued that while nation building might no longer be an appropriate goal nation binding was still a worthwhile objective. In response it was suggested that in the next century this goal might be more readily achieved through electronic highways and use of telecommunications.

The key recommendation – that the user pay the full cost of passenger transportation and that there should be no hidden subsidies – is most likely to hasten the demise of the railways as a form of passenger transportation in Canada. Only high-density commuter transportation, such as that provided by the Ontario government's GO trains, is likely to prove cost effective. It would appear that user-pay is another example of the continuing grip of "Reaganomics" and as long as the Canadian public remains intimidated by the spectre of government deficits it is likely to be a popular recommendation, regardless of its effect on the environment and the safety of travellers. The railways' role in passenger transportation is already seriously eroded.

Though the royal commission's report dealt only with passenger transportation, its free-enterprise philosophy may well be extended to the freight sectors. A suggestion made at the Van Horne symposium was that the Canadian railway infrastructure should be taken over by a separate agency. Such a proposal has already been implemented in Sweden (Hansson and Nilsson 1991). Sweden's Transport Policy Act of 1982 proposed measures similar to those in Canada's NTA and the royal commission's report. Specifically, as Hansson and Nilsson (1991, 153) note, Swedish transport modes and carriers were to operate in a free market, and externalities such as environmental and safety effects were to be considered in all cost calculations. One significant difference was that public transport was to be supplied in an "adequate amount" in all parts of the country. A prominent result of this legislation was that the government separated the nationalized Swedish State Railways, Statens Jarnvagar (SJ), into two publicly held companies: Banverket (BV), responsible for the infrastructure, and a new SJ, responsible only for running traffic. Thus SJ and other companies could buy allocations of time on the railways. The aim was to reintroduce intra- and inter-modal competition. The reborn sj, however, is still in charge of allocating slots on the network, though there are appeal mechanisms to which competing companies can resort. Calculations of marginal costs are complex, especially where they relate to environmental and safety concerns, and at present these costs are not factored into road pricing, thus destroying the principle of a level playing-field between modes. Finally, the Swedish Board of Transport can still subsidize the system.

The solution of separating infrastructure from traffic generation seems attractive for Canada. It will allow for withdrawal of uneconomic lines and real competition within Canada's railway duopoly and possibly with other modes and indeed new railway competitors. Already both railway companies have reduced track (in 1992 CP Rail announced abandonment of its lines east of Sherbrooke, Quebec, thus ending its links from sea to sea – the national icon disappearing without a murmur). The two companies have, in addition, begun sharing track (Gibbon 1993) and reducing labour forces. Divestment of infrastructure seems to be the likely conclusion of this process.

New Trends

The final question to be addressed concerns new trends in transportation. Many of these emerging trends are based on the new information technologies. These new technologies may lead to increased dispersal of jobs away from urban cores, thus allowing for reduction in journey-to-work distances (Garreau 1991). Preliminary work on U.S. work-trip distances by Kumar (1990) supports this hypothesis. Telecommuting may reduce the number of work trips as well as their lengths. Gordon (1986, 43) suggested that by 1990 at least 10 per cent of the office labour force will be working at home or elsewhere offsite for two to four days a week – an overly optimistic prediction. Information on the extent of telecommuting and the socio-economic characteristics of those involved is based largely on anecdotal evidence. The extent of this phenomenon and its impact on travel patterns will have to await more detailed empirical studies (Arnold 1994).

Salomon (1986) provides a typology of the trade-offs between telecommunications and travel demand. He notes three types of interaction – substitution, complementarity, and enhancement. Substitution implies that the need for travel will diminish wherever telecommunications can be used as an alternative. Complementarity can be subdivided into two parts – enhancement effects, in which telecommunications stimulate additional travel between locations, and efficiency considerations, in which the efficiency of transportation systems is improved by reductions in capital and operating costs or improved safety and reduced pollution. Though Salomon is pessimistic concerning reduction in number of trips generated in a telecommunications-based economy (he feels that those journey-to-work trips that are reduced will be replaced by social and recreational trips) he does anticipate a reduction in the temporal concentration of trips resulting from modified travel patterns.

Gillespie and Williams (1988, 1317) discuss the nation-binding effects of telecommunications, though they note that the fact that these systems are usually international in scope acts as a countervailing force. They also state that telecommunications do not reduce the effect of friction of distance but rather remove the influence of distance altogether, concluding: "the denial of any such friction brings into question the very basis of the geography that we take for granted." They dismiss the "electronic highway" metaphor mentioned above, arguing that these telecommunications networks will be essentially proprietary and private and that universal access "would clearly undermine the rationale for their existence which is obviously to gain competitive advantage over others." However, the current Canadian and global philosophy of having the user pay for use of transportation infrastructure would appear to make the highway metaphor perfectly appropriate. Nevertheless, the argument that location on a peripheral node on a telecommunications network carries no locational disadvantage (Gillespie and Williams 1988) is valid and may have a significant effect on the hinterland economies of Canada and on Canada's economy within NAFTA.

DISCUSSION AND CONCLUSION

At the beginning of this chapter we noted several themes in the development of Canada's transportation systems, including governments' financial involvement, the intimate relationship between national policy and strategic importance, the need for redundancy in provision of transportation services, and reinforcement of existing routes when transportation innovations required development of additional infrastructure. We now consider each theme in turn and its probable importance in future.

Government involvement will be reduced if the recommendations of the Final Report of the Royal Commission on National Passenger Transportation (1992) are implemented, as seems likely. However, in an era of global competition and soaring government deficits, can national and provincial governments completely ignore the social ramifications of a purely free-enterprise system? It is improbable, for example, that governments will be totally unresponsive to the transport demands of dispersed rural populations. Demands by various groups that are physically and mentally challenged will also keep the government involved in both subsidization and regulation. Recent U.S. legislation has required provision of transit services for the physically challenged. While similar legislation may not be passed in Canada, many municipalities are likely to provide on a voluntary basis new forms of transportation, such as specially adapted buses, that will meet the needs of such groups. Indeed the Final Report (1992, Summary, 14) states that the physically and mentally challenged should be provided with transportation opportunities similar to those enjoyed by all Canadians.

National and strategic issues are less likely to involve territorial claims in the north and more likely to reflect the need for strategic industries to support a healthy economy, which end can best be achieved through development of high-tech industries. This will include support of the manufacturing of sophisticated transportation vehicles and infrastructure and development of related, information-based transportation systems. The latter include research into Intelligent Vehicle Highway Systems, Automatic Vehicle

Monitoring systems, and use of Transportation Geographic Information Systems (GIS-TS). Significantly, GIS-TS were ignored in the royal commission's report. In defending this omission at the Van Horne symposium, Lou Hyndman stated that computer-based systems were not on the horizon in 1993 – an ignorant oversight. Such systems are now at least a decade old, and some would suggest that their roots go back almost four decades (Waters 1992).

The need for redundancy in Canada's transportation networks is still a primary concern. Many transportations systems rely on routes and infrastructure that have no redundancy, and if one link is removed there is simply no alternative. A link may be damaged by accident, as were bridges in the St Lawrence Seaway system and in Vancouver in recent years. In such cases an entire network may cease to function. A transportation system may also be rendered inoperative by industrial action. Redundancy is today an expensive option and, in an age when cost-efficiency is the rule, an unlikely goal. Moreover, this is a time of relative industrial peace, of weakened unions, when the threat of permanent job loss makes strikes and other forms of industrial action unlikely (in 1980 there were 9.1 million days of work lost due to industrial action; in 1985, 3.1 million days; in 1986, 7.2 million; in 1990, 3.5 million; in 1991, 1.5 million; and in 1992, 1.2 million; and the figure is expected to continue to drop). Indeed, some transportation systems are moving to fewer and fewer players, thus greatly reducing competition and redundancy. In 1987, at the annual meeting of the Canadian Transportation Research Forum, Kevin Peterson, then a senior executive of Pacific Western, showed his "merger-a-month" slide, which portrayed his firm swallowing up numerous competitors in the airline industry. There has been talk of just one Canadian airline, or perhaps none. Redundancy and choice may well be passé (Skene 1994).

Reinforcement of existing routes may once again occur, with introduction of high-speed rail links. These have been proposed for the Quebec City–Montreal–Kingston–Toronto–London–Windsor axis with a possible link to Ottawa (see Paquette 1991; Soberman 1991) and also for Alberta (Smith and Johnson 1978, 125–44). This trend still appears to hold. Only the most populous urban corridors, which already have a variety of transportation links, are likely to benefit from this sophisticated and expensive new technology.

Owram (1992, 4), an historian, notes: "Geography, then, has created particular difficulties for the development of an efficient Canadian transportation system." This situation has affected the public psyche in a deep and enduring manner, conditioning Canadians to accept the need for government subsidy in both construction and maintenance of transportation. Canadian transportation has therefore been seen as a public good, which should not be subject to the vagaries of profit-loss economics. Finally, though most concern has been shown over railway passenger travel, this disquiet could spread to other forms of transportation. What Owram fails to emphasize strongly enough is that this unease has been prevalent throughout the history of Canadian transportation and that it is a deeply geographical problem, which has different effects on the various parts of Canada. The history of the Crow Rate for the transportation of prairie grain is one example of transportation subsidies provided to protect the interests of one region against those of another. Today there is unease in the air transportation industry, and again market forces and national and regional interests appear to be countervailing forces, diametrically opposed to each other. That deregulation will overcome all our difficulties is unlikely, as Dempsey and Goetz (1992) have convincingly shown for U.S. airlines.

In the present economic environment, efficiency and competitiveness, especially global competitiveness, are pre-eminent. But removal of all transportation subsidies, as so stridently and uncompromisingly advocated by the Royal Commission on National Passenger Transportation (1992), will have differing effects across Canada. Densely populated areas will benefit at the expense of sparsely settled regions. Cities will gain, and rural areas will lose. Efficiency will rule, and issues of equity will fall by the wayside. Though this is not new in the annals of Canadian transportation history, governments have tended to fight a rearguard action against these geographical imperatives. What is new in the 1990s is the total capitulation of government policy to market forces. The geographical impact of such policies is likely to be arbitrary and unfair. In the terminology of the politically correct, Canada appears to be entering the era of geographically challenged regions.

NOTES

1 We wish to acknowledge the assistance of Louise Arnold in drawing our attention to the literature on the geographical impact of telecommuting and information technologies.

2 This is noted by Owram (1992), who demonstrates its long-standing importance when he cites the work of MacGibbon (1917): "Transportation policy has not been viewed merely from the standpoint of commercial and economic convenience. In the final outcome these national ideas have dictated the course of development."

3 The details of the initial arrangements for cost sharing for highway construction between the two levels of government were laid out in the Canada Highways Act of 1919.

4 American innovation provided the precedent for the new policy, and by the outbreak of the First World War many provinces had adopted vehicle registration and licensing fees. Fuel taxes were first introduced in the United States by Oregon in 1919. Alberta established them in 1922, and by 1929 all provinces had imposed this tax.

REFERENCES

Arnold, L.J. 1994. "Teleworking: Geographical, Organisational and Social Implications." MSc thesis, Department of Geography, University of Calgary.

Berton, P. 1971. *The Last Spike: The Great Railway 1881–1885.* Toronto: McClelland and Stewart.

David, P.A. 1969. "Transport Innovation and Economic Growth: Professor Fogel on and off the Rails." *Economic History Review,* 2nd series, 22, 506–25.

Davis, L. 1966. "Professor Fogel and the New Economic History." *Economic History Review,* 2nd series, 19, 657–63.

Dempsey, P.S., and Goetz, A.R. 1992. *Airline Deregulation and Laissez Faire Mythology.* New York: Quorum Books.

Flink, J.J. 1975. *The Car Culture.* Cambridge, Mass.: MIT Press.

– 1988. *The Automobile Age.* Cambridge, Mass.: MIT Press.

Fogel, R.W. 1964. *Railroads and American Economic Growth: Essays in Econometric History.* Baltimore: Johns Hopkins University Press.

Garreau, J. 1991. *Edge City: Life on the New Frontier.* New York: Doubleday.

Gibbon, A. 1993. "CN Faces $1 Billion Loss." *Globe and Mail*, 2 March, B1.

Gillespie, A., and Williams, H. 1988. "Telecommunications and the Reconstruction of Regional Comparative Advantage." *Environment and Planning A*, 20, 1311–21.

Goldenberg, S. 1994. *Troubled Skies: Crisis, Competition and Control in Canada's Airline Industry.* Whitby, Ont.: McGraw-Hill Ryerson.

Gordon, G.E. 1986. "Telecommuting: Planning for a New Work Environment." *Journal of Information Systems Management*, 3 no. 3, 37–44.

Guillet, E.C. 1933. *Early Life in Upper Canada.* Toronto: University of Toronto Press.

Hansson, L., and Nilsson, J.-E. 1991. "A New Swedish Railroad Policy: Separation of Infrastructure and Traffic Production." *Transportation Research A*, 25A, 153–9.

Harvey, D. 1989. *The Condition of Postmodernity.* Oxford: Basil Blackwell.

Heaver, T.D. 1985. *A Synopsis of Views and Issues Raised at the Transportation Policy Forum on Freedom to Move – Framework for Transportation Reform.* Centre for Transportation Studies, University of British Columbia, Vancouver.

Innis, H.A. 1956. "Transportation as a Factor in Canadian Economic History." In M.Q. Innis, ed., *Essays in Canadian Economic History*, 62–77. Toronto: University of Toronto Press.

Institute for Transportation Studies (ITS). 1990. "Proceedings of the Alberta Symposium on Future Intercity Passenger Transportation." Institute for Transportation Studies, University of Calgary, Calgary, Alberta.

Jackman, W.T. 1935. *Economic Principles of Transportation.* Toronto: University of Toronto Press.

Johnson, J., and Crowe, D. 1992. "The Effect of Free Trade on Transportation." In R. Lande, ed., *National Transportation Policy*, 155–70. Toronto: Butterworths.

Kansky, K.J. 1963. *Structure of Transport Networks: Relationships between Network Geometry and Regional Characteristics.* Research Paper No. 84, Department of Geography, University of Chicago, Chicago.

Klineman, R. 1992. *The Impact of the Goods and Services Tax on Transportation.* In R. Lande, eds., *National Transportation Policy*, 171–90. Toronto: Butterworths.

Kumar, A. 1990. "Impact of Technological Developments on Urban Form and Travel Behaviour." *Regional Studies*, 24, 137–48.

Lande, R., ed. 1992. *National Transportation Policy.* Toronto: Butterworths.

Lardner, D. 1855. *Railway Economy: A Treatise on the New Art of Transport.* New York. Reprinted 1968 by Augustus M. Kelly Publishing.

MacGibbon, D.A. 1917. *Railway Rates and the Canadian Railway Commission.* Boston: Houghton Mifflin.

Mackintosh, W.A. 1923. "Economic Factors in Canadian History." *Canadian Historical Review*, 4, 1–25.

National Transportation Agency of Canada. 1992. *Annual Review, 1991*. Ottawa: Ministry of Supply and Services.

Nerlove, M. 1966. "Railroads and American Economic Growth." *Journal of Economic History*, 26, 107–15.

O'Callaghan, J.E., and Hartgen, D.T. 1990. *Transportation Impacts of the US-Canada Free Trade Agreement*. Department of Geography, University of North Carolina, Charlotte, North Carolina.

Owram, D.R. 1992. "Icons and Albatrosses: Passenger Transportation as Policy and Symbol in Canada." In Royal Commission on National Passenger Transportation (1992), vol. 3, 1–163.

Paquette, J.R. 1991. "Prospects for High Speed Rail in Canada." In *Evolution in Transportation*, Proceedings of the 26th Annual Meeting of the Canadian Transportation Research Forum, Quebec City, 28–31 May, 1991, 1–14.

Rostow, W.W. 1960. *The Stages of Economic Growth*. Cambridge: Cambridge University Press.

Royal Commission on National Passenger Transportation. 1992. *Directions*, Final Report of the Royal Commission on National Passenger Transportation, 4 vols. Ottawa: Minister of Supply and Services.

Salomon, I. 1986. "Telecommunications and Travel Relationships: A Review." *Transportation Research A*, 20A, 223–38.

Sampson, R.J., Farris, M.T., and Shrock, D.L. 1990. *Domestic Transportation: Practice, Theory and Policy*. 6th edn. Boston: Houghton Mifflin.

Skene, W. 1994. *Turbulence: How Deregulation Destroyed Canada's Airlines*. Vancouver: Douglas and McIntyre.

Smith, P.J., and Johnson, D.B. 1978. "The Edmonton Calgary Corridor." Department of Geography, University of Alberta, Edmonton, Alberta.

Soberman, R.M. 1991. "High Speed Rail and the Real World." In *Evolution in Transportation*, Proceedings of the 26th Annual Meeting of the Canadian Transportation Research Forum, Quebec City, 28–31 May 1991, 15–26.

Taylor, G.R. 1965. Review of Fogel (1964). *American Economic Review*, 55, 890–2.

Trout, J.M., and Trout, E. 1871. *The Railways of Canada*. Toronto: Monetary Times.

Vance, J.E. 1990. *Capturing the Horizon: The Historical Geography of Transportation since the Sixteenth Century*. Baltimore: Johns Hopkins University Press.

Van Horne Institute. 1993. "A Symposium on the Final Report and Recommendations of the Royal Commission on National Passenger Transportation." University of Calgary, Calgary, 26 Feb. 1993.

Waters, N.M. 1973. "The Southern Ontario Railway System: A Network Analysis." MA thesis, Department of Geography, University of Western Ontario, London, Ontario.

– 1992. "Geographic Information Systems in Transportation." *Proceedings of the Transportation Forum*, 6, 112–29.

Whebell, C.F.J. 1969. "Corridors: A Theory of Urban Systems." *Annals, Association of American Geographers*, 59, 1–26.

CHAPTER SEVENTEEN

The Canadian Market and Activities Oriented to Consumers

JAMES W. SIMMONS

While the Canadian economy as a whole continues to rely heavily on trade with foreign markets, a growing number of workers and locations are only indirectly affected by this activity. As in most modern economies, the share of jobs in primary industries and manufacturing has declined over the years: most Canadians now work for the domestic market, and much of their output derives from trade and service activities. This chapter examines the size, spatial distribution, and patterns of change in the Canadian market and then looks at the sectors that directly linked to it: retail, wholesale, consumer services, and finance. While the overwhelming direction of causality runs from changes in the market to changes in distribution activity, the share of services within the economy is now so great that even slight shifts in the relationship between consumers and stores have substantial implications for growth of cities and organization of the urban system. Cities attract consumer activities; consumer activities build cities.

At the same time this huge component of the Canadian economy displays as much concentration of ownership and oligopolistic behaviour as any manufacturing sector. The great department store and supermarket chains are among the largest enterprises in the nation, and the firms that control distribution of automobiles and gasoline are among the largest corporations in the world. Studies of markets and distribution activities in aggregate are a central theme in economic geography, but studies of the location decisions and business strategy of individual enterprises have barely begun.

About 84 per cent of the Canadian GDP goes to personal income; about 33 per cent of personal income goes to retail sales, another 13 per cent goes to commercial (as opposed to community) services, and about 9 per cent is invested in the financial sector. Consumers' decisions are significant in magnitude, but they are even more important in qualitative terms: they specify the requirements of an important component of economic growth. Are households in a mood to save or spend? Do they value convenience or variety and selection? Will they buy cars or clothes, travel or life insurance? And will they choose the big modern shopping plaza or the fancy boutiques? Downtown or the suburban strips? These decisions affect the economy, the industrial structure, and the landscape.

The first section of this chapter looks at the historical trend of relationships between distribution activities and their determinants for the nation as a whole. The second examines the spatial distribution of retail, wholesale, commercial service, and financial activities. Finally, some projections of the market permit speculation about the future of these activities.

THE CANADIAN MARKET

Growth

The Canadian market has temporal, spatial, and social class components, and each one deserves some discussion. The magnitude of the market is measured in market income, defined as the product of population times income per capita. This joint formulation is preferred because the proportion of income allocated to retail sales usually declines as the level of income per capita increases, since the share devoted to taxes, savings, and certain services increases.

The population of Canada has grown relatively smoothly since 1930 (Figure 1), increasing to 2.5 times its size then. Growth was slow for the first two decades, accelerated in the 1950s and 1960s, and has again slowed. As is shown later, population growth is projected to cease by the year 2025. The most important aspect of population growth is the spatial redistribution of economic activity that accompanies it, and the location of population growth depends on the source of growth. Both natural increase (births – deaths) and net immigration (in-migrants – out-migrants) have declined in size relative to net internal migration (relocation of Canadian residents), as measured by interprovincial movement. Each growth component displays a different map. The rate of natural increase is almost uniform across the country. Immigration is the most sensitive to economic growth (Ontario, British Columbia), with internal migration intermediate. The latter is highly responsive to short-distance variations in economic growth but less sensitive to regional differences.

The changes over time in income per capita (Figure 1) have contributed twice as much as population growth to the expansion of the Canadian market. All locations, all social groups, and all economic sectors have participated in this five-fold increase since 1930. The spatial variance in growth, at least at this time scale, is much smaller than for population growth, however. The most notable spatial pattern has been steady convergence of levels of personal income per capita across the country. Restructuring of the rural economy, eliminating many of the discrepancies between rural and urban communities, and redistribution of public-sector funds by equalization payments and personal transfers, have removed much of the variation in disposable income per capita among regions.

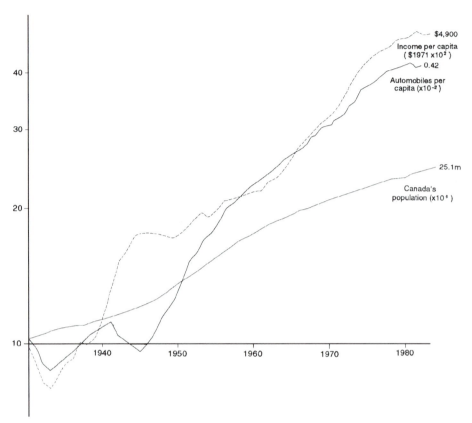

Figure 1
Changes in demand, Canada, 1930–82.

The third item on Figure 1 is a measure of personal mobility: the number of auto-mobiles per capita. The growth of this variable tracks the growth of income per cap-ita, except for the 1940s when the Second World War generated high incomes with no consumer goods. Automobiles per capita is also a surrogate measure for improve-ments in the highway system and widespread reduction in travel time. This variable reflects the greater distances consumers have been able to travel, thus weakening spa-tial ties between place of residence and shopping locations. The result has been a series of restructurings of retail activity – away from the smallest settlements and ageing downtowns toward shopping plazas and retail strips in suburbs of larger cities.

The response over time of the distribution and financial system to all these changes has been growth in employment in retail and wholesale trade, private-sector services, and the financial sector. For the most part these sectors grow at about the same rate – somewhat faster than the total employment – though each sector leads or lags for certain decades. Even with this rapid growth, expansion of employment in these sectors lags behind increases in market income, because of improvements in efficiency over the years resulting from such innovations as self-service, larger stores, automated tellers, and computers. Even so, consumer-oriented activities have increased their share of total employment from 22 per cent to 33 per cent.

Spatial Structure

Distribution activities reach into every community in the country. Even hamlets have gas stations, branch banks, barber shops. Because of the close relationship between the location of these activities and the market, changes in market size and location inevitably require a response by consumer-linked activities.

Three-quarters of the Canadian market is dispersed among the largest 100 urban nodes (Figure 2). The two largest places, Toronto and Montreal, account for one-quarter, and the ten largest, one-half. These nodes are not distributed evenly across the nation but occur in three regions of differing density. The corridor stretching from Windsor to Quebec City includes over 50 per cent of the market in only 2 per cent of the area. In the rest of the ecumene (the settled region) – making up another 12 per cent of the country's area – cities occur at about one-tenth the density of the corridor. Less than one per cent of the Canadian market is found in the 85 per cent of the land area beyond the limits of the ecumene.

Operation of the distribution system is made difficult by the enormous distances among these market concentrations, as well as regional fragmentation. Vancouver and St John's are almost 8,000 km apart: the variance in intercity distance extends over three orders of magnitude, as great as the variation in city size. The distance effects are exacerbated by the physical barriers – the Rockies, the Canadian Shield, the Appalachians, and the ocean inlets of the Atlantic provinces – as well as the language barrier that divides the corridor into two distinct markets for many kinds of goods. The market inevitably imposes a strong regional structure on the organization of distribution activities. At the same time the map reinforces the illusion that Canada is a distinct market, isolated by the national boundary. In fact, this border is less and less effective, with respect to distribution systems, retail chains, and even consumers' behaviour.

Over time this spatial market has changed, just as the overall market has grown. First, the range of urban market sizes has been continually extended as the largest places continue to grow. Cities of all sizes have grown at about the same average rate, except for the very smallest centres (with population of less than 2,000). Even there, the faster growth in income per capita in rural areas partly compensates for the slow growth of population. Much of the variation in market growth over time occurs regionally. From one decade to the next, first one region, then another, leads or lags:

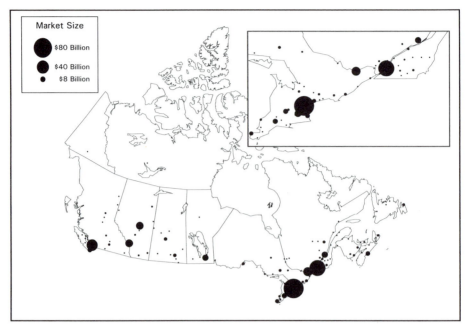

Figure 2
The 100 largest markets in Canada. *Source*: Statistics Canada, *Census of Canada, 1986*.

Ontario in the 1960s, Alberta through to the early 1980s, then Ontario again, and now the west coast. The shifts are dramatic in magnitude and difficult to project. Since 1960 every Canadian province has changed from positive net migration to negative, or vice versa (Simmons 1983). From year to year spatial variations in income per capita shift even more rapidly than population. As a result, retail sales in a small, peripheral city with a resource economy may change as much as 20 per cent from one year to the next.

Because of the size of the largest metropolitan markets, the most important market geography for many retailers lies within the city itself. Toronto and Montreal are each bigger than all the Atlantic provinces together, or any one of the western provinces. A single store in one of these cities can potentially reach three million customers. In the Toronto market (Figure 3a) about 30 per cent of the census metropolitan area (CMA) lives within five miles of downtown and another 30 per cent within ten miles. The variation in income per capita (Figure 3b) is important because the level of consumption is directly related to household income. In 1986, households in the wealthiest census tract in Toronto recorded average incomes of $83,000 – and average expenditures of $48,000. In the poorest tract, households received an average of $7,500 and spent $8,700.

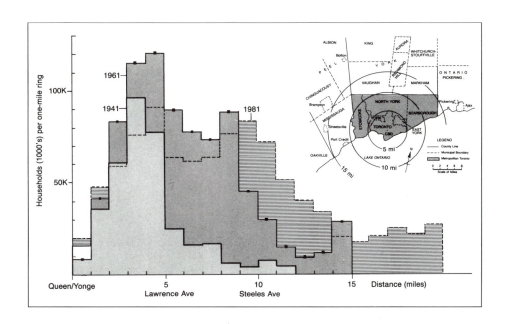

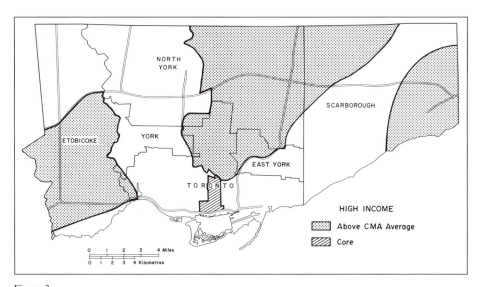

Figure 3
The Toronto market: (a) (top) Households. *Source*: Jones and Simmons (1987).
(b) (bottom) Household income. *Source*: Simmons (1991).

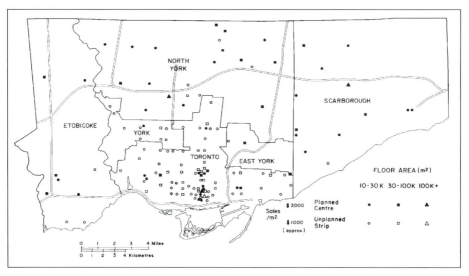

Figure 4
Retail districts in Toronto. *Source*: Simmons (1991).

Within the urban market surface that is shaped by the density of settlement and the income of households, there are nodes of consumption activity analogous to the concentrations of urban settlements across the country as a whole (Figure 4). These retail and service clusters vary in size log-normally, just like the distribution of market size within the national system; but the segmentation of business functions within the distribution system at the metropolitan scale is not related only to size. Because distances are much shorter, customers move easily among a number of nodes, many of which specialize in distinctive goods and services (see Simmons 1991). This is especially true of the high-access locations near the city centre, which have disproportionately more retail activity than the older suburban locations. High-income areas have fewer, larger shopping areas; blue-collar districts have numerous small centres.

The Future

The trends in the domestic market are important aspects of the future of the Canadian economy. While the magnitude and spatial distribution of the market are shaped by external trade and relative sectoral growth, the spatial distribution of the existing market and the processes that govern its rate of change act as powerful constraints on the response of the distribution system. The main elements of the domestic market have their own dynamics and directions of change, which may oppose external pressures or internal policies.

Table 1
The market of the future

(a) National projections

	1986	2001	2011
Population (million) @ current rate*	25.3	28.7	30.2
Minimum	–	28.3	29.0
Maximum	–	29.6	32.3
Household income ($)	35,700	50,000	70,000

* Continued fertility rate of 1.7 child per female; 140,000 immigrants per year.

(b) Provincial population projections for 2011

	Population (000s)									
	Nfld	PEI	NS	NB	Que.	Ont.	Man.	Sask.	Alta	BC
1986	568	127	873	709	6,532	9,102	1,063	1,010	2,366	2,883
Minimum	540	131	931	707	7,032	11,092	1,186	988	2,522	3,274
Maximum	614	153	1,047	819	8,185	12,468	1,355	1,266	3,324	4,004
Mean	578	143	991	761	7,530	11,726	1,261	1,137	2,870	3,600
Range	74	22	116	112	1,153	1,376	169	278	802	730
Range/mean	0.13	0.15	0.12	0.15	0.15	0.12	0.13	0.24	0.28	0.20

Source: Statistics Canada, Population Projections for Canada and the Provinces, Cat. No. 91-520 (1989).

The growth of the overall population is the most predictable element of all. Barring dramatic shifts in immigration policy or fertility rates, Canada can expect a population of about 28.7 million in 2001, and 30.2 million in 2011 (Table 1a), with a well-documented ageing sequence that shifts the share of population aged over 65 from 10 per cent at present to 13 per cent in 2001 and 15 per cent in 2011 (see Sargent 1986).

The projection of economic growth is considerably more uncertain. While the size of the labour force can be anticipated with some accuracy (from population projections), the changes in productivity per capita have proven almost impossible to comprehend ex post, let alone to project with any confidence. The growth in output will depend on such vagaries as the nature of the growth sectors in the economy, which depend in turn on the relative prices of primary commodities and Canada's competitiveness relative to other manufacturing and service-producing nations. The econometric forecasts (Sargent 1986) are based largely on averages over the last decade. Nonetheless they serve to bound the alternative futures. The median projection of 3.2 per cent annual growth in (constant dollars) GDP per year to 1995, dropping to 2.8 per cent thereafter, will increase income per capita by roughly 50 per cent by 2001 and double it by 2011. If past trends continue, consumer activities will both grow and adjust to this change.

But what happens to consumption if incomes increase? Over the years a considerable body of information has been amassed concerning the consumption patterns of

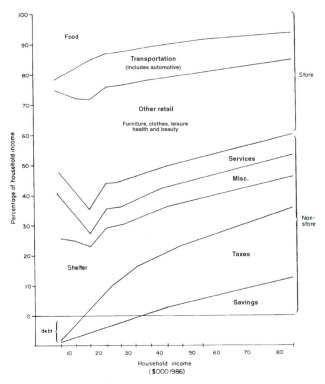

Figure 5
Expenditure patterns by income groups, Canada.

Canadians, and especially the income effects, as summarized in Figure 5. In 1986 the average household income was $35,700; by 2001 it should be the equivalent of $50,000; by 2011, $70,000 in 1986 dollars. By far the most significant effect is the increase in magnitude of retail, service, and financial activity of all kinds. A family income of $22,000 generates $19,800 expenditures and consumes $300 in savings, whereas a household income of $42,000 leads to 1.56 times as much spending and $1,200 in savings. Income growth benefits all distribution sectors, but especially those retail and service sectors that are income-elastic, such as the financial sector.

If per capita incomes increase as projected, the pattern of expenditures should shift toward the right-hand side of the diagram. The level of saving and investment should also increase sharply, and government expenditures should grow as well, as taxes increase. Most of these shifts in the share of expenditure will be achieved at the expense of the housing and food sectors, where the poor spend most of their money. The share of income spent on private transportation will not change greatly, and there may be a modest decline in the share allocated to clothes, furniture, and recreation. Still, most of these latter variations will be unnoticed in the general increase in consumer spending. As households spend more, they will also become more mobile, travelling farther for both recreation and shopping, with continued intensification of specialization resulting in the retail and service roles of different locations, at both the urban system and metropolitan scales.

The major uncertainty revolves around changes in the location of this market. If the federal government maintains its capacity to redistribute income spatially, changes in the spatial pattern of employment or income per capita may not be as significant as the redistribution of the population itself. Part b of Table 1 examines the range of possible population growth rates, based on Statistics Canada projections using the 1986 Census of Canada. The greatest uncertainty occurs in the western provinces, where relatively rapid population growth is possible. At the same time dramatic population decline is extremely unlikely anywhere. Ontario and Quebec grow at a reasonably stable rate, because of their size and complexity (and because no other province could absorb or generate the large number of migrants involved).

To summarize, a number of factors constrain the future development of the space economy and make it more predictable than it has been in the recent past. First, if we treat the space economy as a market rather than as a production system, our concern shifts from value of production to disposable income. The latter varies far less geographically (and temporally) than the former. Second, as the overall rate of population growth declines, the number of persons in the highly mobile age groups, 15–25, declines as well. Thus the economic signals that lead to internal migration are reduced, along with the number of people likely to respond to those signals. The result is a more stable and predictable spatial market.

THE DISTRIBUTION SYSTEM

The Magnitude of Consumer-Linked Activity

By 1986 the four distribution sectors included 4.6 million workers, about 40 per cent of the labour force (Table 2). In order of size, commercial services lead, followed by the retail, finance, and wholesale sectors, in that order; the rate of growth of these activities can be compared to the national growth rate of employment, which was 15.5 per cent. Unfortunately, these activities are the least satisfactorily covered sectors in our national statistical system, so it is difficult to describe the operations in a comparable fashion. Distribution of both firm and establishment size is highly skewed for these activities: median values of outlet size are considerably smaller. From Figure 5, each billion dollars in household income (corresponding to a city of 80,000 people) should generate $500 million in retail sales, 850 retail outlets, and 5,000 retail employees. For commercial services, the numbers are $185 million, 700 outlets, and 7,000 workers, respectively.

Table 2
The magnitude of activities serving the consumer

Sector	Firms* (1984)	Sales* (1984)	Employees[†] (1986)	Sales/firm (1984)	Sales/ employee	Employment[†] growth (1981–6)
Retail	186,300	$133.8	1,529,000	$718,000	$87,500	13.4%
Wholesale	59,400	141.7	568,000	2,386,000	249,500	4.6
Finance	51,100[‡]	–	668,000	–	–	8.8
Commercial services	187,100	48.9	2,082,000	261,000	23,500	16.8
Total	483,900	n.a.	4,847,000	n.a.	n.a.	15.5

* Statistics Canada, *Small Business in Canada*, Cat. No. 61-231, annual. There are an additional 30,000 branch outlets in the retail sector. Figures for sales are given in $billions.
[†] Statistics Canada, *Census of Canada, 1986*. This is a broad definition of employment, counting both part-time and short-term workers.
[‡] These data from Statistics Canada, *Small Business in Canada*, include only real estate and insurance agents. There are another 12,000 banks, trust companies, credit unions, and stockbrokers, with another 10,000 branch locations (Statistics Canada, Cat. No. 61-207).

At the establishment level (as distinguished from the firm) distribution activities are distinguished by their small size and geographical dispersion relative to other economic activities (except farms). The markets they serve are typically very limited in extent, though it is shown below that the small establishments are often linked to-gether into chains and franchise operations of hundreds and even thousands of outlets. The orientation to the consumer is the rationale for this decentralization, which is evi-dent even within the largest market concentrations. Retail margins are high, averaging one-third of the cost of goods purchased, as scale economies compete against the con-sumer's desire for convenience, service, or prestige. Many would say that the retail and service sectors, at least, suffer from severe over-capacity – because there are so few barriers to entry. In fact, neither consumers nor retailers are particularly rational, at least from the point of view of economic logic. This is the world of marketing hype and business strategy, of "selling the sizzle," of the irrationalities of Christmas and Campeau.

Relationships to Markets

The small establishments that dominate these economic sectors adapt easily to small, localized markets. The close relationship between the spatial distributions of markets and the activities that serve them is the most fundamental element in the theory of hu-man geography: it is assumed that knowledge of the size of the market leads inexora-bly to knowledge of the pattern of retail and service activity. This relationship is the starting point for this discussion, as shown in equation 1:

$$\text{distribution activity} = a \, (\text{market size})^b, \tag{1}$$

but it can be specified in many ways, and at various spatial scales. In Table 3a the relationship is fitted at the urban-system scale for the different components of the distribution sectors, using the specification:

$$\text{employment} = a\ (\text{population})^{b_1}\ (\text{income/capita})^{b_2} \qquad (2)$$

At this spatial scale – for 139 urban centres with 10,000 people or more – the fit between employment and the market is very close indeed with an R^2 value of over 0.9 in every instance.

Of particular interest are the two parameters that identify effects of the market size in terms of population (b_1) and income per capita (b_2). Central place theory holds that certain kinds of consumer activities will be more centralized than others: in this case employment in distribution activity increases disproportionately as population increases for all sectors except retail, and very strongly in financial and business services. Some activities are more responsive to higher incomes, as well – financial activities, for example; while others, such as wholesale, may even be negatively related to income.

In the absence of a recent census of business, it is difficult to extend these relationships to other aspects of distribution activity, such as number of stores, sales, or floor area. Nonetheless, there is consistency in the interrelationships among these variables and with the market (Table 3b). Market researchers depend on this consistency: if the number of households and households' income are known, one can work out how many shoes they buy, plus the number of shoe clerks and the floor area to serve them.

Finally, the dynamics of the market relationship – the sensitivity of distribution activity to different kinds of market changes – is captured in the relationship:

$$\text{employment growth rate} = a + b_1\ (\text{population growth rate}) + b_2$$
$$(\text{income/capita growth rate}) + b_3\ (\text{log population}). \qquad (3)$$

First (Table 3c), there is a disproportionate response to population growth ($b_1 > 1.0$), indicating the competitive advantage that a growing city achieves with respect to its neighbours. If Calgary grows quickly and attracts new kinds of business, Winnipeg becomes relatively less attractive to consumers and ultimately loses activities as well. Second, the less-than-proportionate response to income growth reflects the tendency for higher-income households to devote their income to non-store consumption, taxes, and savings. The values of b_3, the coefficient intended to measure hierarchical change, are varied. Most retail activities continue to disperse *down* the urban hierarchy. Service and wholesale activities, however, have grown more rapidly in larger centers, so that in aggregate the total distribution activity has remained just about constant with respect to the urban hierarchy. If the service sector continues to grow more rapidly than the retail sector, a long-standing trend favouring higher growth of distribution activity in smaller places will be reversed. This reversal is confirmed in other studies in Canada (Coffey and Polèse 1988) and the United States (O hUallachain and Reid 1991).

Table 3
Distribution activity and markets

(a) Sectoral relationships, 1986 (139 CMAs and CAs)

Dependent variables	Parameters for independent variables			
	Intercept	Population	Income/capita	R^2
Retail employment	-0.554	0.997	0.332	0.986
Wholesale employment	0.513	1.185	-0.207	0.909
Recreation	-4.789	0.986	1.316	0.963
Personal	-1.952	1.050	0.530	0.981
Financial employment	-4.662	1.183	1.142	0.961
Total distribution	-1.536	1.056	0.639	0.984

(b) Other aspects of the retail hierarchy, 1971 (124 urban-centred regions)

Market income = population * income/capita (by definition)	
Sales by retail chains = 0.149 x (market income) $^{0.920}$	$r^2 = 0.884$.
Retail chain stores = 1.012 x (chain sales, in $million) $^{0.914}$	$r^2 = 0.937$.
Retail employment = 38.8 x (chain sales, in $million) $^{0.927}$	$r^2 = 0.884$ (all stores).
No. of business types = A $(\log_{10}$ retail stores$)^B$ (not calibrated)	

(c) Changes over time, 1971–81 (124 urban-centred regions)

Dependent variables	Parameters for independent variables			
	% growth population	% growth Income/capita	Log_{10} population	R^2
Growth rate of retail employment	1.387	0.386	-0.037	0.692
Growth rate of service employment	1.750	0.102	0.189	0.720
Growth rate of wholesale employment	1.577	0.403	0.022	0.331
Growth rate of total employment	1.558	0.269	-0.020	0.761

Within the metropolitan area the relationship between the market and distribution facilities is more complex (Jones and Simmons 1993). Different retail and service clusters can serve trade areas of different size; specialization is possible; consumers will travel long distances for some things, but not for others. Retail analysts have devoted considerable effort to estimating the size of trade area that a given retail facility can attract. For convenience stores, up to the size of supermarkets, the most appropriate approach is that of the Thiessen polygon: a store serves the part of the market to which it is closer than a competing store. For retail clusters and plazas providing shopping goods the Huff model is widely used. It simply assigns shoppers to competing centres in proportion to the size of the centre and the distance from the consumer. But as Jones (1984) has argued, stores and commercial clusters within the metropolitan region are becoming increasingly specialized, drawing their clientele

Table 4
Concentration in retail activity

(a) Retail chains and department stores, 1986*

	No. of chains		No. of outlets		Sales	
Chain size	No.	%	No.	%	$billion	%
4–9 outlets	769	60.7	5,442	12.0	5,442	9.3
10–49 outlets	353	27.9	6,904	15.2	9,483	16.3
50–99 outlets	75	5.9	8,821	19.4	8,821	15.1
100+ outlets	70	5.5	24,207	53.3	34,530	59.3
Total chains	1,267	100.0	45,375	100.0	58,278	100.0

(b) Share of largest firms (%), 1984[†]

	Retail firms		Wholesale firms		Service firms	
Firm size (sales)	No.	Sales	No.	Sales	No.	Sales
Over $20 million	0.2	39.7	1.3	57.8	0.1	25.5
$2–20 million	3.9	27.5	13.0	28.5	1.2	20.4
$1–2 million	4.6	8.8	10.4	6.2	2.0	10.7
< $1 million	8.7	76.0	24.7	92.5	3.3	56.6

* Statistics Canada, *Retail Chains and Department Stores*, Cat. No. 63-210, annual.
[†] Statistics Canada, *Small Business in Canada: A Statistical Profile, 1984–1986*, Cat. No. 61-231, annual.

from a larger and larger region. Trade areas overlap, so that it is difficult even to isolate a spatial market within the city, let alone identify distribution activities that serve it.

Spatial Organization

Though there are hundreds of thousands of stores, service locations, and financial agencies, most of the activity (measured in sales) is handled by a few, multi-location firms, as indicated in Table 4. Over 40 per cent of all retail sales occur in retail chains, and over half of these sales in chains take place in branches of the 70 largest chains that each control over 100 outlets (Table 4a). In addition, many non-chain retail outlets are bound by franchise contracts to sell common products or operate in a common format. Control of wholesaling and commercial and financial services is also highly concentrated (Table 4b), with 10 per cent of distribution firms accounting for more than three-quarters of all activity.

This concentration of control has several implications for the space economy. First, because of the magnitude of economic activity that is involved in distribution, the spatial structure of multi-location outlets accounts for much of the content – the spatial linkages – of the urban system. Each firm's network generates flows of information, money, and products. Its choice of headquarters (overwhelmingly, Toronto)

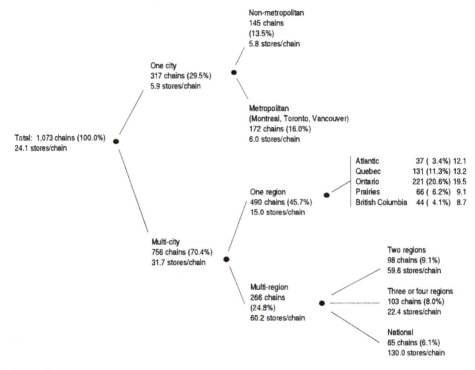

Figure 6
The spatial reach of retail chains, Canada. *Source*: Simmons and Speck (1988).

is related to the urban hierarchy, and the spatial structure of outlets is strongly regional. Together these firms weave a network of spatial relationships among urban
places.

In an analysis of the spatial reach of 1,073 Canadian retail chains (Figure 6), almost
30 per cent operated within a single urban area, and almost half of these were found
within the three largest metropolitan areas. Almost half the retail chains operated in
more than one city, but within the same broad region of the country, meaning that at
least 90 per cent of the outlets were located within the region. These chains were characteristically linked into a regional subsystem, centred on an urban area. There was
little evidence of specialization of chains according to city size within regions, though
some of the multi-region chains focused on the larger urban places. The national
chains were small in number but included a great many outlets and served many cities. They account for the majority of retail chain sales.

Outlets in other consumer-oriented sectors are linked together in the same way. Canadian banks have long shared a national market among a small number of very large
firms, while trust companies are linked into regional and metropolitan networks. A
market shared by a small number of large participants is increasingly characteristic of
distribution sectors in metropolitan markets, and it requires new kinds of theory and

method for analysis – business strategy, instead of neo-classical economics. Such firms expand, merge, specialize, and choose locations according to the logics developed in battles, games, and business schools (Laulajainen 1987). They face enormous uncertainty in their markets – in addition to market growth – because of changes in competitive structure; they may risk all to dominate a market, or to survive a competitor who has overexpanded.

Across the nation these chains now reach into almost every settlement, no matter how small or how remote. Within the metropolitan area, however, where considerable specialization has taken place among retail clusters, the penetration of retail and service chains varies considerably. They completely dominate the larger, planned centres that emerged in the 1960s, because they have formed comfortable alliances with the shopping centre developers that control large amounts of retail space in networks of malls. The older retail strips near the city centre and downtown areas are more likely to retain many independent retail and service activities.

The organization of distribution activities into large multi-location networks of facilities raises the possibility of weakening the relationships between facilities and markets in the future. Interesting trends are emerging; for example, the proportion of employees in these sectors who are actually in contact with customers is declining. Only 45 per cent of retail employees were in sales occupations in 1981, compared with 49 per cent in 1971. The rest of them are managers, typists, accountants, warehousemen, and so on. The major recent innovation in these activities has been computer-based inventory systems, which allow each branch outlet to be directly linked to the head office or warehouse, so that many facets of the operation can be controlled from afar. Half the chain's staff are essentially footloose: they can be located anywhere in the spatial system – in the store or at headquarters, or in some isolated data centre or warehouse in an obscure suburb or small town.

Directions of Change

The distribution sectors described here have been so closely linked to the size and distribution of markets that their futures depend largely on the evolution of the market as outlined above. Nonetheless there is considerable room for variation in the way that the market relationship is specified: the share allocated to different consumption sectors, to settlements of differing sizes, and to the various specialized retail areas within the cities. And there are many ways in which the industry itself can respond to the market, in level of service, store size, and so on – the production function can be changed. Since these decisions will largely reflect consumers' preferences, there is little consensus among retail analysts about directions of change. Nevertheless there are a few key trends, such as the decline in the share of income devoted to retail sales over time and the increased share of retail chains (Table 5). The only thing that can derail the increase in the role of retail and service chains would be a continuation of the recent surge in franchising. The links between the development industry and retail chains are getting stronger and stronger and have been recently expanding into new forms. Both chains and developers are inventing new formats, to suit a variety of market segments, and are

Table 5
Changes in the distribution system

	1961	1966	1971	1976	1981	1986
(a) Retail sales as % of disposable income	59.7	56.9	52.4	45.2	41.7	40.8
(b) Retail chains as % of retail sales	31.4	33.0	37.4	41.8	42.3	41.6
(c) Average sales ($million)						
for retail firms[†]	n.a.	$1,686	1,670	1,517	1,208	n.a.
Average sales ($million)						
for chain outlets	$1,668	1,642	1,694	1,851	1,616	1,346
(d) Average margin (%)						
Supermarkets	n.a.	n.a.	n.a.	19.0	19.0	19.8
Men's clothes	"	"	"	41.1	45.7	46.0
Department stores	"	"	"	37.5	36.9	32.9
(e) Personal savings as % of disposable income	3.3	6.8	6.8	11.6	15.0	10.4

* Statistics Canada, *Retail Chains and Department Stores*, Cat. No. 63-210.

[†] Statistics Canada, *Corporation Financial Statistics*, Cat. No. 61-207; figures in 1986 dollars.

invading new niches. It is now possible to find a retail chain that operates in Italian districts, or upscale boutique areas, or rundown small towns. Even the service industry, long the preserve of "mom and pop" operations, is shifting rapidly toward nationally franchised car repairs, real estate offices, accountants, and beauty salons.

This restructuring need not mean larger, more efficient outlets. On the contrary, if consumers – with two-income families and smaller households – want to spend money for more service or greater choice, there will be a chain or franchise system to fill that niche. After a long period of increasing size of retail outlet, average store sizes have dropped sharply since 1981. If smaller "boutique" units are not taking over, they are at least compensating for the continuing development of superstores and hypermarkets. It may well be that changes in household size and composition carry more weight than potential scale economies. Regardless of the shifts in the size distribution of stores, retail margins have changed very little over time.

How will these changes in distribution activities feed back to alter the settlement geography of the country? The results from various parts of this review suggest, first, that the spatial association between market characteristics and distribution activity will weaken, at all spatial scales, as retail facilities continue to specialize and customers travel farther to obtain the variety and price that they desire. Recreation-linked retailing, for example, will absorb a larger share of all sales. And, as pointed out above, chains are able to separate their customer activities from their backroom managerial activities. Second, the differential growth rate of various consumer-oriented sectors has begun to favour large cities over small. Specialized goods and services, and specialized retail clusters, seek access to larger markets and to other similar activities. While Sears Canada may prefer mass markets, partly isolated from competitors in a suburban mall, Birks Jewellers prefers downtowns or large plazas where there are several other similar stores. So it is with financial institutions and design services, as well.

At the same time, consumer-linked activities are undergoing extraordinary changes that will increase their sensitivity to consumer preferences by an order of magnitude. Retail chains, for example, get daily or weekly reports on sales of each item – permitting them to carry less inventory and to order more frequently. The same chains know the demographic characteristics of their trade areas in exhaustive detail and how to relate these numbers to sales by item. They can switch goods from one region of the country to another, replace a store with any one of a dozen other store formats, or shade prices up or down to match customer response. The same is true of bank branches or supermarkets. This is a far more sophisticated distribution system than it was even a decade ago, and much better positioned to work closely with manufacturers and importers. It may not be particularly efficient in an economic sense, but it performs a vital role from the point of view of the consumer, in a world where style and self-image count more than quantity, quality, and price.

The magnitude of distribution activity within the economy invites intervention by policy-makers. The level of consumption has long been a target for management by Keynesian economists, and there have been intermittent investigations of the level of corporate concentration and competition. But distribution is now is a major determinant of the location of growth – hence regional development – with control over these investments now concentrated within a hundred or so firms, shopping centre developers, and retail chains. Until now, little direct intervention into the distribution system has been observed in Canada; but indirectly the growing role of these sectors has begun to shape the thrust of regional policy. As the share of jobs and income derived from the primary and secondary economic sectors declines, and the share of distribution activities increases, it makes more sense to support local economies directly by providing income supplements that can be spent in the distribution system than by trying to feed the money through the economic base. Why subsidize a failing local industry? And so it has been done: regional policy is now accomplished by mailing cheques – to individuals or local governments – that are then fed through the distribution system to support the local economy. Pension cheques are as effective as salary cheques in supporting stores and services.

The distribution system is often neglected in studies of the space economy; at any rate, it seldom gets the share of attention that it deserves. It is deemed simply to reflect the spatial structure of the market, which is in turn defined by other, more fundamental economic sectors. It is seldom the subject of explicit policy interventions; perhaps because myths of small-town entrepreneurs still obscure the concentration of ownership that exists within these sectors. Yet it is the most visible and frequent link between consumers and the rest of the economy: it displays goods and prices to consumers and carries back to producers their responses. This part of the economy is so large that even subtle shifts in location shape the relative growth of settlements of different size, the health of downtowns, and the rate of deterioration of older parts of cities. These sectors provide an accessible example of how the economy works; and since they have become as oligopolistic as most production activity, they provide a window for the study of business strategy (see Laulajainen 1987) – with all its uncertainty, turbulence, and choices – a different literature for economic geographers to ponder.

REFERENCES

Coffey, W.J., and Polèse, M. 1988. "Locational Shifts in Caandian Employment, 1971–1981." *Canadian Geographer*, 32 no. 3, 248–56.

Jones, Ken G. 1984. *Specialty Retailing*. Monograph No. 14, Department of Geography, York University, Toronto.

Jones, Ken, and Simmons, Jim. 1993. *Location, Location, Location: Analyzing the Retail Environment*. 2nd edn. Toronto: Nelson.

Laulajainen, Risto. 1987. *Spatial Strategies in Retailing*. Dordrecht, Netherlands: D. Reidel.

O hUallachain, B., and Reid, N. 1991. "The Location and Growth of Business and Professional Services in American Metropolitan Areas, 1976–1986." *Annals of the Association of American Geographers*, 81 no. 2, 254–70.

Sargent, John, ed., 1986. *Long-term Economic Prospects for Canada*. Royal Commission on Economic Union and Development Prospects for Canada, Research Studies, Vol. 23. Toronto: University of Toronto Press.

Simmons, James W. 1983. "Forecasting Future Geographies: Provincial Population." *Operational Geographer*, no. 2, 7–12.

– 1986. "The Impact of Distribution Activities on the Canadian Urban System." In J.G. Borchert, L.S. Bourne, and R.G. Sinclair, eds., *Urban Systems in Transition*, 60–9. University of Utrecht, Netherlands Geographical Studies No. 16.

– 1989. "The Turbulent World of Retail Chains." *Operational Geographer*, 7 no. 2, 5–9.

– 1991. "Commercial Structure and Change in Toronto." Research Paper No. 182, Centre for Urban and Community Studies, University of Toronto.

– 1995. *Commercial Activities in Canada, 1994*. Toronto: Centre for the Study of Commercial Activity, Ryerson Polytechnic University.

Simmons, J.W., and Speck, B. 1988. "The Spatial Imprint of Business Strategy." Discussion Paper No. 35, Department of Geography, University of Toronto.

CHAPTER EIGHTEEN

The Role and Location of Service Activities in the Canadian Space Economy

WILLIAM J. COFFEY

One of the major economic phenomena marking the twentieth century has been the growth of service activities in developed countries, where provision of services has displaced production of goods as the principal form of economic activity. In most advanced economies, services now account for well over 50 per cent of both total employment and total economic output, measured as either gross national product (GNP) or gross domestic product (GDP). Using either employment or output measures, Canada ranks as one of the world's most highly "tertiarized" economies. In 1991, employment in service industries constituted 72.7 per cent of total Canadian employment. In terms of output, service industries accounted for 66.4 per cent of Canada's total GDP in 1991, again one of the highest percentages among developed nations (Table 1).

Even more impressive than the absolute size of Canada's service sector is its relative growth. As a proportion of real GDP, service output grew from 57.2 per cent in 1971 to the reported 66.4 per cent in 1991. The share of the service sector in total employment increased from 65.6 per cent in 1971 to 72.7 per cent in 1991, reflecting growth of almost 80 per cent over the two decades, compared to 31.0 per cent for the goods-producing industries. During the period 1971–91, 84 out of every 100 new jobs created were in the service industries. These recent changes are part of the long-term structural evolution of employment, in which the service industries have grown at the expense of the manufacturing and primary industries (Table 2).

In order to comprehend fully the structure and functioning of the Canadian space economy, it is clearly necessary to understand the role played by service activities. Given these impressive statistics, it is not surprising that researchers and policy makers are according more and more attention to this set of activities. In particular, a fundamental public policy issue concerns spatial distribution of service activities. Growth in service output and employment is occurring at a national scale, as we have seen, but what are its spatial consequences? This question is of considerable importance in Canada, a country in which measurement of regional economic disparities and formulation of policies designed to ameliorate them are long-standing elements of the policy environment.

Table 1
The share of services in the labour force and GDP of selected developed
countries, 1991

	Services in labour force		Services in GDP (%)
	%	Growth rate (%) 1971–91	
Canada	72.7	77.4	66.4
United States	71.8	68.7	71.8
Netherlands	69.9	72.0	61.8
Sweden	68.3	42.0	68.0
United Kingdom	70.4	40.9	66.6
France	65.1	47.0	65.6
Japan	58.9	52.2	55.7
Switzerland	60.1	45.4	62.0
Italy	59.2	64.9	60.2
Germany	57.5	43.4	60.6
Spain	56.3	50.8	59.8

Sources: OECD, *Labour Force Statistics, 1972–1992*; The Economist, *Pocket World in Figures*, 1994.

Table 2
Evolution of Canadian labour-force structure (% per sector), 1941–91

	1941	1951	1961	1971	1981	1991
Primary	30.5	20.9	13.9	8.4	6.9	6.1
Secondary	27.9	30.6	28.4	26.0	24.8	21.2
Tertiary	41.6	48.5	57.7	65.6	68.3	72.7

Sources: Statistics Canada, *Census of Canada*, Cat. Nos. 93-152, 93-326.

This chapter identifies and explores the principal issues related to the "where" and the "why" of service activities in Canada. The following section provides empirical evidence of trends in the location of services. While these trends are interesting and significant in themselves, it is the locational factors underlying them that are essential for understanding the dynamics of the Canadian space economy; before these locational factors can be considered, however, it is necessary to examine briefly certain basic notions concerning the nature of service activities. A final section explores the relationship between the growth of service activities and spatial inequalities.

EMPIRICAL TRENDS IN THE LOCATION OF SERVICES

There are a number of spatial frameworks that may be used to examine the location of service employment in Canada.[1] Here, I use two complementary frameworks. The first

Table 3
Service employment location quotients by region, Canada, 1991

	Atlantic	Quebec	Ontario	Prairies	BC	Canadian growth rate 1971–91 (%)
Transport	101	97	86	115	128	38.5
Communication	108	105	96	99	99	80.0
Utilities	106	99	109	99	72	104.8
Wholesale	91	99	99	104	108	76.1
Retail	108	102	99	94	101	99.0
FIRE	68	97	114	86	103	126.4
Producer	56	92	119	84	105	284.4
Education	114	99	98	103	95	70.8
Health	111	108	93	103	96	149.0
Accommodations	94	97	93	102	127	174.4
Public Administration	149	95	98	97	86	73.8
Consumer	96	99	99	98	112	125.6

Note: Location quotients indicate whether a given region is relatively more (LQ>100) or less (LQ<100) specialized in a given sector, using the structure of the national economy as a benchmark.

involves the traditional five-region partitioning of the Canadian space economy, based upon provincial and multi-provincial units (Table 3). Relative to the entire Canadian economy, the Atlantic provinces have a high concentration of public administration and not-for-profit (education and health care) services, but a low concentration of producer services and of finance, insurance, and real estate (FIRE) services; the latter two sectors are concentrated in Ontario, which exhibits a low level of specialization (i.e., a location quotient less than 90) only in transportation services. The strengths and weaknesses of the individual regions of Canada involve a lower level of economic activity in the Atlantic provinces and Quebec, a dynamic Ontario, and economic expansion in the west, where British Columbia's specialization in transport and in accommodations, food, and beverage services is well above the national benchmark.

The variations in sectoral specializations across the urban hierarchy[2] (Table 4) represent a more interesting and important set of information. Producer and FIRE services, in particular, tend to be concentrated in the largest metropolitan areas; retail, consumer, and not-for-profit services are usually more evenly distributed across size categories, except for the two rural classes. A third set of activities, which includes utilities and communications services, tends to be more randomly distributed across the urban hierarchy. It may be argued that the "traditional" regional structure depicted in Table 3 is in fact the result of the spatial dynamics that occur at the level of the urban-rural continuum, as presented in Table 4.

Using both the traditional regional and the urban hierarchical frameworks, analyses of changes in the location of service activities 1971–81 (Coffey and Polèse 1988; Coffey and McRae 1989) have shown that the period may be characterized as one of relative change but absolute stability. At the regional scale, while the prairie prov-

Table 4
Service employment location quotients by urban size category, Canada, 1991

	1000+	500–1000	100–500	50–100	50–100	25–50	25–50	10–25	10–25	<10	<10
				C	P	C	P	C	P	C	P
Transport	104	90	84	78	104	76	88	80	111	94	103
Communication	115	106	102	72	100	72	110	75	86	60	62
Utilities	93	80	94	121	80	119	147	136	95	87	114
Wholesale	119	92	91	102	83	89	83	85	72	83	64
Retail	93	89	101	115	108	126	119	120	115	87	86
FIRE	134	95	91	81	68	77	69	68	64	57	46
Producer	143	109	82	78	60	56	54	64	52	43	28
Education	87	97	111	110	103	98	110	102	103	81	88
Health	88	97	111	107	107	128	111	131	115	85	93
Accommodations	88	92	106	110	127	126	117	120	119	78	80
Public Administration	69	141	109	97	118	79	99	91	125	75	90
Consumer	105	95	96	101	101	104	101	94	89	83	84

Note: Column headings refer to population size (e.g., 25–50 indicates the set of urban centres with populations between 25,000 and 50,000) and whether a given spatial unit lies within (C = central) or beyond (P = peripheral) a 100-km radius of a metropolitan area (population 100,000+).

inces and British Columbia experienced rapid growth in service employment, particularly in FIRE and producer services, Ontario, and to a lesser extent Quebec, saw large absolute growth. The Atlantic provinces experienced slowly increasing concentrations of service employment but continued to lag far behind the other regions. In terms of the urban hierarchy, rural regions and smaller urban centres had the highest growth rates, but the largest metropolitan areas (300+) witnessed very large absolute gains in service employment, reinforcing their position as centres of high-order services (producer services and FIRE activities).

The extreme spatial concentration of high-order or information-intensive services is one of the most prominent characteristics of the Canadian space economy. The majority (51.2 per cent) of all service jobs are located in the country's eight largest metropolitan centres (Table 5); service employment is thus slightly more concentrated than total employment, where the eight largest metropolitan areas account for 48 per cent of all jobs. However, employment in producer services and the FIRE sector is highly concentrated, with half of all Canadian producer-service employment found in the three largest centres (41 per cent in Toronto and Montreal alone); the four largest centres account for half of FIRE employment. As a counterpoint to this information, Table 5 also presents a profile of retail employment, which may be seen to follow closely the distribution of the population (see Simmons, this volume, chap. 17). The remaining sections of this chapter will be devoted primarily to exploring the reasons underlying the spatial concentration of high-order services and the implications of this centralization for the future configuration of the Canadian space economy.

Table 5
The metropolitan concentration of high-order service employment, Canada, 1991

	Population		All sectors		All services		Producer services		FIRE services		Retail services	
	% Canada	Cumul. %	% Canada	Cumul. %	% Canada	Cumul. %	% Canada	Cumul. %	% Canada	Cumul. %	% Canada	Cumul. %
Toronto	14.3	14.3	15.4	15.4	16.0	16.0	26.0	26.0	24.9	24.9	15.1	15.1
Montreal	11.5	25.7	11.4	26.9	11.8	27.8	14.7	40.7	13.5	38.4	11.4	26.5
Vancouver	5.9	31.6	6.2	33.1	6.8	34.6	8.7	49.5	8.2	46.6	6.2	32.7
Ottawa-Hull	3.4	34.9	3.7	36.8	4.4	39.0	5.8	55.3	3.2	49.7	3.3	35.9
Edmonton	3.1	38.0	3.3	40.1	3.6	42.6	3.4	58.7	3.2	53.0	3.3	39.2
Calgary	2.8	40.8	3.1	43.1	3.2	45.8	5.1	63.7	3.5	56.5	2.9	42.1
Winnipeg	2.4	43.2	2.4	45.6	2.7	48.4	2.0	65.7	2.8	59.3	2.4	44.5
Quebec	2.4	45.5	2.4	48.0	2.8	51.2	2.3	68.0	3.0	62.3	2.4	46.9
Hamilton	2.2	47.7	2.2	50.2	2.1	53.4	2.2	70.1	2.3	64.7	2.4	49.3
London	1.4	49.1	1.5	51.7	1.5	54.9	1.3	71.5	1.8	66.5	1.6	50.9

Source: Coffey (1994).

Note: "Cumul."(ative) column is running total of percentages in "% Canada" column.

THE NATURE OF SERVICE ACTIVITIES

A considerable literature exists on the definition, the classification, and the characteristics of service activities. It is neither possible nor appropriate to examine these questions in detail in the present chapter; yet, in order to understand the location of these activities, a certain basic understanding of the nature of "the service sector" is essential.[3]

Definition and Classification

A service activity results in a product of an immaterial or ephemeral nature, such as a sale or a consultation; the typical service is produced and consumed simultaneously, requiring the close collaboration of the producer and the consumer, and generally can neither be stored nor owned. It tends to be labour intensive rather than capital intensive, though there are degrees of variation. Further, a service is usually defined in a negative or residual sense: it is an activity that does not produce or modify physical goods. Obviously, the distinction between goods and services often becomes blurred. Do restaurants, book publishers, and software producers provide goods or services? Do construction and repair fall within the secondary or tertiary sectors?

Starting with the simple Fisher-Clark (c. 1940) partitioning of the economy into three sectors (as in Tables 1 and 2) – a typology based on the output of economic activities – the classification of services has become highly complex, with the "quaternary" (information-related) and "quinary" (administrative) sectors sometimes being separated from the tertiary. More recent typologies have been based on user sectors (for example, producer or intermediate-demand services versus household, or final-demand services; public v. private services; marketed v. non-marketed services); on occupational classifications (such as white-, grey-, and blue-collar occupations); on role or function in the production system (for example, circulation, distribution, regulation); and on what is being processed or handled (such as goods, information, persons, or public security). A further distinction that will be treated below involves that between basic and non-basic services.

The major point to be retained from the myriad of typologies available is that perhaps the principal attribute of "the service sector" is its vast heterogeneity. The notion of the rise of the service sector as a monolithic shift within the economy is an unwarranted simplification. In fact, the very notion of "a service sector" represents a profound ecological fallacy. Different service activities are characterized by different rates of growth (see the right-hand column of Table 3), by different levels of use of capital and human resources, by different growth stimuli and growth constraints, by different levels of remuneration, and by different capacities to contribute to economic development. In addition, and perhaps most importantly in a geographic context, various service activities are characterized by different locational tendencies and/or constraints. Thus it is necessary to subdivide "the service sector" in some logical way: Tables 3 and 4 use a typology based on the "product" or output of sectors (or, more precisely, on the Standard Industrial Classification system).[4] It is clear from these tables that such a typology captures important differences in locational tendency, as do the distinctions of public versus private services, market v. non-marketed services, producer v. household services, and basic v. non-basic services, and as do classifications based on occupation.

Service Growth: Fact or Illusion?

One of the fundamental questions involving service activities is whether their recent rapid growth in all developed countries is more apparent than real. It is generally acknowledged that service growth is a function of six factors: (1) shifts in household consumption preferences combined with an income elasticity of demand greater than unity; (2) shifts away from in-house service production by firms and toward external purchase of services from specialized, free-standing establishments; (3) the increasing tendency of households to purchase external services rather than to produce them internally (in part a function of increased female participation in the labour force); (4) the lower level of productivity in services than in the goods-producing sector; (5) lower and more flexible wages due to the disproportionate number of small and medium-sized firms in the service industry; and (6) increased consumption of public services. Certain authors (such as Gershuny and Miles 1983) argue that the net effect of these factors is that the growth of employment and output in the service industries is a statistical artifact; "taking in each other's washing" does not constitute economic growth. This argument is vigorously contested by other researchers who cite creation of new services as well as real growth in service output and rising productivity in service activities. The issue is far from being satisfactorily resolved. In the specific case of the second factor cited, however, the empirical evidence is categorical. Not only has the growth of producer services been found to be real, but growth in output and employment has been achieved in in-house and external producer services simultaneously; a simple displacement of internal producer services to free-standing establishments is clearly not occurring ((McCrackin 1985; Tschetter 1987; Kutscher 1988; Beyers 1989; Illeris 1989).

In a related controversy over competing views of structural change, the "post-industrial" or "information" society perspective interprets the growth of services in a positive manner, arguing that in the natural evolution of the economy growth is sustained more by knowledge, skills, and information than by physical capital. In contrast, the "deindustrialization" perspective holds that the growth of services is a negative phenomenon related to the inability of goods-producing industries to develop or to the internationalization of capital and the shift of production activities to newly industrializing countries. According to the former view, the information presented in Table 1 would indicate that Canada is a highly advanced country; according to the latter view, Canada's economy is exceedingly weak. Again, this is a debate that will not be easily resolved.

Basic versus Non-Basic Services

One of the most important distinctions involving services is furnished by economic base theory, which posits that the economic base of a region (or of a city) has two principal elements: basic, or export-oriented activities and non-basic, or residentiary activities (those serving the needs of the local population). Obviously, the distinction is often difficult to establish; rare is the economic activity that produces exclusively for export or for local consumption. According to the theory, it is the former set of activities that serve a "propulsive" or "engine of growth" function; they create injections into the local economy, which, through the multiplier mechanism and the circular flow of income, stimulate local economic growth. Residentiary activities, in contrast, are induced activities; not being export-oriented, they do not promote growth but, rather, follow from it.

Economic base theory has traditionally considered that the principal basic activity or propulsive force of a local (regional or urban) economy is manufacturing and that the residentiary sector consists of services. For this reason, economic geography and regional science long ignored the potential role of services in economic development. This neglect has been reinforced by the original Fisher-Clark typology, which relegates services to a residual category composed of "non-productive" activities, and by central place theory, which emphasizes the role of residentiary services in structuring a space economy.

It is now widely recognized, however, that not all services may be justifiably regarded as residentiary activities; a significant proportion are basic, in that they are not only exportable (tradable) but also responsive to external demand. Producer services, in particular, have emerged as one of the fastest-growing components of both inter-regional and international trade.[5] Canadian-based consulting services alone accounted for $817 million worth of international exports in 1989, representing a real increase (calculated in constant dollars) of over 400 per cent relative to the 1969 value. Similarly, for the same period, Canadian exports of insurance and administration services more than tripled, reaching $746 million and $519 million, respectively, in 1989 (see also, in this volume, Norcliffe, chap. 2, and Britton, chap. 14). The importance of producer services may be further emphasized by citing the case of New York City, where approximately two-thirds of all export earnings are now generated by the producer service sector (Drennan 1987). In addition to the producer service category, however, certain other types of services may create major economic injections: transportation, communications, and financial or consumer services. One subset of the latter category, services related to tourism, is particularly important for many regions, both rural and urban; rather than the services being "transported" across the boundaries of the local region, however, it is the tourists who transport themselves into the region in order to receive the services.

The Role of Producer Services

Producer services[6] are intermediate-demand functions that serve as inputs into production of goods or of other services (and, as such, are perhaps more correctly characterized as indirect elements of the production process[7]), enhancing the efficiency of operation and the value of output at various stages in the production process. All firms, whether specializing in fabrication of goods or in provision of services, employ a range of producer services as inputs into their production functions. The more advanced or modern the "value chain" (i.e., the production process) of a firm, the more numerous and more complicated will be the links in the chain and the greater will be the importance of producer service inputs. Indeed, a wide range of empirical evidence has demonstrated that producer services occupy a major and expanding role both within firms and within national, regional, and metropolitan economies.

A fundamental decision that each firm (and, indeed, each establishment within a firm) must face, irrespective of the sector of activity to which it belongs, concerns whether to "make" or to "buy" a specific producer service input; that is, whether to provide a given service internally by assigning its own personnel to production of the service, or to contract out provision of the service to experts in the employ of external (either affiliated or free-standing) specialized establishments. The decision to internalize or to externalize given service inputs is one of the most important strategic decisions that an establishment must make, as this choice ultimately affects its cost structure, its modes of operation and organization, and very possibly its location. The "make or buy" decision of individual establishments also affects the structure of an economy, as measured by official statistics on employment or output by sector of activity. A lawyer, for example, hired in-house by a manufacturing establishment will increase the level of employment in the manufacturing sector, whereas the same lawyer working in the context of a law firm, even though performing the same functions for the same organization, will augment employment in the producer services sector.

Among all service activities, producer services have the greatest potential for stimulating economic development. Why is this so? First, producer services comprise the most rapidly growing sector in the majority of developed countries. In Canada, over the period 1971–91, employment in producer services grew by 284.4 per cent, more than one hundred percentage points more than their closest competitor, the accommodations, food, and beverage sector (see Table 3). By way of comparison, the growth rate for manufacturing was 22 per cent over the period, and that for total employment was 65 per cent. This increased demand for producer services is in turn a function of three factors. (1) Because of the changing organizational structures of both goods- and service-producing firms, the quest for the enhanced organizational flexibility and the external economies of scale needed to compete on national and international markets has led to the increased purchase of factor inputs, including services, from external establishments. (2) The increased complexity of modern society and, in particular, increased government intervention and regulation require firms to use a whole range of specialized services; as the demand for many of these services is often sporadic, it is generally more economical to purchase them externally than to provide them internally. (3) The expanding role of product innovation and of market differentiation on the part of both goods- and service-producing industries generally requires highly specialized, knowledge-based inputs.

In sum, all these phenomena require the intervention of specialists – engineers, lawyers, accountants, management consultants, advertising professionals, and so forth – who can analyse situations, process information, produce required documentation, and assist in decision-making. The rise of producer services thus reflects important transformations in modern economies in terms of both what types of goods are produced and how these goods are produced. In addition, the amount of information that a firm must process (gather, store, analyse, distribute) is rapidly increasing. The opportunities and constraints presented by a firm's social and economic environment, as well as those imposed by its internal structure and functioning, need to be continually evaluated, with certain interventions or adjustments frequently being initiated as a result of this process of evaluation.

Second, producer services can constitute a particularly important element of the economic base of a region. Many types of producer services have a particularly high propensity to be exported (Coffey and Polèse 1987a; 1987b; Marshall 1988; Illeris 1989). However, even those that are not directly exported, but that provide intermediate inputs for firms producing goods and services for export, may be considered as "indirectly basic" activities.

Third, and perhaps most importantly, through their role in investment, innovation, and technological change, producer services play a key role in economic development, particularly in facilitating overall economic change and adjustment. In an age of rapid technological change, certain producer services provide the source and the vehicle of that change. Producer services are thus an important part of the supply capacity of an economy: they influence its adjustment in response to changing economic circumstances, and they may help to adapt skills, attitudes, products, and processes to changes or to reduce the structural, organizational, managerial, and informational barriers to adjustment.

LOCATION

What factors underlie the varying locational patterns of service sectors that are observed in Tables 4 and 5? While this question is highly complex, certain valid generalizations can be made. Above all, the ability to explain the location of service activities rests on the distinctions made above between classes of services.

In this section, locational tendencies related to the traditionally defined regions shown in Table 3 will generally not be considered. While differences in service concentration between Toronto and a medium-sized city such as Guelph, Ontario, can be addressed using a coherent body of geographical and economic theory, the fact that Canada's largest city is in Ontario rather than in Nova Scotia is a separate issue and requires a different form of analysis. Further, the present analysis will remain at the inter-urban scale, defined broadly so as to include rural settlements; the question of intra-metropolitan location – the central business district versus the suburbs – is an entirely distinct matter, involving different factors, and will not be examined.

Public, Mixed, and Final-Demand Private Services

It is first necessary to distinguish between public and private services. The former may be further subdivided into two types. Not-for-profit services (education, health, and welfare) are generally provided to households by various levels of government. (This is the case in Canada, but there are considerable variations between countries in terms of the level of privatization of "public" services and, consequently, in the degree to which they really are "not-for-profit"). It is therefore logical that these activities are located near the households that they serve. Reality reflects this logic, except for over-representation in the 100–300 city class (many of which are regional university and medical centres), very slight under-representation in the 300+ cities, and under-representation in rural areas (fewer than 10,000 people); the latter under-representation involves virtually all service classes and may be attributed to the low population density of rural areas.

A second type of public service consists of public administration at the federal, provincial, and municipal levels and national defence at the federal level. Once again, these activities generally follow the distribution of the population, particularly in the case of municipal administration. The location of this set cannot be entirely explained by the distribution of the population, however, even if there is a tendency to concentrate many administrative functions in relatively large national, provincial, and regional capitals. On the one hand, certain functions may have very particular locational requirements; for example, the maritime branch of the Canadian armed forces generally needs to be located in an ocean port. On the other hand, many decisions concerning the location of public administration functions reflect neither distribution of population nor considerations of economic efficiency; in many instances these activities may be regarded as indirect elements of regional development policy. Thus we find the philatelic branch of Canada Post in Antigonish, Nova Scotia, and a major facility of the Department of Veterans' Affairs in Charlottetown, PEI. Specific physical factors may also be involved, as in the case of the location of Parks Canada activities.

Transport, utilities, and communication services represent a mixture of government-operated activities (airports and ports; municipal transport), crown corporations (Ontario Hydro or CNCP telecommunications), and private services (private truck or bus firms; Bell Canada). These subsectors are very irregularly distributed (Table 4), as a result of both the wide diversity of activities and the varied patterns of ownership. Some of these activities are in the final-demand sector and located near the population (transport of persons). Others may be located hear a physical feature (marine transport) or a natural resource (natural gas), or in proximity to primary or secondary activities (grain storage, transport of goods). In still other cases, the pattern may reflect certain "strategic" locations (Canadian National Railway facilities, located until recently in Moncton, the "hub" of the Atlantic provinces' land transport routes).

For the remaining private services, two distinct classes may be identified. First, the location of retail and consumer services (final-demand or residentiary activities) can be largely explained by the distribution of households, perhaps weighted by purchasing power; this is shown in Table 4 and even in Table 3. The absence of these activities in rural regions and the corresponding over-concentration in small and medium-sized cities may of course be explained by reference to central place theory; the sparsely distributed rural population must travel to the closest central place in order to avail itself of these services. Consumer services, especially personal services, tend to follow the distribution of the population even more closely than do retail services (see Simmons, this volume, chap. 17). Note the relatively high concentration of retail services in the Atlantic provinces (see Table 3). This has been made possible by federal transfer payments, which, rather than promoting economic development, have artificially inflated the capacity for consumer spending.

The Location of Private, Intermediate-Demand Services

It is the second group of private services, those such as FIRE and producer services, principally serving intermediate demand, that are the main contributors to the spatial concentration of tertiary employment in the Canadian space economy. From an intellectual perspective, this group provides the most fertile ground for inquiry since its locational patterns tend not to be explained by distribution of population, natural resources, or physical features.

A number of complementary frameworks can be used to explain the location of private, intermediate-demand services. A first involves what Illeris (1989) refers to as the "structural explanation" – the structural effect in shift-share analysis. The basic reasoning is that since employment is growing more rapidly in high-order, intermediate-demand services and less rapidly in routine service activities, the logical result is increased concentration of employment in large cities such as Toronto and Montreal, where the most rapidly growing classes of activities are over-represented. (See right-hand column of Table 3 for an indication of differences in growth rates by sector). This is not highly satisfactory as an explanatory framework, as the argument is largely circular. Further, it only partially explains the increasing concentration of these activities, while neglecting the question of initial concentration. The frameworks reviewed below more directly contribute to our understanding of both initial and increasing concentration of intermediate-demand services.

A second explanation is based on the concept of agglomeration economies, or externalities. Here, the argument is that the concentration of intermediate-demand services in a few large cities minimizes transaction costs associated with their production and delivery. In particular, maintaining face-to-face contact between producers and consumers is potentially the most expensive element of intermediate service production, and the one that can be most significantly reduced by spatial agglomeration. A well-developed literature indicates that telecommunications technologies may be successfully substituted for face-to-face contact only when the information to be transmitted is relatively standardized or if the individuals know one another intimately and trust one another fully (Gottmann 1977). In the case of negotiations, strategic discussions, and other dialogical situations, face-to-face contact remains absolutely essential (Pye 1979).

Thus in large cities such as Toronto and Montreal, these forces of agglomeration generally produce what has been termed a "complex of corporate activities": the spatial clustering and mutual symbiosis of the head offices of primary-, secondary-, and tertiary-sector firms; high-order financial establishments; and the high-order producer services that serve the first two types, as well as other producer-service firms (approximately 50 per cent of all producer services serve as intermediate inputs into other service firms [Illeris 1989]). Such forces of agglomeration produce corporate complexes at the regional (Edmonton) and national (Toronto) scales, in which accessibility to the corresponding market area is maximized. In addition, the growth of international trade in producer services has encouraged their location in certain "world cities" (a status to which Toronto, Montreal, and Vancouver aspire), which, due to both the convergence of airline routes and cultural factors, have the greatest accessibility to other international centres.

A third framework – one of the most comprehensive explanations formulated thus far – is furnished by the simple locational model proposed by Coffey and Polèse (1987a; 1987b). This model, which adopts a micro-economic approach and is the result of both inductive and deductive lines of inquiry, posits that in its locational decision a high-order service firm (i.e., producer service or FIRE activities) will seek to minimize a production or cost function involving three factor inputs: complementary intermediate-demand services, human resources, and the cost of communicating (delivering) its output. The first two elements represent externally purchased and internally provided factor inputs, respectively, and the possibility for factor substitution exists. The decision concerning this substitution between externally purchased services and internal human resources will be affected by certain of the economic factors introduced above (for example, the advantages of vertical disintegration); in particular, the propensity to purchase services externally will be increased where non-standardized service inputs with relatively unpredictable demand are involved.

The cost of each factor input and thus the value of the production function will vary with location. In sum, each high-order service establishment is subjected to three locational pulls: toward urban centres characterized by the availability of diversified, complementary producer services, toward centres with specialized pools of skilled labour, and toward the market for its output. Certain locations, such as Toronto, Montreal, and Vancouver, may combine two or more of these attributes and thus enjoy a major advantage in attracting and retaining high-order service firms; a location such as Halifax, in spite of its highly skilled labour force, will be at a disadvantage with respect to these larger centres.

In terms of the third element of the model, the market for high-order, intermediate-demand services, the spatial pattern of corporate ownership and control tends to centralize the location of producer-service and FIRE activities. Headquarters and divisional head offices of large corporations have a high propensity to purchase from nearby sources most of those high-order services consumed by their widely dispersed branches. As corporate control and its associated spatial division of administrative functions tend to be highly concentrated in a few large metropolitan areas, demand for these intermediate-service inputs will be similarly concentrated. As noted above, face-to-face contact between producer and purchaser of such services reflects qualitative characteristics that cannot be reproduced by long-distance communication. Thus firms supplying intermediate services to administrative offices will usually reflect the locational pattern of the latter. In non-metropolitan areas, independently owned firms rather than branch establishments tend to purchase their intermediate-service inputs from local firms.

A final point concerning location of private, intermediate-demand services poses a question rather than furnishing an explanation. The potential impact of telecommunications technology on the location of these services promises to be a crucial issue for development of the Canadian space economy. The likely effect of technological advances in this area on location of high-order services remains unclear, as indicated by the existence of two schools of thought on the matter. Conventional wisdom holds that these new technologies will permit decentralization of high-order service activities by making it possible to transact business efficiently over vast distances. Some argue, however, that new technologies will further centralize intermediate-demand services in a few large metropolitan centres by permitting these activities to "decouple" from the production establishments that they direct.

Though there is little empirical research on this issue, that which does exist, along with the major portion of conceptual work, tends to support the second view. Telecommunications technology seems a necessary but not sufficient condition for locational change (Hepworth 1989). For certain services, particular locational constraints may be relaxed, but it does not follow that these services can operate equally well everywhere. The notion that telecommunication technology will diminish the advantages of agglomeration may be relevant in the case of standardized "back office," intermediate services such as data processing. For higher-level, "front office" functions, however – those requiring face-to-face contact – decentralization is far less likely. Hepworth (1986) and Lesser and Hall (1987) have furnished empirical evidence that firms in Canada are using telecommunications technology to maintain and to increase spatial concentration.

The principal factor underlying this failure of intermediate-demand services to behave in the manner suggested by the conventional wisdom relates to the pattern of technological diffusion: there is generally a time lag in adoption of telecommunications technologies and provision of the necessary infrastructure, with the process of diffusion following the urban hierarchy. Thus establishments in metropolitan areas enjoy an initial advantage in acquiring new technologies and, because of availability of skilled human resources, generally benefit from the options presented by these technologies. The evolution of telecommunications technology is thus a two-edged sword. It can enable intermediate-demand services to free themselves from certain of the locational constraints that have ruled other types of economic activity; but it enables firms to centralize their high-level management and scientific and technical functions far from their production facilities.

In summary, service activities as a whole are often perceived as being "footloose" or free of the types of locational constraints typically cited to explain, for example, the relative absence of manufacturing in the Atlantic provinces. While many service activities do have locational constraints, certain classes of them are characterized, at least in theory, by greater locational flexibility than many other types of economic activities. It is precisely for this reason that services have begun to generate a lot of interest as vehicles for stimulating economic growth in the peripheral regions of Canada.

CONCLUSION: SERVICES AND SPATIAL INEQUALITIES

Any evaluation of the ability of service activities to influence economic development in a given region of Canada depends above all on the definition of development employed. If it is defined modestly, in terms of job creation, residentiary functions such as consumer services, retailing, or not-for-profit activities may be just as effective development instruments as high-order basic services. If development is measured, however, in a more rigorous manner, involving structural change, productivity increase, market-earned income, and technological change, intermediate-demand, private services stand virtually alone as possible vehicles for development. In a regional development context, therefore, the principal issue is whether those services that are most capable of contributing to the development of a given area are sufficiently footloose to locate in peripheral regions or, at least, outside large metropolitan centres. The data and the conceptual arguments reviewed above, as well as those developed in more detail elsewhere (Coffey and McRae 1989; Coffey and Polèse 1989), suggest that the more optimistic views concerning the capacity of service activities to enhance the economic development prospects of Canada's peripheral regions are largely unjustified.

The locational trends of all services, and intermediate-demand services in particular, are most fruitfully examined in the framework of the urban-rural continuum. In this context, we have seen that the high-order services capable of stimulating economic development are highly centralized in a small set of large cities. The disparities observed between the traditionally defined regions of Canada, those that have preoccupied Canadians to such an extent over the past 50 years, may thus in large measure be viewed as a function of the inequalities in the spatial distribution of metropolitan centres. The ten largest metropolitan areas displayed in Table 5, containing approximately 70 per cent of all Canadian FIRE and producer-service employment, are distributed across only five provinces; six of the ten are in Ontario and Quebec, along the Windsor-to-Quebec axis that has long been Canada's principal agglomeration of population and economic activity. Such concentration of the most rapidly growing sectors of the economy does not bode well for the development prospects of the remaining provinces and regions, with few or no large cities. On the basis of the evidence examined, it is unrealistic to expect that this spatial concentration of high-order services will reverse itself significantly in the near future. In a regional development context, then, the principal question is whether the spatial concentration of these activities will increase, thus exacerbating the economic inequalities between Canada's regions.

In terms of the economic geography of its service industries, Canada is not significantly different from most other developed countries; its space economy is subject to the same phenomena and processes that are currently found elsewhere among advanced economies (Marshall 1988; Coffey and Polèse 1989; Illeris 1989). Further, the same explanatory frameworks that apply to Canada may be employed in these other nations. In Canada, as in other developed countries, the impact of service growth on the configuration of the space economy, and particularly on spatial inequalities, will be a major issue of the coming decades.

NOTES

1 Employment rather than output must generally be used to examine location of services; locationally specific data on output are simply not available.

2 Table 4 employs a set of urban hierarchy categories that are defined using two criteria: population size and relative distance (i.e., within or beyond a 100-km radius from a metropolitan centre with more than 100,000 residents). The relative-distance criterion permits us to distinguish those activities that locate in remote areas from those on the edge of a metropolitan area – i.e., to distinguish true decentralization from extended suburbanization. See Coffey and Polèse (1988) for a more detailed discussion of this distinction.

3 General discussions of service activities may be found in Stanback et al. (1981); Gershuny and Miles (1983); Daniels (1985); and Illeris (1989).

4 Use of occupation to define service activities represents an important complement to this sectoral definition, which tends to underestimate service employment in an economy. For example, using a sectoral definition, service functions such as those provided by a lawyer or by an accountant working in the head office of a manufacturing firm would appear as employment in the goods-producing sectors. Between 25 and 35 per cent of all employment in Canada's "goods producing" sectors consists of service functions (Coffey and McRae 1989).

5 Some researchers, however, argue that even producer services are of limited importance in regional economic growth. In the Canadian context, a lively debate concerning the ability of services to function as a propulsive force in the economy was generated by the Economic Council of Canada's (1984) *Western Transition* report, which presented an optimistic view of the role of services in generating growth and in offsetting declines in the resource sectors of the western provinces.

6 There exists no absolute definition of the concept "producer services"; each researcher is free to employ an operational definition that is derived from the particular objectives of the study undertaken. While certain authors have employed a broad definition, including transportation, communications, and public utility services, it is much more conventional to restrict the set of producer services to a more narrowly defined group of high-order office activities. The latter approach is favoured in the present context, where the term "producer services" is synonymous with division M of Statistics Canada's 1980 Standard Industrial Classification: employment agencies and personnel suppliers (class 771); computer and related services (772); accounting and bookkeeping services (773); advertising services (774); architectural, engineering, and other scientific and technical services (775); offices of lawyers and notaries (776); management consulting services (777); and other business services (779).

7 The separation between goods- and service-producing activities, as reflected in the Fisher-Clark typology and its derivatives, is largely artificial. One of the principal characteristics of a modern economic system is the increasing interdependence between these two types of activities; the growing tertiarization of goods production and the corresponding increase in the capital intensity of services are widely acknowledged.

REFERENCES

Beyers, W.B. 1989. *The Producer Services and Economic Development in the United States: The Last Decade*. Washington, DC: Economic Development Administration.

Coffey, W.J. 1994. *The Evolution of Canada's Metropolitan Economies*. Montreal: Institute for Research on Public Policy.

Coffey, W.J., and Bailly, A.S. 1991. "Producer Services and Flexible Production: An Exploratory Analysis." *Growth and Change*, 22, 95–117.

Coffey, W.J., and McRae, J.J. 1989. *Service Industries in Regional Development*. Montreal: Institute for Research on Public Policy.

Coffey, W.J., and Polèse, M. 1987a. "Intra-firm Trade in Business Services: Implications for the Location of Office-Based Activities." *Papers of the Regional Science Association*, 62, 71–80.

– 1987b. "Trade and Location of Producer Services: A Canadian Perspective." *Environment and Planning A*, 19, 597–611.

– 1988. "Locational Shifts in Canadian Employment, 1971–1981: Decentralization v. Decongestion." *Canadian Geographer*, 32, 248–56.

– 1989. "Producer Services and Regional Development: A Policy-Oriented Perspective." *Papers of the Regional Science Association*, 67, 13–27.

Daniels, P.W. 1985. *Service Industries: A Geographical Appraisal*. London: Methuen.

Drennan, M. 1987. "New York in the World Economy." *Survey of Regional Literature*, no. 4, 7–12.

Economic Council of Canada. 1984. *Western Transition*. Ottawa: Ministry of Supply and Services.

Gershuny, J.I., and Miles, I.D. 1983. *The New Service Economy*. London: Frances Pinter.

Gottmann, J. 1977. "Megalopolis and Antipolis: The Telephone and the Structure of the City." In I. de Sola Pool, ed., *The Social Impact of the Telephone*, 303–17. Cambridge, Mass.: MIT Press.

Hepworth, M.E. 1986. "The Geography of Technological Change in the Information Economy." *Regional Studies*, 20, 407–24.

– 1989. *Geography of the Information Economy*. New York: Guildford Press.

Illeris, S. 1989. *Services and Regions in Europe*. Aldershot: Gower.

Kutscher, R.E. 1988. "Growth of Service Employment in the United States." In B.R. Guile and J.B. Quinn, eds., *Technology in Services: Policies for Growth, Trade and Employment*, 47–75. Washington, DC: National Academy Press.

Lesser, B., and Hall, P. 1987. *Telecommunications Services and Regional Development*. Halifax: Institute for Research on Public Policy.

McCrackin, B. 1985. "Why Are Business and Professional Services Growing So Rapidly?" *Federal Reserve Bank of Atlanta Economic Review*, Aug., 14–28.

Marshall, J.N. 1988. *Services and Uneven Development*. Oxford: Oxford University Press.

Pye, R. 1979. "Office Location: The Role of Communications and Technology." In P.W. Daniels, ed., *Spatial Patterns of Office Growth and Location*, 239–76. Chichester: Wiley & Sons.

Stanback, T.M., Bearse, P.J., Noyelle, T.J., and Karasek, R.A. 1981. *Services: The New Economy*. Totawa, NJ: Rowman & Allanheld.

Tschetter, J. 1987. "Producer Services Industries: Why Are They Growing So Rapidly?" *Monthly Labor Review*, Dec., 31–40.

CHAPTER NINETEEN

Quaternary Places in Canada

R. KEITH SEMPLE

Geography has a rich literature and tradition of explaining the location of economic activities in the primary, secondary, and tertiary sectors and providing an understanding of the influence of these activities on the evolution of places with which they have been closely associated. In the 1970s, as the service sector grew in complexity, interest developed in explaining office location, especially offices located in large towers clustered in major urban places, which provide a multitude of simple and complex services (see Coffey, this volume, chap. 18). Most office clusters locate in the central cores of cities, though others are in suburban centres or at the sites of industrial production or resource extraction. In so far as these offices are free standing and provide routine services, central place and industrial location theory adequately explain their location (Daniels 1979; Gad 1985; 1991; Huang 1989). The location of the most important office towers, however, especially those associated with corporate decision making, does not find an explanation in any of these ideas and requires a theory based on the quaternary activities of command, control, decision making, and coordination (Semple 1985).

The location of these activities defines quaternary places, which rely on the generation, processing, and movement of information within a system of urban centres – in this case, the quaternary place system. The central principle of quaternary analysis is that head office managerial functions are separate from operational units, since corporations locate their elite management teams in central or suburban office nodes in order to maximize their effectiveness. The core question is whether it is possible to identify a hierarchy of quaternary places reflecting the organizational structure of business enterprises or whether the locational pattern of quaternary decision making is more complex.

Using these concepts, this chapter examines the spatial distribution of quaternary places in Canada, taking 1989 as the base year. It seeks answers to several related questions. Why do corporations choose specific cities to locate their corporate headquarters? Why do locations decline in effectiveness? Do cities specialize in the headquarters of corporations that are associated with specific subsectors of the economy?

Do distinct corporate hierarchies exist in quaternary place systems much like they do in central place systems? Do foreign corporations locate their Canadian head offices in a specialized set of places or do they emulate the existing head office locations of domestic companies?

These questions are particularly pertinent at this time because the 1980s witnessed globalization of international business and for Canada this meant deregulation of the banking industry (see Dobilas, this volume, chap. 5) and the transport sector (see Lea and Waters, this volume, chap. 16), privatization of state-controlled corporations (see Davis, this volume, chap. 20), and freer trade internationally (see Norcliffe, this volume, chap. 2) and interprovincially. As industries are increasingly deregulated and free, or induced, to compete internationally it is important to consider likely changes in the location of their corporate headquarters; this chapter considers these sources of change after establishing the origins of Canada's corporate head offices.

QUATERNARY PLACE AND CORPORATE EVOLUTION

The large corporation emerged at the end of the nineteenth century in Canada, soon after the United States, when business units could operate more profitably through a centralized managerial hierarchy than by means of decentralized market mechanisms. This hierarchy itself became a source of power because it created profits through co-ordination. It substituted the visible hand of management for the invisible hand of the market (Chandler 1962; 1977). This was particularly important for corporations whose operational units were spatially dispersed. In Canada, as corporations grew larger and more complex and financial intermediaries matured, pools of capital began to accumulate in the nation's competing quaternary centres.

Two especially important tendencies in the accumulation of capital are industrial concentration and financial centralization. Concentration of industrial capital is the direct result of the reinvestment of profits, which increases the mass of wealth controlled by individual corporations. Centralization of financial capital, in contrast, is a form of competition among corporations in which blocs of capital lose their independent existence and are merged into fewer large financial concerns. Here the money capital of individual corporations becomes centralized in banks and financial institutions as loan or financial capital (Carroll 1986).

Throughout this century very large enterprises have held the possibility of cooperation and coordination among small groups of competing corporations, which could restrict price competition and increase corporate power in key centres of the Canadian quaternary place system. The same process of concentration and monopolization characterizes the banking sector, which is also dominated by a few large corporations. Most of these banks have their corporate headquarters in Montreal and Toronto (Carroll 1986), which serves to integrate regional economies into a national economy. Since the late 1960s, Toronto has dominated the national economy simply because the largest and most powerful corporations have their key decision makers located there in close proximity and Montreal has relinquished its title "metropole du Canada" to a more dynamic group of internationally oriented corporations located in Toronto.

Corporate Development

The formation of the various financial and non-financial subsectors of the economy is linked directly to the historical roots of the largest corporations, as is the corporate geography of Canada. In essence, the cities that encouraged the emergence and development of corporate headquarters form a system of financial and non-financial decision-making places.

In the period prior to Confederation, small-scale manufacturing grew, supporting emerging cities. Large businesses such as Molson (founded 1785) and the Bank of Montreal (1817)[1] provided crucial impetus to the concentration of capital in important quaternary places and in fewer hands (Naylor 1975). Charters were given to major financial intermediaries, and[2] by 1867 a basis existed for formation of Canadian and foreign blocs of financial and industrial capital.[3] After Confederation, the era of Macdonald's National Policy was one of rapid growth for blocs of capital, especially in Montreal and Toronto. New forms of financial intermediaries emerged, such as trusts and credit unions,[4] modern stock exchanges were born,[5] and the concentration of industrial capital occurred in resources, communications, transportation, and utilities,[6] as well as in the primary manufacturing sectors and capital equipment such as heavy machinery.[7] New corporations also developed in service industries – retailing, wholesaling, business services, and construction.[8] Montreal dominated the quaternary place system in 1918. Though Toronto was in second rank, the industrial centres of southern Ontario were beginning to mature, and control centres of the west were beginning to emerge.

Canada's financial system developed in the interwar period 1919–39; the large investment banks underwrote major issues, and investment firms were able to pyramid their shareholdings and exercise operational power over multiple companies (Wheeler 1986). By 1930, a wave of mergers further centralized industry while many new domestic and foreign firms of present significance were formed. The cooperative movement which marked the beginning of a new commercial base in the country outside the traditional heartland.[9]

The era of growth after 1945 introduced a tremendous flow of American and other international manufacturing capital (see MacPherson, this volume, chap. 4) to Toronto in particular. This great inflow might have signalled capitulation of Canadian manufacturing capital to stronger monopoly interests coming in from the south and elsewhere, but Niosi (1981; 1985) characterizes the late 1960s and 1970s as a period of declining American hegemony and increasing Canadian nationalism. He shows that between 1970 and 1978 foreign control of non-financial assets in Canada fell from 36 to 29 per cent as crown corporations and Canadian businesses purchased control of numerous internationally owned companies (see MacPherson, this volume, chap. 4).

Carroll (1986) suggests that during this period a large amount of financial and industrial capital was controlled in Canada despite the massive foreign flows into Toronto. This indigenous position was consolidated by means of take-overs and mergers, by the shifting of investments to expanding industries, and by repatriation of control of specific firms from foreign to Canadian interests. Many domestic manufacturing firms begun in this period were headquartered outside Toronto, and there was substantial development of the resource sector, especially petroleum, centred in Calgary. The service sector experienced massive corporate development, the financial sector was augmented with a number of diversified consumer finance companies, and corporate real estate developers emerged during the building booms of the period.

A stable component of monopoly capital, based largely in Toronto and Montreal until the late 1970s, integrated under the control of corporate executives through interlocking directorates. By way of change, globalization, which emerged in full force in the 1980s, stimulated restructuring of leading investment companies.[10] Though the recession of the early 1980s initiated a period of corporate restructuring that was intensified by the recession of the early 1990s, some investment empires expanded (for example, Peter and Edward Bronfman consolidated a major collection of assets), but the collapse of some major holding companies was triggered by dramatic declines in commercial property values.

One of the most interesting blocs of financial capital to emerge in the 1980s was the Montreal-based Quebec pension fund (the $36-billion Caisse de dépôt et placement du Québec). By the end of the decade the company ranked among Canada's top ten financial institutions and supported the development of Quebec corporations long headquartered in Montreal. Nevertheless, as noted above, deregulation was a major force at work during the 1980s, an example being that over fifty foreign banks opened schedule-II banks in Canada (see Dobilas, this volume, chap. 5).

The Quaternary Place System

The current resolution of the development process that linked cities and corporations in mutual support now involves over 3,000 corporations. To organize these by activity and location I assigned them to one of the eight financial subsectors or 20 non-financial subsectors,[11] selecting the largest financial corporations on the basis of their assets, whereas for manufacturing, service, and resource companies I used revenues. Every firm, classified by subsector, was assigned to the city of its corporate headquarters, and a set of urban places has been identified that contains the nation's major corporations. These cities form the Canadian quaternary place system.

The top 25 financial centres in Canada in 1989 contained almost all the 460 largest financial intermediaries in the country (Table 1), while firms in Toronto controlled 48 per cent of the system's assets, ensuring that it forms the apex of the financial decision-making hierarchy. Montreal ranks second, with 28 per cent of the assets while the next ten cities, located between Quebec City and Stratford, each has more than 1 per cent of the corporate assets of the major financial institutions. Banks account for the

Table 1
Total assets ($million) of financial corporations: Canadian headquarters, 1989

Rank	City	Top 62 banks	Top 35 life insurance	Top 20 trusts	Top 70 diversified finance	Top 33 utilities	Top 42 real estate developers	Top 88 credit unions	Top 40 general insurance	Top 460 total	Percentage of total
1	Toronto	285,881	129,254	50,130	41,618	43,502	43,897	4,959	21,017	620,258	47.98
2	Montreal	239,109	6,468	24,242	47,748	35,025	1,180	–	5,512	359,282	27.79
3	Quebec City	–	11,580	619	–	–	–	37,282	519	50,000	3.87
4	Ottawa	–	4,794	–	29,809	312	–	511	–	35,426	2.74
5	Calgary	–	–	4,002	695	16,878	12,458	1,003	186	35,222	2.72
6	London	–	–	32,665	1,028	–	159	168	102	34,112	2.64
7	Vancouver	6,138	–	–	430	12,529	3,505	6,898	2,857	32,357	2.50
8	Winnipeg	–	16,866	–	2,051	5,528	1,521	1,492	948	28,406	2.20
9	Halifax	–	–	17,431	75	1,594	171	–	231	19,502	1.51
10	Edmonton	341	–	813	10,235	3,563	2,794	1,417	281	19,444	1.50
11	Kitchener	–	16,486	–	–	–	–	93	955	17,534	1.36
12	Stratford	–	–	14,051	–	–	–	–	–	14,051	1.09
13	Regina	–	–	–	–	3,539	–	546	421	4,506	0.35
14	St John's	–	–	–	–	2,724	–	–	64	2,788	0.22
15	Fredericton	–	–	–	–	2,730	–	–	–	2,730	0.21
16	Hamilton	–	–	–	487	134	–	658	248	1,527	0.12
17	Saskatoon	–	–	1,016	–	–	–	454	41	1,511	0.12
18	Guelph	–	–	–	–	–	–	–	1,487	1,487	0.12
19	Stellarton	–	–	–	~	–	1,355	–	–	1,355	0.10
20	Barrie	–	–	–	1,260	–	–	–	–	1,260	0.10
21	Kingston	–	1,125	–	–	–	–	–	–	1,125	0.09
22	Brandon	–	–	–	–	–	–	81	626	707	0.05
23	Victoria	–	–	–	–	–	–	698	–	698	0.05
24	St Catharines	–	–	–	–	–	–	688	–	688	0.05
25	Cambridge	–	–	–	–	–	–	–	604	604	0.05
	Subtotal top 25	531,469	186,573	144,969	135,434	128,280	67,040	56,948	36,099	1,286,590	99.53
	Total	531,469	186,573	145,336	135,505	128,288	67,040	62,323	36,154	1,292,688	100.00
	% of total	41.12	14.43	11.24	10.48	9.92	5.19	4.82	2.80	100.00	

Table 2
Total revenues ($million) of non-financial corporations: Canadian headquarters, 1989

Rank	City	Top 1,350 service	Top 1,100 manufacturing	Top 200 resource	Top 2,650 total	Percentage of all cities
1	Toronto	122,485	147,334	36,484	306,303	44.14
2	Montreal	91,571	61,573	976	154,120	22.21
3	Calgary	13,591	8,420	24,581	46,592	6.72
4	Vancouver	20,667	16,132	6,560	43,359	6.25
5	Winnipeg	20,774	1,521	567	22,862	3.30
6	Hamilton	4,285	7,892	–	12,177	1.76
7	Windsor	299	11,454	–	11,753	1.68
8	London	2,130	8,845	–	10,975	1.58
9	Ottawa	5,991	4,341	–	10,332	1.49
10	Edmonton	7,087	1,185	405	8,677	1.25
11	Saskatoon	3,815	834	974	5,623	0.81
12	Kitchener	1,711	2,288	–	3,999	0.58
13	Saint John	774	1,601	1,500	3,875	0.56
14	Regina	2,952	499	321	3,772	0.54
15	Halifax	1,606	1,871	–	3,477	0.50
16	St Catharines	220	2,552	–	2,472	0.36
17	Sarnia	–	2,163	–	2,163	0.31
18	Quebec City	867	1,290	–	2,157	0.31
19	Florenceville	–	2,107	–	2,107	0.30
20	Stellarton	1,767	–	–	1,767	0.25
21	Kamloops	617	1,057	–	1,674	0.24
22	St John's	1,061	439	–	1,500	0.22
23	Guelph	153	1,231	–	1,384	0.20
24	Granby	853	237	–	1,090	0.16
25	Brantford	101	984	–	1,085	0.16
	Subtotal top 25	305,377	287,550	72,368	665,295	95.89
	Total	317,724	302,975	73,121	693,820	100.00
	% of total	45.79	43.67	10.54	100.00	

largest bloc of financial power (41 per cent of the total of all blocs), and Toronto is the dominant place for six of eight sectors. Nevertheless, Montreal is the most important diversified finance place and Quebec the major credit union centre. Ottawa's position derives from its many federal crown corporations that specialize in diversified finance.

A summary of the revenues of non-financial corporations headquartered in the top 25 centres in the subsectors (Table 2) establishes Toronto (44 per cent of revenues) as the undisputed leading decision-making centre, with Montreal (22 per cent) a distant second. Services, with 1,350 top-ranking firms, is the largest of the non-finance sectors (46 per cent of revenues) and the top 25 places account for 96 per cent of the revenues of this sector (see Coffey, this volume, chap. 18). Calgary ranks third in this table, due to its specialization in resources. When the three major non-financial sectors are considered in detail (Tables 2.1, 2.2 and 2.3), the two merchandising subsectors,

Table 3
Total revenues ($millions) of service corporations: Canadian headquarters, 1989

Rank	City	Top 125 retail	Top 340 wholesale	Top 75 communication	Top 70 transport	Top 150 construction	Top 230 misc. service	Top 20 cooperatives	Top 140 business service	Top 1,350 total	Percentage of total
1	Toronto	54,887	19,840	15,946	4,086	6,321	11,642	745	9,018	122,485	38.55
2	Montreal	14,450	14,097	22,827	22,292	7,835	3,624	4,705	1,641	91,571	28.82
3	Winnipeg	1,731	11,398	740	2,957	145	1,005	2,594	204	20,774	6.54
4	Vancouver	1,562	10,096	1,886	1,870	1,747	1,922	385	1,199	20,667	6.50
5	Calgary	4,565	951	–	3,319	1,054	500	2,301	901	13,591	4.28
6	Edmonton	1,651	414	1,539	179	2,767	110	427	–	7,087	2.23
7	Ottawa	50	586	4,125	103	123	850	–	154	5,991	1.89
8	Hamilton	40	1,283	67	1,856	240	799	–	–	4,285	1.35
9	Saskatoon	483	623	21	70	445	248	1,877	48	3,815	1.20
10	Regina	167	123	566	15	–	32	1,837	212	2,952	0.93
11	London	310	265	55	–	1,294	206	–	–	2,130	0.67
12	Stellarton	1,596	–	–	28	–	–	143	–	1,767	0.56
13	Kitchener	1,265	182	–	105	38	20	–	101	1,711	0.54
14	Halifax	–	584	455	219	80	72	116	80	1,606	0.51
15	Moncton	52	–	–	299	–	483	409	–	1,243	0.39
16	St John's	191	180	267	221	101	78	23	–	1,061	0.34
17	Ouebec City	–	250	55	209	215	113	25	–	867	0.27
18	Granby	55	16	–	–	–	–	782	–	853	0.27
19	Saint John	–	251	362	20	116	–	25	–	774	0.24
20	Barrie	644	35	–	–	–	–	–	–	679	0.2i
21	Kamloops	–	–	–	–	–	554	63	–	617	0.19
22	Victoriaville	45	395	–	–	121	–	42	–	603	0.19
23	Charlottetown	–	–	48	487	–	12	13	–	560	0.18
24	Rimouski	–	–	212	–	–	–	328	–	540	0.17
25	Drummondville	–	457	–	–	–	–	–	–	457	0.14
	Subtotal top 25	83,774	62,026	49,171	38,335	22,642	22,270	16,840	13,558	308,586	97.21
	Total	84,599	64,135	49,201	39,817	24,071	23,398	18,853	13,650	317,724	100.0

Table 4
Total revenues ($million), manufacturing corporations: Canadian headquarters, 1989

Rank	City	Top 100 transport machinery	Top 120 food	Top 115 forestry	Top 90 metal	Top 120 chemical	Top 105 electrical	Top 160 machinery	Top 150 textiles	Top 80 misc. mfg.	Top 60 pharmaceutical	Top 1,100 total	Percentage of total
1	Toronto	48,621	18,805	11,582	7,676	11,915	18,249	15,343	5,102	3,999	6,042	147,334	48.63
2	Montreal	4,961	14,779	9,394	15,426	5,409	1,813	1,618	4,514	1,526	2,133	61,573	20.38
3	Vancouver	883	705	9,936	399	515	755	121	140	2,622	56	16,132	5.32
4	Windsor	11,294	–	–	88	–	–	49	–	–	23	11,454	3.78
5	London	1,377	4,856	163	1,503	4,840	–	295	–	651	–	8,845	2.92
6	Calgary	201	1,532	145	25	390	168	111	26	1,372	–	8,420	2.78
7	Hamilton	1,183	159	233	4,190	241	871	653	85	128	–	7,892	2.60
8	Ottawa	–	–	144	780	–	1,604	1,375	58	–	139	4,341	1.35
9	Kitchener	354	719	88	69	427	314	157	160	–	–	2,288	0.76
10	St Catharines	786	220	545	160	97	–	–	–	411	33	2,252	0.74
11	Sarnia	–	–	–	–	2,163	–	–	–	–	–	2,163	0.71
12	Florenceville	–	2,107	–	–	–	–	–	–	–	–	2,107	0.70
13	Halifax	255	1,148	–	–	354	114	–	–	–	–	1,871	0.62
14	Saint John	312	548	741	–	–	–	–	–	–	–	1,601	0.53
15	Winnipeg	242	287	89	169	129	–	54	306	189	56	1,521	0.50
16	Quebec City	–	–	940	–	–	46	44	84	176	–	1,290	0.43
17	Guelph	25	–	–	71	–	403	424	308	–	–	1,231	0.42
18	Edmonton	–	607	–	24	–	–	62	187	305	–	1,185	0.39
19	Kamloops	–	–	1,021	36	–	–	–	–	–	–	1,057	0.35
20	Brantford	120	–	–	17	192	379	276	–	–	–	984	0.32
21	Saskatoon	–	452	–	19	–	46	307	10	–	–	834	0.28
22	Cambridge	–	–	–	216	–	120	418	12	–	–	766	0.25
23	Prince George	–	–	736	–	–	–	–	–	–	–	736	0.24
24	Kingsley Falls	–	–	681	–	–	–	–	10	–	–	691	0.23
25	Brockville	–	–	–	341	–	213	48	–	–	–	602	0.20
	Subtotal top 25	70,614	46,924	36,438	31,209	26,672	25,095	21,355	11,002	11,379	8,482	289,170	95.44
	Total	72,146	49,462	39,843	32,427	27,540	25,148	23,087	12,691	12,124	8,507	302,975	100.0
	% of sector	23.81	16.33	13.15	10.70	9.09	8.30	7.62	4.19	4.00	2.81	100.	

Table 5
Total revenues ($million), resource corporations: Canadian headquarters, 1989

Rank	City	Top 115 petroleum	Top 85 mining	Top 200 total	Percentage of total
1	Toronto	14,974	21,510	36,484	49.90
2	Calgary	24,581	–	24,581	33.62
3	Vancouver	1,196	5,370	6,560	8.97
4	Saint John	1,500	–	1,500	2.05
5	Montreal	214	762	976	1.33
6	Saskatoon	26	948	974	1.33
7	Winnipeg	–	567	567	0.78
8	Edmonton	50	355	405	0.55
9	Regina	182	139	321	0.44
10	Sydney	–	191	191	0.26
11	Red Deer	125	–	125	0.18
	Other	–	431	431	0.58
	Total	42,848	30,273	73,121	100.0

retailing and wholesaling, account for a large share (47 per cent) of the service sector. Toronto (see Simmons, this volume, chap. 17) is the largest headquarters centre (39 per cent of revenues) and, with Montreal, dominates national service control functions, while Winnipeg and Vancouver coordinate service activities in the west. These four cities have a balanced representation of corporate head offices in all service subsectors. Other cities are smaller and more specialized (see Table 3).

Toronto dominates manufacturing – every industry except metal – with Montreal a considerable distance behind (Table 4). Vancouver's third-place standing reflects the size of its forestry industry, and other cities are also specialized – Hamilton (steel), Ottawa (electrical), and Windsor in transportation equipment, which is the largest manufacturing subsector (24 per cent of revenues). These findings accord with other measures of the importance of the auto industry and aircraft (see Holmes, this volume, chap. 13).

The hierarchy of resource-based decision-making centres (Table 5) is focused on Toronto (mining) and Calgary (petroleum), but Vancouver is the dominant mining centre in the west, though Saskatoon, with the world's largest uranium and potash corporations (Cameco and Potash Corp.), is a regional mining decision centre.

Foreign firms control 12 per cent of the assets of financial corporations (Table 6), and Toronto, as Canada's international financial centre, is the location of 59 per cent of the assets of these corporations. London, Ontario, is the second choice for foreign financial headquarters, primarily because of BAT PLC and its holdings in CT Finance, and Montreal ranks third, as the choice of French-based banks. Vancouver ranks fourth, with the Hongkong Bank, and Ottawa fifth, due to Metropolitan Life. In the non-financial sectors foreign control is over 38 per cent (Table 7) but reaches its maximum (54 per cent) in manufacturing, while within the service sector there is 43 per cent

Table 6
Assets ($million), foreign-controlled financial corporations: Canadian headquarters, 1989

Rank	City	Banks	Life insurance	Trusts	Diversified finance	Utilities	Real estate developers	Credit unions	General insurance	Total	Percentage of total	Percentage foreign
1	Toronto	41,600	7,338	689	20,696	–	4,530	–	15,937	91,190	59.11	14.70
2	London	–	–	32,665	1,028	–	–	–	–	33,693	21.84	98.77
3	Montreal	6,345	5,576	–	–	–	–	–	1,462	13,383	8.68	3.72
4	Vancouver	6,138	–	–	–	–	–	–	–	6,138	3.98	18.97
5	Ottawa	–	4,749	–	–	–	–	–	–	4,749	3.08	13.41
6	Kitchener	–	3,321	–	–	–	–	–	–	3,321	2.15	18.94
7	Calgary	–	–	–	–	733	–	–	111	844	0.55	2.40
8	Hamilton	–	–	–	487	–	–	–	–	487	0.32	31.89
9	Halifax	–	–	–	–	–	–	–	231	231	0.15	1.18
10	Trail	–	–	–	–	117	–	–	–	117	0.08	100.00
11	Winnipeg											
	Total top 11	54,083	21,384	33,354	22,211	850	4,530	–	17,855	154,267	100.00	11.93
	% foreign	12.06	11.46	23.01	16.40	0.66	6.76	–	49.46	11.93		
	% of sector	35.06	13.06	21.62	14.40	0.55	2.94	–	11.57	100.00		

Table 7
Revenues ($million), foreign-controlled non-financial corporations: Canadian headquarters, 1989

Rank	City	Service	Manufacturing	Resource	Total	Percentage of total	Percentage foreign
1	Toronto	35,936	103,995	15,226	155,157	58.73	50.65
2	Montreal	13,868	21,410	–	35,278	13.36	22.89
3	Calgary	5,401	–	15,287	20,688	7.83	44.40
4	Vancouver	6,843	5,835	1,165	13,843	5.24	41.93
5	Windsor	–	11,251	–	11,251	4.26	95.72
6	Hamilton	998	3,416	–	4,414	1.67	36.34
7	Winnipeg	2,614	520	495	3,588	1.36	15.69
8	London	145	2,851	–	2,996	1.13	27.30
9	Sarnia	–	2,163	–	2,163	0.82	100.00
10	Ottawa	–	1,760	–	1,760	0.67	17.03
11	St Catharines	–	1,,547	–	1,547	0.59	62.58
12	Kitchener	–	1042	–	1,042	0.39	26.06
13	Kamloops	–	1,021	–	1,021	0.39	60.99
14	Guelph	–	588	–	588	0.22	42.49
15	Brockville	–	581	–	581	0.22	90.50
16	Prince George	–	569	–	569	0.22	75.26
17	Peterborough	–	557	–	557	0.21	84.91
18	Cambridge	–	511	–	511	0.19	57.42
19	Charlottetown	487	–	–	487	0.18	81.84
20	St Thomas	–	389	–	389	0.15	89.22
	Subtotal top 20	66,292	160,000	32,132	258,430	97.83	40.75
	Total	67,092	164,752	32,310	264,154	100.00	38.07
	% foreign	21.12	54.38	44.18	38.07	21.12	54.38
	% of total	25.40	62.37	12.23	100.00	25.40	62.37

foreign control in wholesaling but only 25 per cent in retailing (Table 8), and little or no control in communications, transportation, cooperatives, and business services. Toronto is the primary headquarters location for 59 per cent of the revenues of all foreign headquarters, and locally this amounts to 51 per cent of the revenues of all companies. Montreal and Calgary are distant second and third choices, but together the top four cities include headquarters responsible for 85 per cent of the total of foreign revenue, and this locational concentration is much greater than for domestic firms.

In manufacturing (Table 9) 63 per cent of the revenue total is concentrated in Toronto, and only 13 per cent shows up in Montreal. Windsor ranks number three due to the presence of Chrysler Canada. Transportation equipment (83 per cent foreign controlled) accounts for 37 per cent of the revenues generated by foreign corporations. High levels of control also occur in chemicals, electrical, machinery, and especially pharmaceuticals, sectors closely associated with Ontario and its quaternary places. Southern Ontario and its industrial quaternary places appear vulnerable with respect to any negative changes in the international economic environment. By contrast, in the resource sector (Table 10) foreign firms control 44 per cent and over 94 per cent of this

Table 8
Revenues ($million), foreign service corporations: Canadian headquarters, 1989

Rank	City	Retail	Wholesale	Communication	Tranportation	Construction	Misc. service	Coops	Service	Total	% all cities	% foreign
1	Toronto	11,964	15,084	1,115	1,229	1,189	3,373	–	1,982	35,936	53.56	29.34
2	Montreal	265	4,902	2,126	85	4,898	1,592	–	–	13,868	20.66	15.14
3	Vancouver	601	4,491	–	451	814	486	–	–	6,843	10.20	33.11
4	Calgary	4,198	251	–	270	246	190	–	246	5,401	8.05	39.74
5	Winnipeg	–	2,501	–	–	–	91	–	22	2,614	3.90	12.58
6	Hamilton	–	791	–	–	–	207	–	–	998	1.49	23.29
7	Charlottetown	–	–	–	487	–	–	–	–	487	0.73	100.00
8	Edmonton	–	30	–	–	159	45	–	–	234	0.35	3.30
9	London	–	145	–	–	–	–	–	–	145	0.22	6.81
10	Sept lles	–	–	–	100	–	–	–	–	100	0.15	100.00
	Subtotal top 10	17,028	28,195	3,241	2,622	7,306	5,984	–	2,250	66,626	99.31	
	Total	17,028	28,633	3,241	2,622	7,306	6,012	–	2,250	67,092	100.00	21.12
	% foreign	20.13	44.64	6.59	6.59	30.35	25.69	–	16.48	21.12		
	% of sector	25.38	42.68	4.83	3.91	10.89	8.96	–	3.35	100.00		

Table 9
Revenues ($million), foreign manufacturing corporations: Canadian headquarters, 1989

Rank	City	Transport machinery	Food	Forestry	Metal	Chemical	Electrical	Machinery	Textiles	Misc. mfg.	Pharma-ceutical	Total	Percentage of total	Percentage foreign
1	Toronto	43,305	14,703	3,081	2,198	9,889	7,098	14,099	931	2,715	5,936	103,955	63.10	70.56
2	Montreal	1,627	5,126	2,758	1,598	4,671	1,173	1,168	671	687	1,931	21,410	13.00	34.77
3	Windsor	11,225	–	–	–	–	–	26	–	–	–	11,251	6.83	98.23
4	Vancouver	828	–	4,084	163	–	599	–	40	121	–	5,835	3.54	36.17
5	Hamilton	1,183	51	233	118	331	847	653	–	–	–	3,416	2.07	43.28
6	London	292	110	–	1,503	–	–	295	–	651	–	2,851	1.73	32.23
7	Sarnia	–	–	–	–	2,163	–	–	–	–	–	2,163	1.31	100.00
8	Ottawa	–	–	–	35	–	608	1,059	58	–	–	1,760	1.07	40.54
9	St Catharines	494	–	545	–	97	–	–	–	411	–	1,547	0.94	68.69
10	Kitchener	288	–	40	41	427	135	111	–	–	–	1,042	0.63	45.54
11	Kamloops	–	–	1,021	–	–	–	–	–	–	–	1,021	0.62	96.59
12	Guelph	–	–	–	71	–	252	265	–	–	–	588	0.36	47.77
13	Brockville	–	–	–	341	–	213	27	–	–	–	581	0.35	96.51
14	Prince George	–	–	569	–	–	–	–	–	–	–	569	0.35	77.30
15	Peterborough	135	342	80	–	–	–	–	–	–	–	557	0.34	97.37
16	Winnipeg	242	–	–	–	89	–	–	–	189	–	520	0.32	34.19
17	Cambridge	–	–	–	–	–	93	418	–	–	–	511	0.31	66.71
18	St Thomas	–	–	–	95	43	–	192	–	59	–	389	0.24	89.22
19	St Jean	80	–	–	235	61	–	–	–	–	–	376	0.23	87.23
20	Halifax	–	–	–	–	354	–	–	–	–	–	354	0.21	18.92
	Subtotal top 20	59,699	20,332	12,411	6,398	18,125	11,018	18,313	1,700	4,833	7,867	160,696	97.54	55.57
	Total	60,197	20,453	13,141	6,457	18,917	11,191	19,097	2,360	5,047	7,892	164,752	100.00	54.38
	% foreign	83.43	41.35	32.98	19.91	68.69	44.50	82.72	18.60	41.62	92.77	54.38		
	% of sector	36.53	12.41	7.98	3.92	11.48	6.79	11.59	1.43	3.06	4.79	100.00		

Table 10
Revenues ($million), foreign resource corporations: Canadian headquarters, 1989

Rank	City	Petroleum	Mining	Total	% all cities	% foreign
1	Calgary	15,287	–	15,287	47.31	62.19
2	Toronto	12,764	2,462	15,226	47.12	41.73
3	Vancouver	669	496	1,165	3.61	17.76
4	Winnipeg	–	454	454	1.41	80.07
5	Saskatoon	–	178	178	0.55	18.28
	Total top 5	28,720	3,590	32,310	100.00	
	% foreign	67.02	11.86	44.18		
	% of sector	88.88	11.12	100.00		

is associated with Calgary and Toronto, a combination that produces a stronger western orientation in control.

The Foreign Component

The geography of foreign parents of Canadian corporations (Table 11) demonstrates a high concentration of assets controlled, even within the most important 25 cities. London alone controls 38 per cent of all assets, followed by New York (14 per cent), Detroit (10 per cent), and Chicago (6 per cent); only ten cities control over 86 per cent of international intermediary financial assets in Canada. In the three non-financial sectors (Table 12), Detroit changes place with London, and the big three control 40 per cent of revenues. The origins of investment in Canada are an extremely dispersed international group of cities, indicating the true globalization of non-financial sectors, especially manufacturing. That more than 100 foreign cities have a substantial stake in the Canadian economy indicates a broad base to its attractiveness and represents an element of stability. It also reflects a very real element of competition, though the geographic concentration of these cities in the United States is very strong (see MacPherson, this volume, chap. 4).

Since foreign control is such a significant component of the Canadian system of quaternary places, the foreign component of each domestic place has been assigned to the relevant international centre. The assets associated with financial corporations based in New York, for example, were subtracted from the domestic totals of their Canadian locations and assigned directly to New York, as if it were a domestic decision centre. When undertaken for over 150 foreign cities for the financial sector (Figure 1), and the top 35 domestic and foreign quaternary places, five levels in the decision-making hierarchy may be distinguished.[12] The number associated with each city in Figure 1 represents the proportion of all Canadian assets controlled by that place; for example, 41 per cent of all Canadian corporate assets are controlled directly from Toronto. This total excludes those foreign assets controlled by foreign subsidiaries, which are assigned directly to the foreign parent's locations. In this manner a true picture of

Table 11
Financial assets ($million) of foreign-controlled corporations: parent
locations, 1989

Rank	City	Total financial assets	Percentage of total
1	London	59,029	38.26
2	New York	20,939	13.57
3	Detroit	15,862	10.28
4	Chicago	9,111	5.91
5	Hong Kong	6,138	3.98
6	Zurich	6,082	3.94
7	Paris	5,844	3.79
8	Tokyo	5,588	3.62
9	Stamford	2,747	1.78
10	Basel	2,376	1.54
11	Hartford	2,164	1.40
12	San Francisco	2,128	1.38
13	Osaka	1,772	1.15
14	Frankfurt	1,729	1.12
15	Los Angeles	1,490	0.97
16	Rotterdam	1,144	0.74
17	Milan	848	0.55
18	Amsterdam	794	0.51
19	Omaha	689	0.45
20	Boston	654	0.42
21	Philadelphia	585	0.38
22	Nagoya	538	0.35
23	Pittsburgh	535	0.35
24	Tampa	519	0.34
25	Bloomington, Ill.	478	0.31
	Subtotal top 25	149,783	97.09
	Others	4,484	2.91
	Total	154,267	100.00

decision making appears, with international centres taking their place with the Canadian system.

The financial quaternary place system of Canada is dominated by blocs of financial capital controlled from Toronto and Montreal, with smaller blocs associated with third-level financial centres such as London, England, Quebec City, and Calgary.[13] When comparable transformations are undertaken for the revenues of non-financial activities (Figure 2), Toronto and Montreal emerge as decision-making centres with almost equal power, Toronto no longer dominating the system when its foreign fraction is subtracted. Though the two largest blocs of industrial and commercial capital are still domestically controlled, the five second-tier centres are foreign cities. The foreign penetration of the hierarchy increases from level 2 to level 5, indicating the vulnerability of Canada to decisions made abroad (Wheeler 1988). As noted above, the positive feature of this pattern is the urban diversity of foreign control, which can

Table 12
Revenues ($million) of foreign non-financial corporations: parent locations, 1989

Rank	City	Service	Manufacturing	Resource	Total revenues	Percentage of all cities
1	Detroit	2,210	46,812	–	49,022	18.56
2	New York	8,571	17,898	1,473	27,942	10.58
3	London	3,996	19,201	4,288	27,485	10.40
4	Chicago	6,679	8,371	3,374	18,424	6.97
5	Tokyo	11,183	4,288	–	15,471	5.86
6	Dallas	667	1,626	10,007	12,300	4.66
7	Paris	7,142	1,387	2,624	11,153	4.22
8	San Francisco	5,066	1,637	1,547	8,250	3.12
9	Philadelphia	626	4,229	1,215	6,070	2.30
10	Stamford	3,420	1,638	–	5,058	1.91
11	The Hague	–	–	4,917	4,917	1.86
12	Minneapolis	2,160	2,402	–	4,562	1.73
13	Osaka	3,840	503	–	4,343	1.64
14	Pittsburgh	30	3,197	–	3,227	1.22
15	Buffalo	–	2,849	–	2,849	1.08
16	Houston	546	1,685	488	2,719	1.03
17	Los Angeles	653	1,209	830	2,692	1.02
18	Wellington	40	2,619	–	2,659	1.01
19	Seattle	282	2,111	–	2,393	0.91
20	Boston	645	1,621	–	2,266	0.86
21	Koln	195	1,839	–	2,034	0.77
22	Midland, Mich.	–	2,033	–	2,033	0.77
23	Stockholm	579	1,220	–	1,799	0.68
24	Frankfurt	284	1,002	492	1,778	0.67
25	St Louis	–	1,734	–	1,734	0.66
	Subtotal top 25	58,814	133,111	31,255	223,180	84.49
	Total	67,092	164,752	32,310	264,154	100.00
	% sector	25.40	62.37	12.23	100.00	

be interpreted as a stabilizing element for the Canadian quaternary system (Palmer and Friedland 1987).

GLOBAL ADJUSTMENTS FOR THE 1990S

Over the past decade the rapid internationalization of flows of capital has transformed the relationship between capital accumulation and national states. In the first place, cross-penetration of capital has made the advanced economies more dependent on one another (Berry 1989). Among the seven major capitalist countries, for example, there has been a dramatic increase in the extent to which business cycles are synchronized. Prior to the late 1970s this occurred within western Europe and within North America, but much less so across continents. In the 1980s, however, the North

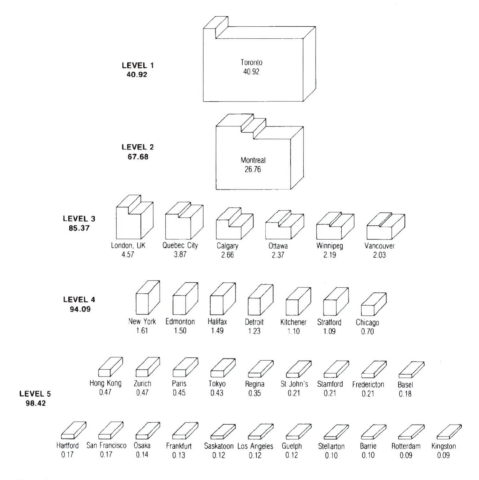

Figure 1
Canada's quaternary place system, 1989, for top 35 centres for financial sector (domestic versus foreign component as percentage of total assets).

American economy became more closely linked with the economies of Japan and Europe (MacEwan 1984), and the coherence of temporal patterns of accumulation is a result in large part of decisions by executives of large corporations, consumers' groups, and governments (Erickson and Hayward 1991). The globalization of capital has important implications for the actions of both corporations and governments. Increasingly the dynamics of the world economy cannot be understood with reference to a single nation or groups of nations, since productive decisions are made on a global scale.

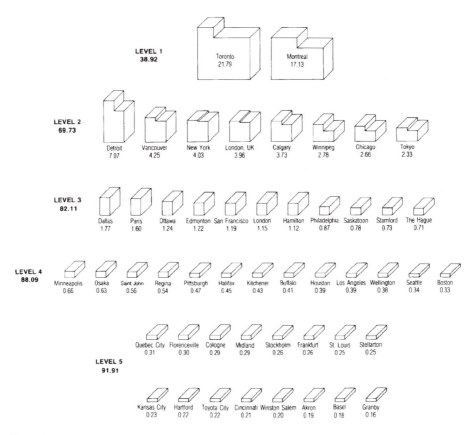

Figure 2
Canada's quaternary place system, 1989, for top 50 centres for non-financial sectors (domestic versus foreign component as percentage of total revenue).

Privatization

Canada has developed through a mixed economy, split between large, state-owned enterprises and private firms. The historical attachment to the crown corporation began with the railways, which were built after 1867 as one of the conditions of Confederation. Utilities, telephone systems, aircraft manufacturing companies, a national air carrier, and oil companies, among others, have all been created at some time when there has been a need for government participation. Usually these initiatives were taken because the private sector was unable to assume the risks and the financial community lacked the resources to capitalize major projects. What started as a measured response to the practical economic and political realities of the day became a common practice, with successive federal and provincial governments adding to the inventory with the best of intentions and in the national interest. After 125 years of accumulating commercial holdings the federal government alone had amassed 53 parent and 113 subsidiary crown corporations employing over 170,000 people. In the face of mounting annual deficits and limits imposed by tough expenditure restraints, however, privatization has emerged as a favoured policy option.[14] Since there is growing public demand for increasingly scarce government resources, privatization has been viewed by a variety of governments as a necessary step (McDermid 1990).

Privatization is the divestiture of state-owned assets or enterprises to the private sector, and the key word is "divestiture" – transfer of ownership and authority for that business to private entrepreneurs. The additional arguments that have been made in favour of government divestiture centre on increased competition, greater efficiency and lower costs of production. Efficiency can lead to innovation and new products and better methods of production, which are necessary if there are to be new jobs and a larger market share. This process can lead to a larger tax base, reducing the strain on state treasuries and permitting provision of more and better business infrastructure.

Deregulation and Free Trade

Federal economic policies and Canadian business strategies have converged on a continental process that has strengthened indigenous firms in their home market and pushed them toward international expansion, particularly in the United States. Today the maturity of the Canadian business structure has reduced the permanence of foreign direct investment, especially in manufacturing, mining, oil and gas, and pipelines. Because Canada possesses an independent financial power that accrues to advanced nations, there has been repatriation of foreign-controlled companies through share purchases and expanded investments abroad. The importance of u.s. investments is partly that of an external pressure contributing to fast concentration and centralization. This insight was not lost on Canada's Royal Commission on Corporate Concentration (Gorecki and Stanbury 1979), which concluded that the growth of domestic firms, even at the cost of an increase in concentration, will offset both the proportion of the economy under foreign control and some of the undesirable consequences of foreign investment (Carroll 1986).

Deregulation of the Canadian transportation system (see Lea and Waters, this volume, chap. 16) puts the pressure of competition on hundreds of small firms, cuts margins, and improves efficiency. The anticipated effect is rationalization in the industry, resulting in a restructured, reorganized transport sector. Deregulation forces trucking companies to become larger and more efficient and the railways to become smaller and more effective, while for air carriers it has resulted in a round of mergers that two carriers have survived. Each has been forced to set up a hub-and-spoke air network so that the remodelled system can compete when and if an "open sky" policy is adopted for the continent. The two railways and a rationalized trucking system are adapting to the stronger north-south orientation of commerce in the 1990s.

Free trade is part of the process of privatization, deregulation, and globalization, and firms are rationalizing in the hope of competing in global markets. For the Canadian quaternary place system this rationalization will remove many inefficient firms. As others specialize and globalize their production and services, some will transfer corporate decision making to other places. As globalization continues, places lose some of their competitive edge and major corporations seek alternate locations for headquarters. This trend was illustrated quite clearly in the 1970s, when corporations moved their headquarters from Montreal to Toronto during a period of political unrest. In the late 1980s, companies demonstrated that they will move from expensive locations in Ontario; for example, Varity Corp. moved to Buffalo, Crown Life to Regina, Trans Canada Pipe Lines to Calgary, Cameco to Saskatoon, McLeod-Stedman to Winnipeg, and Shell to Calgary. For quaternary places in southern Ontario, Toronto in particular, these relocations may signal an era of rapid decentralization of decision making, and Toronto may become a "humbler metropolis" (Mittelstaedt 1991).

NOTES

1 Other examples are the Champlain and St Lawrence Railroad (1836), now part of CNR, and Canadian Steamship Lines (1845), all of Montreal, and Consumers Gas, Toronto (1848). All firms in this and other notes are listed by their current corporate name, headquarters city, and date of incorporation or charter. The *Blue Book of Canadian Business* provides corporate profiles for most of the firms.

2 These include the Bank of Nova Scotia, Halifax-Toronto (1832); Royal Insurance, Toronto (1835); the biggest life insurer, Sun Life, Montreal-Toronto (1847), and major accounting firms such as Richardson Greenshields, Kingston-Winnipeg (1857).

3 The material on the evolution of independent, indigenous fractions of financial capital in this and subsequent sections draws heavily on Carroll (1986).

4 For example, Montreal Trust (1889) and Caisse populaire Desjardins, Quebec City (1900).

5 Montreal (1874), Toronto (1878), Vancouver (1907), and Alberta, in Calgary (1914).

6 Imperial Oil, Toronto (1880), and Cominco, Vancouver (1906); Bell Canada, Montreal (1880), and Sasktel, Regina (1906); Canadian Pacific, Montreal (1881); and Ontario Hydro, Toronto (1906).

7 Varity, Toronto (1891); General Electric, Toronto (1882); Quaker Oats, Peterborough (1901); Stelco, Hamilton (1910); MacMillan Bloedel, Vancouver (1911); and Ford, Toronto (1911).

8 For example, Weston, Toronto (1882); United Grain Growers, Winnipeg (1911); Marsh and McLennan, Toronto (1914); and Lundrigan-Comstock, Corner Brook (1916).

9 Manitoba Pool, Winnipeg (1925); Federated Coop, Saskatoon (1928); Coop Atlantic, Moncton (1928); and Agropur, Granby (1938). The Credit Union Central of Saskatchewan, Regina (1941), is also associated with the movement.

10 Companies such as Canadian Pacific Investments, Power Corporation, Argus, and BCE Inc.

11 The author maintains a permanent computerized data file for some 5,000 Canadian public, private, and crown corporations for the period 1954 to the present. The data are gathered from well-known Canadian and international business services, including Financial Post Corporation Services, Dun's Key Businesses (Top 20,000 Canadian), the Globe and Mail *Report on Business*, and World Scope Corporate Profiles. All data are cross-checked for accuracy.

12 See Wheeler and Brown (1985) on the quaternary place corporate hierarchy in the U.S. south.

13 See Wheeler (1986; 1988) for definitions of dominant quaternary centre.

14 Recent examples of major privatizations include, at the federal level, Air Canada, Montreal; De Havilland, Toronto; Teleglobe Canada, Ottawa; Petro-Canada, Calgary; and Canadair, Montreal; and, at the provincial level, Cameco, Saskatoon; AGT, Edmonton; B.C. Gas Corp., Vancouver; and Potash Corp., Saskatoon.

REFERENCES

Berry, B.J.L. 1989. "Comparative Geography of the Global Economy: Cultures, Corporations, and the Nation State." *Economic Geography*, 65, 1–18.

Carroll, W.K. 1986. *Corporate Power and Canadian Capitalism*. Vancouver: University of British Columbia Press.

Chandler, A.D. 1962. *Strategy and Structure*. Cambridge, Mass.: MIT Press.

– 1977. *The Visible Hand*. Cambridge, Mass.: Harvard University Press.

Daniels, P.W. 1979. *The Spatial Pattern of Office Growth and Location*. New York: Wiley.

Erickson, R.A., and Hayward, D.J. 1991. "The International Flows of Industrial Exports from U.S. Regions." *Annals, Association of American Geographers*, 81, 371–90.

Gad, G.H.K. 1985. "Office Centralization: The Locational Dynamics of Suburbanization in Toronto." *Urban Geography*, 6, 331–51.

– 1991. "Office Location in Canada's Major Urban Centres with Emphasis on Toronto and Suburbs." In T. Binting and P. Fillion, eds., *Canadian Cities in Transition*, 432–59. Toronto: Oxford University Press.

Gorecki, P.K., and Stanbury, W.T. 1979. *Perspectives on the Royal Commission on Corporate Concentration*. Toronto, Butterworths.

Huang, S.S.L. 1989. "Office Suburbanization in Toronto: Fragmentation, Workforce Composition and Labour Sheds." PhD dissertation, University of Toronto.

McDermid, T., 1990, "Canada's Privatization Experience." In *International Privatization: Global Trends, Policies, Processes, Experiences*, 57–62. Saskatoon: Institute for Saskatchewan Enterprise.

MacEwan, A. 1984. "Interdependence and Instability: Do the Levels of Output in Advanced Capitalistic Countries Increasingly Move Up and Down Together?" *Review of Radical Political Economics*, 16, 57–79.

Manderson, K. 1990. "The Global Insiders." *Financial Post*, Summer, 169–83.

Mittelstaedt, M. 1991. "Humbler Metropolis." *Globe and Mail*, 26 Oct. 1991.

Naylor, T. 1975. *The History of Canadian Business 1867–1914: Vol. 1, The Banks and Financial Capital*. Toronto: Lorimer.

– 1975. *The History of Canadian Business, 1867–1914: Vol. 2, Industrial Development*. Toronto: Lorimer.

Niosi, J. 1981. *Canadian Capitalism: A Study of Power in the Canadian Business Establishment*. Toronto: Lorimer.

– 1985. *Canadian Multinationals*. Kitchener, Ont.: Dumont Press.

Palmer, D.A., and Friedland, R. 1987. "Corporation, Class, and City System." In M.S. Mizruchi and M. Schwartz, eds., *Intercorporate Relations: The Structural Analysis of Business*, 145–84. Cambridge: Cambridge, University Press.

Semple, R.K. 1985. "Toward a Quaternary Place Theory." *Urban Geography*, 6, 285–96.

Wheeler, J.O. 1986. "Corporate Spatial Links with Financial Institutions: The Role of the Metropolitan Hierarchy." *Annals, Association of American Geographers*, 76, 262–74.

– 1988., "Spatial Ownership Links of Major Corporations: The Dallas and Pittsburgh Examples." *Economic Geography*, 64, 1–16.

Wheeler, J.O., and Brown, C.L. 1985. "The Metropolitan Corporate Hierarchy in the U.S. South." *Economic Geography*, 61, 66–78.

PART FIVE

Political Economic Questions

It has been argued throughout this book that the "openness" of the Canadian economy helps explain its locational development. The concept of openness has changed in significance over time; it also takes on different meanings for different sectors, as illustrated by the reliance of Canada on access to foreign markets to absorb its massive exports of resources while historically it placed tariffs on imports of highly manufactured goods in order to assist the development of secondary industry. The locational legacy of the policy of protection but with open borders to investment capital has been quite striking. Central Canada, especially southern Ontario, not only gained from its location in the Canadian urban system but throughout much of this century attracted a substantial share of u.s. corporate investment. Today, southern Ontario, especially the Toronto area, is Canada's prime industrial and advanced-service location.

With the liberalization of trade, in terms of tariffs and non-tariff barriers under the General Agreement on Tariffs and Trade (GATT) and the North American Free Trade Agreement (NAFTA), capital investment is free to choose locations of maximum corporate advantage. Other Canadian economic policies have sought to stimulate or support development of most economic sectors; various forms of infrastructure investment in transportation, communications, and urban development have involved the three levels of government, while some federal programs have covered part of exploration costs associated with new mineral developments. New coal, oil, and gas developments ("megaprojects") have been brought into production by public participation of provinces and the federal government. In agriculture (see also Found, this volume, chap. 9) price support mechanisms (supply management) have been a costly economic and social mainstay. Duty remission for imported machinery has encouraged use of new equipment, and in manufacturing there have been programs that provide direct support (and tax concessions) for modernization, including adoption of more efficient techniques.

Some of these policies have been delivered without obvious reference to location, while others reflect Canada's resolve to combat sharp and lasting regional differences

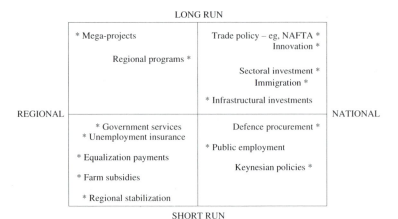

Figure 1
Locational and temporal classification of sample public-sector programs
(after Simmons 1981).

in per-capita earned income (Britton 1988; Cannon 1989; Davis, this volume, chap. 20). As a political consequence of the constitutional form of Canada, "equalization" payments are made by the federal government to several provinces to assist them in providing otherwise-unaffordable social, health, educational, and other public services. In addition, "stabilization" programs offset significant falls in provincial income because of seasonal conditions or world price shifts. An evolving stream of regional economic development initiatives funded by the central government has also attempted to create new jobs in the Atlantic region and Quebec and to diversify the employment structure of western Canada. These programs do not have a strong reputation for long-term effectiveness, and their share of the federal budget has declined to about 3 per cent of annual expenditures (Britton 1988). Nevertheless, the response of provinces to these programs is one of considerable rivalry and dissatisfaction (Savoie 1990).

A simple view of the complex range of such programs is provided in Figure 1. The combination of spatial scale and time horizon (Simmons 1981) differentiates policies that address the nation (trade) from either federal programs that redistribute funds or provincial and local initiatives that affect more limited areas. Initiatives such as regional development with a long-run redistributive goal are distinguished from those that aim at reducing short-run economic fluctuations. Though the purpose of many policies is to influence the location of private-sector production, other programs are designed to maintain consumption, especially in low-income regions, and to secure provision of public services at a reasonable standard wherever there is sufficient demand. About 25 per cent of GDP (collected through taxation) is redistributed to reduce the discrepancies between the spatial patterns of consumption and income earned by the Canadian population.

The scale of the public part of the Canadian space economy, as outlined by Davis (this volume, chap. 20), may be gauged in terms of measures of public spending, revenue, and public employment. Unlike 25 years ago, the role of the public sector as employer is large (30 per cent), though it varies greatly between provinces – high in Atlantic Canada and low in Ontario and British Columbia. These variations reflect the location pattern of federal ministries and enterprises and differences in the need for public provision of services across the country. There has been a significant decline in the role of the federal government as a source of public-sector expenditure. Now, the federal and provincial governments each spend more than 40 per cent (and local governments, 16 per cent), reflecting the increasing devolution of effective authority over public expenditure to the provinces. Despite this trend, Davis shows that by the mid-1980s federal fiscal inputs into Atlantic Canada supported levels of final demand up to 45 per cent higher than provincial output.

Davis argues that because of transfers within and between provinces, the spatial distribution of final demand is, to a high degree, politically determined. The net fiscal support (input) provided to provincial demand, which includes transfers, contributes to accumulated debts. Despite the history of regional development initiatives in response to the limited industrialization of the poorer regions, transfers have been pared down by Ottawa. Until recently the contributions to transfers came from two resource economies, British Columbia and Alberta, and only Ontario contributed from its base of secondary manufacturing. As a result of the economic crisis of the 1990s, however, the weakness of Ontario's manufacturing has been exposed, and the core of the Canadian industrial economy is now subjected to the type of restructuring that many northeastern U.S. states experienced during the 1970s (Britton 1993). Continuing problems of productivity growth, slow restructuring to meet international competition, and a low rate of innovation have left Ontario susceptible to the ravages of the recession: provincial unemployment rose to over 11 per cent during 1993, but it was 8.4 per cent (seasonally adjusted) by December 1994.

Despite substantial recovery from the recession, with 467,000 new jobs being generated in 1994, technological change and innovation are the basis for an improvement in the performance of the Canadian industrial economy. There are, however, substantial impediments to innovation by Canadian firms. Britton, Gilmour, Smith, and Steed (this volume, chap. 22) argue that Canadian companies must overhaul current innovation practices in all sectors so as to generate competitive strength through new or redesigned and improved products, services, and processes. Unfortunately, Canada has few large "flagship" companies with business networks that stimulate innovation by smaller companies. There are also only a few young Canadian, technology-based companies of intermediate size ("threshold companies") that are successful in highly specialized, international markets. In contrast to the world's leading firms, which favour incremental innovation, the dominant approach among even the leading Canadian firms, in many industries, is acquisition and minor modification of technology developed elsewhere.

In reviewing Canada's options, Britton and his colleagues (this volume, chap. 22) suggest that although trade liberalization has encouraged competition, this is likely to be only part of the solution. The answer, however, is not more policy measures that embody subsidies, which are liable to be judged grounds for countervailing trade actions under NAFTA.

The most advantageous locations for manufacturing are the strong spatial concentrations of firms that require and supply a variety of industrial, business, and financial services, equipment, and components. This conclusion applies also to the producer and financial services and head office functions. In these locations both small and large firms can benefit from improved accessibility to inputs, flexible supply systems, and savings in communication and delivery times. While Toronto and, to a lesser extent, Montreal provide Canada's best-developed range of these benefits, these are not islands of superior industrial performance. Rather, small and medium-sized firms, which are concentrated in these locations, suffer from inefficient contact and inadequate linkage with other firms and a slow rate of innovation. Thus in all locations new initiatives are needed to improve the flow of industrial information between firms about inputs and markets.

Canada has consistently undervalued the evidence that advisory service networks are needed to increase the rate in all industries at which firms install new processing technology and get access to sources of technical and business information and venture capital. If available, these would go some way toward increasing design inputs and re-engineering and speeding the rate at which workers are trained and new product concepts and production techniques are adopted or adapted. There has been, however, little real policy action in response, despite strong and widespread international agreement that small firms require assistance in acquiring information about new technology and implementing new methods of work. Traditionally, much of Canadian policy making has overlooked structural and behavioural weaknesses in Canada's major industrial locations in favour of policies to generate jobs, especially in less-favoured regions. Now, flawed and incomplete innovation systems in Canada's industrial regions also are of pressing concern for the innovation policies of all levels of government.

Methods to increase productivity and restructure industry involve not only new activities by firms but also new skills for workers if Canada's trading performance is to improve. The importance of labour skills for continuous improvement in design, manufacture, and marketing of products provides the context for Rutherford's chapter (this volume, chap. 21). He argues that training programs have been weak and substantial structural unemployment has led to a two-tier labour market. Labour market policies have begun to change: training is now a stronger focus, there has been a definite shift from welfare to workfare programs, and responsibility has shifted from Ottawa to provincial governments. Thus the increased demands that industrial innovation places on labour skills and labour mobility have been acknowledged as a burden to be shared by industry, labour, and government. Nevertheless, Canada's ability to respond to labour market changes has been woefully inadequate given recent recessions, trade liberalization, and the pace of technological evolution. According to Rutherford, the mismatch of labour to industrial needs continues, institutional adjustment is critical, and only limited progress has been made.

REFERENCES

Britton, John N.H. 1988. "Economic Change and the Regional Question." *Canadian Journal of Regional Science*, 11, 125–36.

– 1993. "A Regional Industrial Perspective on Canada under Free Trade." *International Journal of Urban and Regional Research*, 17, 559–77.

Cannon, J.B. 1989. "New Directions in Regional Policy." *Canadian Geographer*, 33, 230–9.

Savoie, Donald J. 1990. *The Politics of Public Spending in Canada.* Toronto: University of Toronto Press.

Simmons, J.W. 1981. "The Impact of Government on the Canadian Urban System: Income Taxes, Transfer Payments, and Employment." Research Paper No. 126, Centre for Urban and Community Studies, University of Toronto.

CHAPTER TWENTY

Canada's Public Space Economy

J. TAIT DAVIS

In Canada you are reminded of the government every day. It parades itself before you. It is not content to be the servant but will be the master.   Henry Thoreau 1886, IX, 103

The 30 years between 1956 and 1986 witnessed a significant broadening of the scope and scale of governmental intervention in the economic affairs of individuals, businesses, and regions in Canada. To a much greater extent than is commonly appreciated, the structure and spatial organization of Canada's economy are shaped by its public institutions. In 1986 the public sector accounted for about 30 per cent of all employment and owned about 40 per cent of the nation's non-residential capital stocks, and governments spent more than one-quarter of gross domestic product (GDP) on production of public goods and services. The sheer magnitude of this physical presence in the economy and its structurally significant regional variations constitute the first of two perspectives on the public space economy considered in this chapter.

The second perspective focuses on the role of governments in interception and redirection of factor payments, or money flows. Governments at all levels currently determine where and by whom at least half of Canada's GDP will be spent. Just over one-quarter of GDP, appropriated through taxation, is redistributed among individuals, business enterprises, and regions. Such interventions reduce the factor incomes accruing to some individuals and regions, increase those of others, and are the principal policy mechanism employed to compensate for the persistent geographic discrepancy between production and consumption capacities that characterize the Canadian space economy.

In the theory of economic geography the term "spatial organization" implies that economic structures and associated flows of commodities and factor payments are organized to serve some objective purpose (Gore 1984, 3–8), usually expressed in terms of efficiency: to use resources (production factors) so as to realize the greatest profit or utility, and to achieve the highest possible level of spatial interaction or interregional exchange consistent with the principles of comparative advantage (Morrill

1970). Where, as in the Canadian case, the objectives of public-sector intervention reflect political rather than economic criteria, a degree of inefficiency can be expected. Difficulties will be created for structural adjustment in provincial economies and for national economic policy. These concerns are the focus of the concluding section.

THE PUBLIC SECTOR IN CANADA

The federal structure of Canada's public sector and the allocations of responsibility and authority within that framework derive from the way in which two imperatives were reconciled at the time of Confederation: the need to create a system of governance that could assume the tasks of defence and economic development as Britain withdrew and the American Civil War came to an end and the need to accommodate strong regional loyalties that made establishment of a unitary state impracticable (Norrie, Simeon, and Krashick 1986). Canada's governments have played an active, interventionist role in the spatial and structural characteristics of economic development and maintained a political commitment to the demographic, cultural, and economic viability of the constituent provinces. In policy terms, Canada's economic regions are identified with provincial entities that vary greatly in area, population, and natural resource endowment, and there is a tradition of resolving income and welfare-distribution issues at a provincial rather than an individual scale.

Estimates of the size of Canada's public sector, and of the relative magnitudes of its several components – federal, provincial, local, and municipal governments, education and health services, and public enterprises – vary considerably, reflecting numerous problems of classification and measurement. Public enterprises present particular difficulties. The principal criterion for inclusion of an enterprise in the public sector is government ownership and control, a status that may change. De Havilland Aircraft, for example, was a public enterprise from 1974 to 1987 but a private-sector company before and since; Canada Post has been a crown corporation since 1981 but operated previously as a federal government department. The criteria developed by Foot (1978) are largely followed here but yield estimates of employment that are less than those of other sources – for example, the Macdonald Commission (1985).

Employment

Direct public-sector employment more than doubled between 1960 and 1986, rising from 1.2 million to 2.6 million jobs and from 19 to 23 per cent of total employment. The indirect employment generated by government purchases of goods and services from the private sector approximates 7 per cent of total national employment (Bucovetsky 1979).

Government employment has increased from a mid-1960s low of 39 per cent to 46 per cent of the public-sector total, and by 1986 governments and their wholly owned enterprises had become dominant, accounting for more than 60 per cent of public-sector and 14 per cent of all jobs (Table 1). Federal-government employment

Table 1

Public-sector employment, non-residential capital, and enterprises in the national economy, 1960–86

	1960		1966		1976		1986	
	Number	%	Number	%	Number	%	Number	%
Employment								
Federal government	203,013	3.4	221,345	3.1	417,980	4.4	369,087	3.2
Provincial government	139,434	2.3	211,809	3.0	375,600	4.0	480,000	4.2
Local government	149,403	2.5	168,439	2.4	253,668	2.7	349,820	3.0
Education	260,496	4.4	406,650	5.7	530,827	5.6	554,750	4.8
Hospitals	183,189	3.1	272,433	3.8	364,770	3.8	432,281	3.7
Public enterprises	218,280	3.7	253,816	3.5	315,158	3.3	406,737	3.5
Total public sector	1,153,815	19.3	1,534,492	21.5	2,258,003	23.8	2,592,675	22.5
Total employment	5,965,000	100.0	7,152,000	100.0	9,476,000	100.0	11,545,000	100.0
Public sector in (net) non-residential capital ($million, 1971 prices)								
Government sector	20,869	19.4	27,974	18.8	43,410	19.8	54,925	17.4
Federal	8,378	7.8	9,281	6.2	11,362	5.2	12,241	3.9
Provincial	7,155	6.6	10,580	7.1	17,967	8.2	22,875	7.2
Municipal	5,336	5.0	8,113	5.4	14,081	6.4	19,809	6.3
Schools	3,075	2.9	5,547	3.7	9,683	4.4	9,934	3.1
Universities	642	0.6	1,754	1.2	3,671	1.7	3,716	1.2
Hospitals	2,082	1.9	3,039	2.0	4,478	2.0	6,,240	2.0
Utilities/urban transit	5,623	5.2	17,777	11.9	32,614	14.9	52216	16.5
Total public sector	32,291	30.0	56,091	37.6	93,856	42.9	127,031	40.2
Total non-residential capital	107,748	100.0	149,046	100.0	218,869	100.0	315,912	100.0
Public enterprises (number)								
Federal	41		45		65		87	
Provincial	85		113		208		257	
Total federal and provincial	126		158		273		344	
Local/municipal	n.a.		n.a.		n.a.		500	

Sources: On employment: Statistics Canada, *Federal Government and Enterprises*, Cat. No. 72-004; *Provincial Government and Enterprises*, Cat. No. 72-007; *Municipal Government*, Cat. No. 72-009; *Capital Stock*, Cat. No. 13-211. On state enterprises: Harry M. Kitchen, "Local Government Enterprise in Canada," Economic Council of Canada, Discussion Paper No. 300, 1986, 329.

grew rapidly from the mid-1960s through the mid-1970s but has increased little since then. The decline to 1986 reflects the transfer of Canada Post and its 57,000 or so employees to the ranks of public enterprise in 1981. Provincial- and municipal-government employment continued to grow rapidly through the 1970s and 1980s, and by 1986 provincial governments employed more people than the federal government and municipal governments almost as many. This is one of the most important trends of the past 25 years and reflects a significant redistribution of political authority.

In education, the strongest employment growth occured during the 1950s and 1960s, coincident with the maturing of the "baby boom" generation; hospitals lagged behind education by about a decade. Both education and hospitals represented a smaller fraction of total public-sector employment in 1986 than 20 years earlier.

Non-residential Capital Stock

Over the past 25 years the public-sector share of total capital has increased by a third, from 30 to 40 per cent. This share peaked in the 1970s and has declined moderately since then. Net capital stocks comprise buildings and engineering works, such as roads, dams, machinery, and equipment, but exclude housing (Table 1). These productive capital stocks may be partitioned into industrial capital (manufacturing and non-manufacturing industries, such as transportation, electric power, gas distribution, and urban water systems) and capital employed in all non-manufacturing commercial activity. Social capital includes governments, education, health, churches, and other non-governmental institutions. Federal and provincial state enterprises accounted for $162.4 billion (40.5 per cent) of the net fixed assets of all Canadian corporations (1985), but public ownership dominates in only a few areas, mainly capital-intensive sectors such as electric power, transportation, and communications.

There has been rapid growth in the capital stocks owned by provincial and municipal governments. Since the mid-1970s each has owned capital stocks greater than those of the federal government, indicative of a significant shift in control of public-sector capital assets. The tripling over 25 years of the proportion of capital stocks in utilities and urban transit reflects development of economic infrastructure in the 1960s and 1970s, a trend that noticeably slackened in the 1980s. The share of universities in institutional capital rose from 8 to 17 per cent (1956–86), while that of hospitals declined through the 1960s and 1970s before regaining its 1956 level of 28 per cent by 1986. Over 30 years the share of the public sector in Canada's non-governmental institutional capital resources has risen from two-thirds to 90 per cent, reflecting absorption by the state of educational and health responsibilities performed in significant part by the private sector as recently as the 1950s.

Public Enterprises

In 1986 more than 400,000 people – 3.5 per cent of total employment – worked in some 844 public enterprises. Of these, 500 were the creatures of local government, 257 were established by provincial governments, and 87 by federal authority (Prichard 1983, Economic Council of Canada 1987). At least 226 of 344 federal and provincial enterprises were established or acquired after 1966. Some of the largest firms in the country are public enterprises: 10 of the top 100 revenue earners in 1988 and 11 of the top 50 in terms of assets.

The use of state enterprises or crown corporations as instruments of public policy varies greatly among the OECD countries. Canada, like Japan, has relatively few in-

Table 2
Relative size of government in GNP/GDP (national accounts basis), 1926–86

Indicator	1926	1950	1960	1970	1980	1986
Expenditures of total government sector as % of GNP/GDP*	15.7	22.1	29.7	36.4	41.8	45.1
Government current expenditure on goods and services as % of consumer plus government current expenditures on goods and services	10.0	13.4	17.2	24.8	25.9	25.1
Government fixed-capital formation as % of total fixed-capital formation	13.0	13.5	18.4	17.6	12.0	12.5
Government transfer payments to persons as % of total personal income	1.8	7.2	10.0	10.2	12.3	14.3
Government wages and salaries as % of total wages and salaries [†]	10.1	11.7	16.0	22.9	24.2	25.7

Sources: Statistics Canada, Historical Statistical Compendium, prepared in 1985 for the Royal Commission on the Economic Union and Development Prospects for Canada, Tables 2.1, 2.16, 2.9, and 5.2; *Provincial Economic Accounts*, Cat. No. 13-213; *National Income and Expenditure Accounts, Vol. 1*, Cat. No. 13-531.

* All levels, net of intergovernmental transfers.

[†] Wages, salaries, and supplementary labour income (including military pay and allowances).

dustrial sectors dominated by state enterprises, as compared to France, Italy, or the United Kingdom (Economic Council of Canada 1986). Canada is distinctive in its inconsistency over time and among the provinces about the role of the state in the manufacturing and natural resource sectors. Public-sector utilities, urban transit services, alcoholic beverage distribution, and lottery corporations were often created to displace private-sector businesses in a public interest and have become virtual monopolies in their sector or region. Other enterprises were established as symbols of nation building or sovereignty – Canadian National Railways (founded 1913), Air Canada (1937), and Petro-Canada (1976). Still others entered the public sector as a result of efforts to preserve businesses of "critical" regional or strategic significance, part of governments' efforts to manage processes of structural adjustment at a regional or national scale – Canadair (1976), Clairtone Sound (1967), and Fisheries Products International (1983) – or as the by-product of federal-provincial conflicts – B.C. Hydro (1962). Recently, the federal government in particular has privatized selected public enterprises, such as Air Canada and Petro-Canada.

Government Revenue and Expenditure in GDP

The importance of governments in financial flows of the Canadian economy increased sharply over the 1960s and has continued to grow in several important categories since the 1970s (Table 2). In 1986, total government expenditures represented 45 per cent of GDP and 25 per cent of all expenditure for current goods and services. Wages and salaries – the major part of government current expenditure for goods and services – have risen from 16 to 26 per cent of all wages and salaries paid in Canada. Transfers to individuals currently approximate 14 per cent of personal income – double the level of the 1950s.

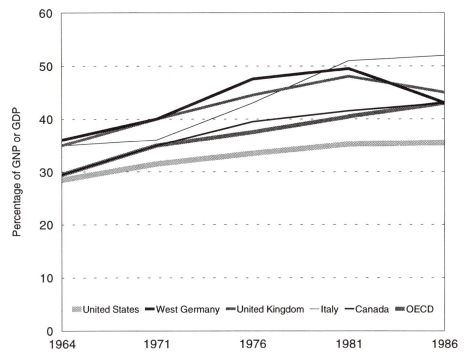

Figure 1
Government expenditure as percentage of GNP/GDP, selected countries, 1964–86.

The greater part of the increase in government expenditure has been on transfers, which have risen from about 10 per cent of GDP (1960) to 24 per cent (1986). The total includes transfers to individuals and businesses as well as interest on the public debt. Some of the sums involved are very large in the context of the Canadian economy; for example, the $42.4 billion paid in interest on the public debt in 1986 was more than the total interest and investment income ($40.7 billion) of Canadians.

Over the past 25 years, the part played by Ottawa in redirecting flows of factor income has declined, while provincial governments have become increasingly interventionist. In the early 1960s about 60 per cent of all government expenditure was at the federal level. By 1986 the federal and provincial governments each accounted for somewhat more than 40 per cent of expenditure (after intergovernmental transfers), and local governments for 16 per cent.

Comparisons with other OECD countries indicate that the size of Canada's public sector falls between the extremes. Government expenditures as a percentage of GDP have tended to rise more rapidly in Canada and have ranged above the OECD average since the early 1970s (Figure 1). Canada is distinctive among Western industrialized countries in its relatively large government budget deficit (Figure 2).

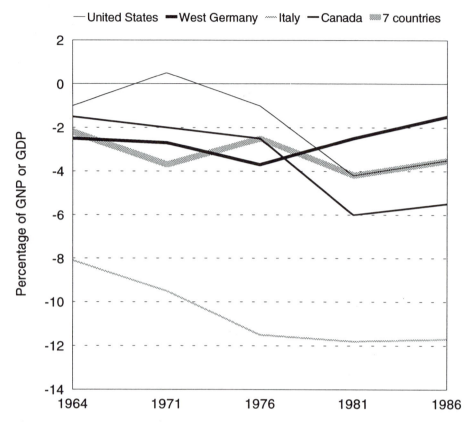

Figure 2
General government (all levels) fiscal balance, selected countries, 1964–86.

THE PUBLIC SECTOR IN THE STRUCTURE
OF PROVINCIAL ECONOMIES

Provincial economies vary greatly in size and structure, and so national aggregates tend to reflect the characteristics of the largest provinces – Ontario and Quebec – and to conceal substantial departures from the national averages in the smaller provincial economies.

Employment and Wages

Over the last 20 years the relative importance of the public sector in total employment has increased in almost all provinces, as has interprovincial variation. By 1986, three distinct groups of provinces were evident. In three Atlantic provinces (Newfoundland,

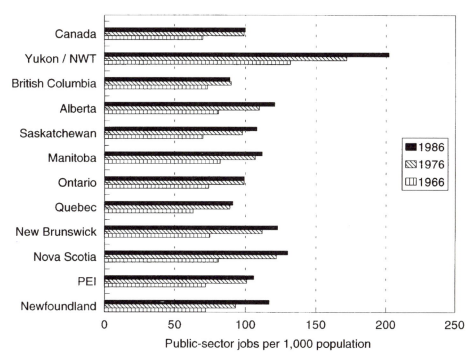

Figure 3
Public employment trends, by province, Canada, 1966–86.

Nova Scotia, and New Brunswick) the public sector provided one job in three, substantially more than the one in four of 1966. In the three prairie provinces and Prince Edward Island one job in four is in the public sector, higher than the one job in five in 1966. In Ontario, Quebec, and British Columbia the public sector provided one job in five in 1986, a modestly smaller proportion than in 1966 for Ontario and British Columbia and a somewhat larger proportion in the case of Quebec. Public-sector employment has grown more than twice as fast as total employment in Atlantic Canada and one and a half times as fast in Quebec, but more slowly than total employment in Ontario and British Columbia. Overall, the relative importance of the public sector as a source of employment has increased in eight of ten provinces and diminished only slightly in two.

Provinces allocate responsibilities between provincial and local governments in different ways, but employment by provinces and municipalities tends to fall in a narrow range about the 1986 national average of 32 per cent of the public sector workforce. Federal government employment, in contrast, does vary considerably by province. In 1986, federal departments and enterprises accounted for more than 30 per cent of public-sector and 11 per cent of total employment in Nova Scotia and Prince Edward Island, in contrast to Saskatchewan, where they provided 15 per cent of public-sector and 4 per cent of total employment.

Table 3
Provincial distribution of public-sector employment and wages ($000), 1986

Province	Government				Other public		Total		Public % total	
	Fed.	Prov.	Munic.	Enter-prises	Educ.	Health	Public	Total empl.	Empl.	Wages
1 in 5 jobs										
Ontario	139.6	126.8	152.6	133.1	195.4	143.7	891.2	4,555.0	19.6	21.4
British Columbia	36.0	53.7	29.1	39.7	51.8	40.8	251.1	1,274.0	19.7	24.1
Quebec	66.8	98.6	60.7	101.4	151.2	125.6	604.2	2,866.0	21.1	26.1
Regional %	13.9	16.0	13.9	15.7	22.8	17.8	1,746.5	8,695.0	20.1	23.3
1 in 4 jobs										
Saskatchewan	10.0	23.2	18.2	20.6	22.0	16.7	110.6	457.0	24.2	35.2
Manitoba	17.2	18.4	14.8	28.9	23.9	18.9	122.0	493.0	24.8	31.0
Alberta	25.0	70.7	56.1	43.2	52.0	43.6	290.6	1,146.0	25.4	28.2
Prince Edward Island	3.4	4.4	0.3	1.2	2.5	1.7	13.4	52.0	25.8	40.6
Regional %	10.34	21.74	16.66	17.49	18.69	15.08	426.1	1,691.0	25.2	30.3
1 in 3 jobs										
Nova Scotia	32.2	21.1	10.5	11.5	22.2	17.5	115.1	344.0	33.5	40.2
New Brunswick	13.1	33.2	3.9	12.0	15.1	12.6	90.0	267.0	33.7	42.7
Newfoundland	7.6	22.6	2.4	8.1	15.8	10.3	66.8	181.0	36.9	44.6
Regional %	19.46	28.29	6.20	11.63	19.54	14.89	272.0	792.0	34.3	42.0
Yukon	1.1	2.5	0.2	0.6	0.5	0.2	5.1	n.a.	n.a.	57.8
Northwest Territories	2.0	4.9	0.9	0.7	1.1	0.6	10.3	n.a.	n.a.	
Outside Canada	15.1	0.0	0.0	5.7	0.0	0.0	20.8	n.a.	n.a.	
Totals	369.1	480.0	349.8	406.7	553.4	432.3	2,591.4	11,635.0	22.3	25.7
National %	14.24	18.52	13.50	15.70	21.36	16.68				

The trend (Figure 3) is to a widening difference among the provinces in the ratio of public employment to population about a rising national average. The public sector's contribution to total wages and salaries (Table 3) averaged 26 per cent in Canada as a whole in 1986 but ranged widely, from 21 per cent in Ontario to nearly 45 per cent in Newfoundland. The public sector occupies an important place in the employment and income structure of all the provinces and plays a dominant role in several.

Large differences among the provinces in the magnitude of the public sector in employment, wages, and salaries suggest that the boundary between public- and private-sector components of the economy is not drawn with any consistency. In Atlantic Canada and the territories, one-third of employment and 40 per cent of wages are provided by the public sector. In the prairie provinces, notwithstanding Alberta's supposed private-sector orientation, the public sector accounts for nearly one job in four and pays upward of 30 per cent of all wages and salaries. At these magnitudes the public-sector wage bill represents a substantial cost to the provincial and national economies.

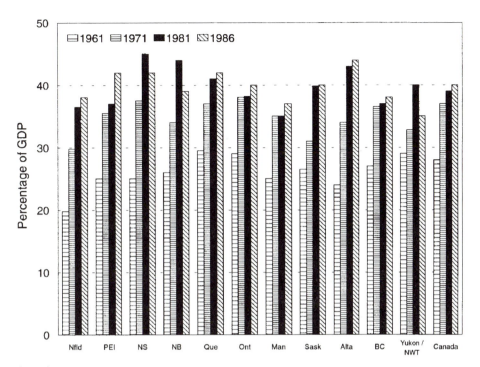

Figure 4a
Government revenues as percentage of provincial GDP, 1961–86.

Government Revenue and Expenditure in Provincial GDP

Through taxation, governments at all levels intercept factor payment and private expenditure flows and use the revenues to produce public goods and services (exhaustive expenditure) and to redistribute incomes among individuals, businesses, and regions (transfer expenditure). For Canada, government revenues rose between 1961 and 1986 from 27 to 39 per cent of GDP; expenditures (exhaustive and transfer) increased from 30 to 47 per cent (Figure 4). The difference between government revenues and expenditures constitutes a net fiscal input (or drain) to (from) a provincial economy.

Government expenditures have converged over the past 25 years on the rising national average and by 1984–86 were – at more than 50 per cent of GDP in seven provinces and nearly 40 per cent in the other three – the dominant magnitude in provincial expenditures. In contrast, differences among the provinces in the magnitude of government revenue drains from the economy have widened. In 1986 all provinces enjoyed a net fiscal input (government expenditure greater than revenues), a situation possible only through deficit financing. In those Atlantic and prairie provinces most

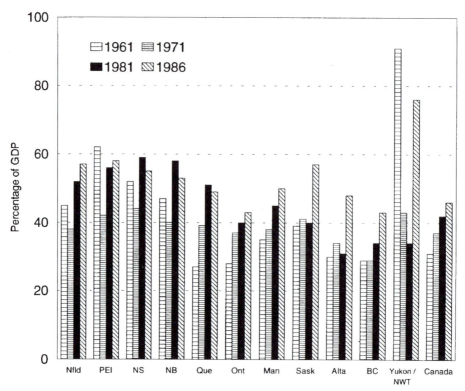

Figure 4b
Government expenditures as percentage of GDP, by province, 1961–86.

dependent on net fiscal inputs, revenues have increased more rapidly than expenditure, to narrow the financing gap. In central Canada and the two westernmost provinces, the reverse has been true, as government expenditures have increased more rapidly than revenues.

In all provinces, governments' role in non-residential fixed-capital formation was substantially less in 1986 than 25 years earlier, averaging 18 per cent but ranging above 24 per cent in Atlantic Canada (39 per cent in Prince Edward Island) to a low of 15 per cent in Ontario. The role of government assistance to fixed-capital formation by business has expanded greatly over the past 25 years and by 1986 was of major importance in the territories, Newfoundland, Nova Scotia, New Brunswick, and Quebec.

The distinction between exhaustive and transfer expenditure is important to an explanation of the impact of government expenditure in the provincial economies. The extent to which a provincial economy is skewed toward production of public goods and services (Figure 5) is more accurately captured when transfer payments to individuals are removed from consideration (Simmons and Bourne 1982; Simmons 1984). In the three-year period 1961–63 government expenditures, defined this way,

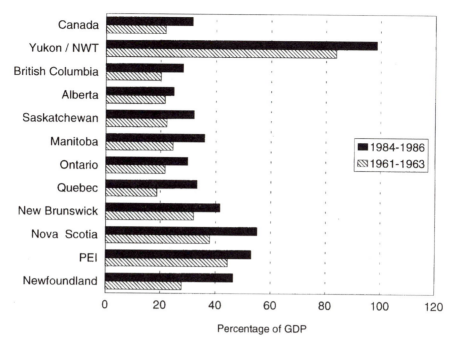

Figure 5a
Total government expenditures as percentage of GDP, by province, 1961–63 and 1984–86.

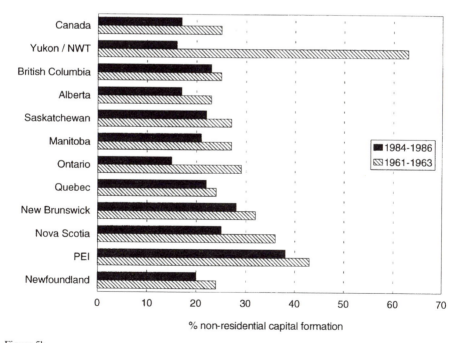

Figure 5b
Government capital formation as percentage of non-residential capital formation, by province, 1961–63 and 1984–86.

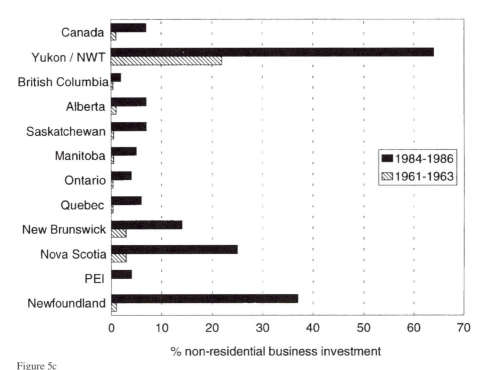

Figure 5c
Government capital assistance as percentage of non-residential business investment, by province, 1961–63 and 1984–86.

represented a third or more of provincial GDP in only the three Maritime provinces. By 1984–86 there were seven provinces where government expenditures exceeded this threshold, averaging 45 per cent, compared with 31 per cent in the earlier period. It is difficult to exaggerate the extent to which production in these provinces is biased toward provision of politically determined, public goods and services and highly dependent on government expenditures. In the three remaining provinces – Ontario, Alberta, and British Columbia – government expenditures averaged 23 per cent of GDP in 1961–63 and 29 per cent in 1984–86. Differences in the demand orientation of these two sets of provinces have widened considerably.

Classification of Canada's Provincial Economies

The public sector's presence, as measured by employment and expenditure in GDP, has increased since the 1960s in all provinces. Over the same period the provinces also became much more disparate. British Columbia, with the lowest public-sector presence in 1984–86, and Nova Scotia and Prince Edward Island, with the highest, are much more dissimilar than the corresponding provinces in 1961–63 (Quebec and Nova Scotia). Twenty-five years ago there was very little variation among the prov-

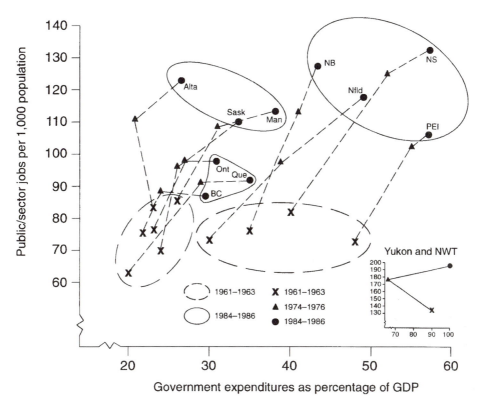

Figure 6
Trends in public-sector employment and government expenditures, by province, 1961–63 to 1984–86.

inces in the ratio of public-sector employment to population; distinctions related mainly to differences in government expenditure in GDP. By 1984–86, however, the provincial economies were differentiated in terms of both public-sector employment and government expenditure.

Broadly speaking, the three-region categorization suggested in Figure 6 for 1984–86 is reinforced by other indicators of public-sector activity (Table 4). First, Atlantic Canada has high levels of public-sector employment and government expenditure in GDP. Total government expenditures range between 50 and 80 per cent of GDP, and government exhaustive expenditure between 30 and 45 per cent. The public sector represents more than 40 per cent of all wages and salaries. Second, the prairie region experiences intermediate levels of public-sector and public-enterprise employment, but also a higher fraction of publicly owned, non-residential capital stocks than either central or Atlantic Canada. Total government expenditures range above 50 per cent of GDP in Manitoba and Saskatchewan, in contrast with 39 per cent in Alberta, and government exhaustive expenditure ranges between 19 and 25 per cent. The public sector repre-

Table 4
Summary characteristics of three Canadian regions, 1984–86

	Canada	Central Canada + BC	Prairie region	Atlantic region
General				
Percentage of Canadian population	100%	73%	18%	9%
Percentage of Canadian employment	100%	75%	18%	7%
Per-capita GDP in 1986	$20,029	$20,511	$21,218	$13,457
Employment and wages				
Public-sector employment per 1,000 population	101	94	117	125
Public-sector in total employment	22%	20%	25%	34%
Public-sector in total wages and salaries	26%	21–26%	28–35%	40–45%
Public enterprises				
Number: provincial enterprises/all	255/844	108/558	78/133	67/88
Enterprise employment per 1,000 population	16	15	21	14
Non-residential capital stocks				
Public sector percentage of total	38%	37%	42%	39%
Government expenditure in GDP				
Current, excluding transfers to individuals	32%	28–33%	25–36%	42–55%
Total expenditure	46%	42–49%	47–57%	52–57%
Government directed in final demand				
In total final demand	47%	43–51%	48–53%	53–58%
In private final demand	40%	33–44%	40–51%	43–51%

sents between 28 and 35 per cent of all wages and salaries. Third, central Canada and British Columbia are positioned at the Canadian means and generally display the lowest levels of public-sector employment and government expenditure. Total government expenditure falls between 40 (Ontario) and 51 (Quebec) per cent of GDP, and government exhaustive expenditure ranges from 20 per cent of GDP in Ontario to 24 per cent in Quebec. The public sector represents 20 to 26 per cent of the total wage bill.

Both Alberta and Quebec appear to be evolving differently from other provinces in their class – Alberta because of a relatively smaller public-sector presence in expenditure terms, Quebec because of a larger. Expansion of Quebec's public sector has moved that province toward the intervention levels of Atlantic Canada and away from its historic similarity to Ontario and British Columbia.

GOVERNMENT INTERVENTION
IN THE FLOWS OF FACTOR INCOMES

One of the most persistent problems in the Canadian space economy is the inability of several provincial economies all the time, and of others some of the time, to generate enough income to support a level of provincial consumption comparable to the average of all the provinces, let alone to that of the most productive. Though not always

Table 5
Provincial GDP and final demand, 1986

Province	Final demand ($)	GDP ($)	Country ranking next	
			below	above
Newfoundland	16,065	11,937	New Zealand	Italy
Prince Edward Island	16,130	11,611	Italy	United Kingdom
Nova Scotia	19,277	14,404	Austria	Netherlands
New Brunswick	16,453	13,841	Belgium	Austria
Quebec	18,329	18,262	Japan	Sweden
Ontario	21,054	22,429	Norway	United States
Manitoba	18,657	17,829	Denmark	Japan
Saskatchewan	17,985	17,021	Finland	Denmark
Alberta	21,607	24,531	United States	Switzerland
British Columbia	19,625	19,554	Sweden	Canada
Yukon and NWT	40,013	29,538	Norway	United States
Canada	19,747	20,029	Sweden	Norway
Ratio: highest to lowest province	1.34	2.11		

Note: GDP comparisons are from World Bank, *World Development Report*, 1988, with Canadian GDP data adjusted to a comparable basis.

the same province, the most productive province has been at least twice as productive in per-capita terms as the least over the entire 25-year period. In geographic terms there is a chronic mismatch between distribution of productive economic activity and that of the population of consuming individuals and institutions (Davis 1980).

Differences in Provincial Production and Productivity

Canada's total GDP is divided very unevenly among the provinces. More than 60 per cent of national GDP originated in Quebec and Ontario over the entire period. Though relatively faster growth in the two westernmost provinces somewhat eroded the dominance of the central provinces, the range in scale among the provincial economies is so great that even major shifts in the smaller economies have little influence on the national averages. In terms of productivity, Canada is a federation of very unequal provincial economies. In 1986, as in most years, only two provinces had per-capita GDP above the national average; the most productive province (Alberta) generated a per-capita GDP more than twice that of the least productive (Prince Edward Island), a difference comparable to that between New Zealand (per-capita GNP = U.S.$7,460 in 1986) and Switzerland (per-capita GNP = U.S.$17,680) (Table 5).

Discordance between Provincial Product and Expenditure

Money flows from consumers to producers and from producers to suppliers of labour, capital and materials link the production and consumption modes of an economy. In a closed system the total value of goods and services produced in a regional economy

(GDP) must equal total spending for consumption and investment (final demand). In most provinces final-demand expenditure differs greatly from provincial GDP. The larger the difference, the greater the separation of productive capacity from consumption and investment expenditure of individuals, businesses, and governments. Such discordance is illustrated in Figure 7 by the ratio of per-capita final demand to GDP: a ratio of 100 represents final demand = GDP.

Throughout the 25 years per-capita final-demand expenditure has exceeded GDP by a wide margin in all provinces of Atlantic Canada and the territories and, in most years and by smaller margins, in Manitoba and British Columbia. Saskatchewan, reflecting the size of its agricultural sector, is the most variable of the provinces, but there, in most years, final demand has been less than GDP. In Quebec since the mid-1970s provincial GDP and final demand have been in near balance. Post-1970s economic growth has reversed the excess of final demand over GDP in Alberta. Only in Ontario has final demand been substantially less than GDP over the entire period.

Canada's space economy is characterized not only by large and persistent differences in the productivity of the provincial economies but also by large and persistent differences between provincial GDP (the measure of total productivity in the economy) and provincial final demand (the sum of current consumption and investment expenditure by its governments and population). Were the spending in a provincial economy limited to the value of goods and services it produces, substantial differences in provincial welfare would result. The provincial distribution of final demand is significantly different from that of GDP, and provincial variation much less. The persistence of this discrepancy is attributable to offsetting financial flows, largely the result of governmental intervention to redirect flows of factor incomes from provinces in which they are earned to provinces in which they are spent (Davis 1983; Britton 1988; Cannon 1989).

Differences between final demand and GDP reflect the net financial flows resulting from both government and private-sector transactions. To what extent is discordance between final demand and GDP linked to net fiscal input (or drain) by governments, particularly Ottawa? Since the mid-1970s net non-governmental flows appear to have had only minor significance in Atlantic Canada; by the mid-1980s federal fiscal inputs dominated net financial flows, supporting levels of final demand 25 to 45 per cent greater than provincial productivity (Figure 7). Dependence on government intervention was greater in the 1980s than in the 1960s, when only half the excess demand could be associated with a fiscal input by governments. Atlantic Canada has become, increasingly over the past 25 years, a de facto ward of Ottawa.

In all other provinces except Alberta, net fiscal inputs by governments, mainly federal, appear to have partially offset the effect on final demand of substantial outflows through non-governmental transactions. Alberta alone experienced a net fiscal drain by government in the 1980s (approximately 4 per cent of GDP). In nine of ten provinces and the territories the level of final demand is either dependent on or sustained at higher-than-market-determined levels through fiscal intervention by governments.

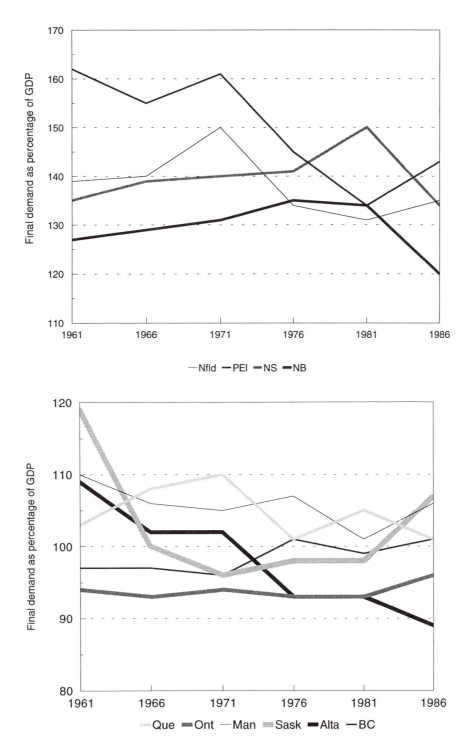

Figure 7
Final demand as percentage of GDP, by province, 1961–86.

Table 6
Federal transfers as % of provincial revenues, 1961–86

Province	1961–63	1974–76	1984–86
Newfoundland	60.8	51.2	47.8
Prince Edward Island	56.2	59.1	49.0
Nova Scotia	45.6	47.6	41.0
New Brunswick	46.4	47.7	41.8
Quebec	25.4	24.1	23.7
Ontario	21.0	21.1	16.1
Manitoba	35.4	36.1	30.6
Saskatchewan	27.8	24.0	20.4
Alberta	26.2	16.4	11.4
British Columbia	22.9	18.3	18.2
Yukon and NWT	57.1	85.2	78.7
Canada	26.8	24.9	21.4

Sources: Computed from provincial revenue and expenditure, national-accounts basis, in Statistics Canada, *Provincial Economic Accounts*, Cat. No. 13-213, Table 5.

Transfers to Provinces

Provincial dependence on federal transfers has generally declined over the past 25 years from 27 to 21 per cent of provincial revenue in Canada as a whole as provincial governments have come to raise more of their revenues directly (Table 6). In this respect the level of government revenues in a province is notably less dependent on the federally directed system of intergovernmental transfers than it used to be. Even so, dependence on these transfers remains high in Atlantic Canada and the territories (over 40 per cent of provincial revenues) and intermediate in Manitoba and Quebec, where their relative magnitude has changed least. The relatively higher levels in these provinces reflect mainly distribution of federal equalization payments.

The implications of low productivity in several provinces for the fiscal capacity of governments have been recognized since the time of Confederation. The Constitution Act of 1982 affirms a long-standing practice in committing Parliament and the government of Canada to the "principle of making equalization payments to ensure that provincial governments have sufficient revenues to provide reasonably comparable levels of public services at reasonably comparable levels of taxation." These equalization payments are intended to compensate for a province's lack of a tax base to support provision of public goods and services at a level comparable to that of a nationally average province (Macdonald Commission 1985).

In 1985–86, equalization payments were estimated at $5 billion – about one-quarter of total federal transfers to the provinces and about 2 per cent of all current expenditure by governments. All provinces except British Columbia, Alberta, Saskatchewan, and Ontario qualified for such payments in 1985–86. In 1980–81 there were only three exceptions, as Saskatchewan's economy was relatively depressed. In 1979–80 both British Columbia and Ontario qualified, though Ontario waived its entitlement.

Intraprovincial Transfers

Most provinces now pursue the type of redistributive policy more traditionally associated with federal equalization payments, so that the level and quality of government services offered in different parts of a province should not be compromised by any productivity-related deficiency in the fiscal capacity of local governments.

In Canada intraprovincial transfers rose from about one-third to one-half of local government revenues between 1961 and 1976, a level maintained through the mid-1980s (Table 7). As local governments have become more dependent on provincial transfers as a source of revenue, their responsibility for expenditure has declined. They were responsible for 22 per cent of total government expenditure after intergovernmental transfers in the early 1960s. By 1984–86 this portion had fallen to 16 per cent, a decline most evident in the more populous provinces – Ontario, Quebec, British Columbia, and Alberta. In New Brunswick, education employees were transferred from local to provincial responsibility (Table 7).

Trends of the past 25 years have increased the importance of provincial governments in control of government revenue and expenditure, making them more independent of Ottawa. Provinces have further consolidated their power by increasing the dependence of their local governments on transfer revenues and by reducing the scope of local government responsibility for expenditure. Clearly the provincial governments have emerged from the changing structure of the public sector as the dominant force shaping the public space economy.

Government in the Structure of GDP and Final Demand

On a national-accounts basis, government expenditures, exclusive of transfers to individuals, currently average 32 per cent of GDP in Canada as a whole. The provincial economies, however, display considerable variation, with government expenditure for goods and services ranging in 1984–86 from a low of 25 per cent of GDP in Alberta to a high of 55 per cent in Nova Scotia. In all provincial economies, government expenditures, exclusive of transfers to individuals, were at least half again as large in 1986 as 25 years earlier. Between 1961 and 1986 total government expenditure, inclusive of transfers to individuals, rose from 30 to 47 per cent of GDP in Canada as a whole and as high as 75 to 80 per cent in Newfoundland and Prince Edward Island. When governments determine where and by whom half or more of GDP will be spent it is difficult to exaggerate the extent to which the structure and spatial distribution of final demand has become politicized.

Both within and between provinces a politically determined spatial distribution of final demand differs markedly from distribution of production and productivity. The public sector's presence has thus expanded beyond the fiscal capacity of several pro-

Table 7

Federal and provincial transfers in local government revenues
(percentage of total revenue), 1961–86

Province/Source	1961–63	1974–76	1984–86
Newfoundland			
Federal	0.0	5.4	8.4
Provincial	29.2	42.2	36.0
Prince Edward Island			
Federal	0.0	1.8	1.1
Provincial	52.6	88.1	85.8
Nova Scotia			
Federal	4.2	2.0	2.9
Provincial	31.6	58.2	64.6
New Brunswick			
Federal	5.2	3.3	3.8
Provincial	34.9	52.3	63.8
Quebec			
Federal	0.8	1.1	1.2
Provincial	28.9	56.4	55.7
Ontario			
Federal	1.7	1.5	1.6
Provincial	36.9	48.8	44.4
Manitoba			
Federal	1.7	1.7	1.6
Provincial	27.9	36.6	53.2
Saskatchewan			
Federal	0.9	0.9	0.8
Provincial	33.6	46.1	48.6
Alberta			
Federal	1.0	0.9	0.6
Provincial	38.7	49.5	51.7
British Columbia			
Federal	1.3	1.2	1.6
Provincial	36.6	39.6	47.2
Yukon and NWT			
Federal	0.0	3.6	6.7
Provincial	71.4	69.1	67.7
Canada			
Federal	1.5	1.3	1.5
Provincial	34.4	49.7	49.8

Sources: Computed from provincial revenue and expenditure, national-accounts
basis, in Statistics Canada, *Provincial Economic Accounts*, Cat. No. 3-213, Table 6.

vincial economies and of many provincial subregions as well. The doubling of government transfers to individuals from 10 to 21 per cent of total personal expenditure, together with the increase in government current expenditure for goods and services from 19 to 26 per cent of total current expenditure, has politicized the structure as well as the spatial distribution of final demand. These discordances between consumption on the one hand and the productivity and fiscal capacity of the provincial economies on the other have created a significant structural rigidity that impedes structural adjustment at both provincial and intraprovincial scales.

CANADA: DIVIDED IN THE NAME OF UNITY

The increasing structural importance of the public sector became evident in all provinces over the period 1961–86, though the provincial economies are on diverging courses in this respect, becoming more unlike in the degree and nature of their dependence on the public sector. By 1986 three distinct regional groupings can be distinguished, each differentiated from the others in the degree to which its economic structure is oriented toward production of public or politically determined goods and services rather than private, market-determined ones and, in final demand, dependent on intergovernmental transfers to sustain current levels of public and personal expenditure. On a national scale, governmental interventions have had major consequences: geographic separation of production and productivity from consumption; spatial distribution and structure of final demand that are more politically than economically determined; erosion of political accountability in the system, as revenues raised in one jurisdiction are disbursed in another; and balkanization of economic policy, as provincial priorities displace national ones.

Structural differences of the magnitude present in Canada reflect a considerable divergence among the provinces on the boundary between public and private sectors (Mathias 1971, 58–9). The proportion of a provincial economy that corresponds with the public sector is not, by and large, a significant source of revenue for governments. Structural divergence means that the effective private-sector tax base is, relative to GDP, much smaller in some provinces than in others. Changes in boundaries between the public and private sectors configure provincial tax bases in different ways, making formulation of a truly national taxation policy almost impossible. As well, the basic principle that "crown" does not tax "crown" means that certain activities constituted in one province as public enterprises are not subject to federal taxation, while those in other parts of Canada are. Electricity generation and mining interests have both figured prominently in federal-provincial disputes over tax jurisdiction.

In recent years several provinces have either indicated or acted on their intention to withdraw from Tax Collection Agreements with Ottawa, reflecting the difficulties in formulating a national taxation policy. Differences in the way the boundary is drawn between public and private sectors are an important part of this problem, though certainly not the whole of it. Introduction in 1990 by the federal government of a goods and services tax (GST) ultimately paid at the point of final demand may have been in part a consequence of the inability to reach agreement with the provinces on a uniform policy for taxing factor incomes – wages, corporation profits, rent and interest receipts, and so on. The GST, which targets the expenditure side of the economy, after government-directed redistribution of factor incomes, may be a more equitable way to raise additional revenues.

Other factors also operate to configure provincial tax bases in different ways. Resource revenues, for example, represented 45 per cent of the Alberta government's revenues in 1981 but were insignificant in several other provinces. The fact that most natural resources remain under provincial ownership means in principle that economic rents accruing to provincially owned resources are outside the federal tax base, despite bilateral federal-provincial agreements under which resource rents are shared. The proliferation of such arrangements is one indication of the difficulty of establishing a national taxation policy and the need to define the tax base differently in different provinces.

While studies such as those carried out by the Macdonald Commission generally point to the flexibility of such agreements as a fundamental strength of the Canadian system, such measures conceal serious problems in the management of Canada's space economy. The diverging economic structures of the provinces, and the substantially enlarged role played by provincial governments negotiating bilaterally with Ottawa on such important economic issues as tax and expenditure policy, have notably eroded the ability of the federal government to articulate national, as opposed to provincial, priorities.

Established Program Financing (EPF) is a case in point. EPF is the largest single transfer program, involving more than half of all federal transfers to the provinces. Historically, these transfers were made to support national programs and standards in such areas as health and post-secondary education. At present, provinces are able to use them in accord with their own priorities. Increasingly, ad-hoc "executive federalism" has been substituted for a national policy framework and clear, legally constituted allocations of responsibility. Courchene (1984) characterized executive federalism as "decision-making conferences of senior executives from federal and provincial governments." This bilateral approach to making policy results in political agreements on such matters as revenue sharing that are not exposed to any judicial test of their legality. Such practises further erode accountability in the system and contribute to the breaking down of any coherent framework for national economic policy.

The Macdonald Commission described Canada's system of equalization and program transfer payments as the reflection of a long-standing commitment by the federation to inter-regional sharing, suggesting that this principle lies at the heart of the idea of Canada. There are, however, grounds for challenging the sense of this strategy for coping with persistent, productivity-related, below-average fiscal capacities in several provinces. In 1985–86, six provinces, containing nearly 10 million people, almost 40 per cent of the Canadian total, had below-average fiscal capacity and were by definition incapable of providing an average level of government services. Quebec, the second-most populous province, is in most years among those qualifying for equalization payments. Provision of an "average" level of public goods and services thus exceeds the fiscal capacity of governments in at least half the country. Is the public sector too large, and is the "average" level of government services too expensive for the economy to support?

By the mid-1980s federal intervention provided a net fiscal input to nine provinces and the territories and a net fiscal drain from only one – Alberta. Ninety per cent of Canadians, including those in the populous and relatively prosperous provinces of British Columbia, Ontario, and Quebec, were benefiting from a net federal fiscal input. There is not much reality left in the Canadian myth of inter-regional sharing. Rather, the support afforded by governments' net fiscal inputs to provincial final demand owes more to the deficit financing of government expenditure – a resource transfer from the future to the present – than to a current transfer from the more to the less prosperous regions.

Difficulties for national policy formulation are also evident in government policy and program responses to large differences in the productivity of the provincial economies. The dominant strategy in terms of costs to the economy has been to compensate for low productivity in several provinces through a complex system of government-directed fiscal transfers, including equalization payments, to provincial and local governments and to individuals and business enterprises. Another strategy, pursued with varying intensity over the years, has been the attempt to influence investment decisions of business enterprises to favour locations in low-productivity provinces through a range of measures, including operating subsidies to private and public enterprises and capital assistance to specific investment or development projects.

The long-pursued policy objective of ensuring that all Canadians, regardless of their province of residence, should have roughly equivalent access to public goods and services suggests that in the poorer regions the public sector would be larger than it might be otherwise. It is also to be expected that the ratios of public employment to total population for such provinces would tend to cluster in a relatively narrow band about the national average, as it did in the 1960s. This is manifestly not the case now, as these ratios have come over the past 25 years to vary over an extremely wide range and to strengthen further the already strong destabilizing and decentralizing tendencies in Canada's space economy.

Nowhere are the problems more evident than in efforts to formulate economic policies to address the need for massive structural adjustment in the economies of the least-productive provinces. Were these provinces Third World countries, structural adjustment would probably begin with a reduction in the size of the public sector and the economic burden that it represents. The process would continue with policies that would increase the mobility of capital and labour and encourage their investment in activities and regions where their productivity is enhanced. A policy framework that assumes that labour ought to be able to find employment in its province of birth denies, particularly to those resident in the smaller and weaker provincial economies, the possibility of employment in the broad and growing range of occupations that those economies do not and often cannot provide. This high opportunity cost is an essential flaw in the existing policy framework and a major contributing factor to persistent low productivity.

The fundamental problem of relatively low productivity in several provinces remains essentially the same as 25 or even 50 years ago. Though the Canadian approach to this problem has been one of compensating for the welfare implications of low productivity by subsidizing government and private consumption at a provincial scale, resulting separation of production from consumption has destabilized the organization of Canada's space economy. The tension between spatially different distributions of production and consumption has been sustained only by means of increasing political interventions in the flows of factor incomes and in such far-reaching decisions as where to locate new private as well as public investment, how the future workforce will be educated and trained, and what incentives or disincentives will be placed in the way of interprovincial trade and migration. Decisions in these areas have come to reflect political concerns rather than economic efficiency.

While there are always trade-offs to be made between economic and non-economic policy objectives, the broadening, increasingly politicized, interventionist role played by the public sector in the structure and spatial organization of Canada's economy goes beyond the making of such trade-offs to the substitution of political for economic criteria. Richard Gwyn, quoting an anonymous American official, stated the basic issue posed by this trend: "Your problem is that you've politicized your economy. Your government is almost incapable of taking actual economic decisions as opposed to political ones, to support this or that group or industry or region" (Gwyn 1985).

Given the structural and organizational significance of the public sector in the economy, a reasonable question would seem to be whether Canada's system of public-sector institutions and interventions meets the basic tests of an efficient spatial organization and thus increases and enriches the economic opportunities and potentials of Canadian individuals and businesses. This is the geographic variant of the central, historically rooted question addressed periodically by a succession of royal commissions: does our system of public-sector institutions operate to maximize economic well-being? It has been the frequent conclusion of these inquiries that Canada's public system does not do so.

Two characteristically Canadian responses to this negative finding are typically advanced. First, where the failure to maximize economic opportunity can be traced to some defect in the design and operation of public institutions, then appropriate adjustments to that system may be indicated. Periodic adjustments to revenue and cost-sharing arrangements and the growth of "executive federalism" have been the response. Second, the system's failure may be rationalized in terms of non-economic criteria; one may argue that economic efficiency must be compromised by the need, in Canada's federal system, to preserve regional, linguistic, and cultural communities and to compensate for large regional disparities in resources and productive capacities. Any resulting diminution in economic opportunities is seen to be inherent in the assumptions and trade-offs inherent in our federal system – part of the price of being Canadian.

One might reasonably ask whether this confederation is not fundamentally flawed in entrenching the dependence of some on others and in placing no limits on preservation of regional communities. Where, as in Atlantic Canada, governments spend between 50 and 80 per cent of provincial GDP, it should be embarrassing to pretend that this distorted economic structure preserves historic regional communities. The political commitment to sustain the political, cultural, and economic viability of provinces and regions that lack the capacity to sustain themselves constitutes the greatest single impediment to mobilization of Canada's human and capital resources in occupations and regions where their productive potential can best be realized. It is this characteristic of Canada's public space economy, reinforced over the past 25 years by the rise to prominence of interventionist and introspective provincial governments, that is most worrisome for the continued viability of Canada as a nation in a more open and competitive world economy.

REFERENCES

Britton, John N.H. 1988. "Economic Change and the Regional Question." *Canadian Journal of Regional Science*, 11, 125–36.

Bucovetsky, Meyer W. 1979. "Government as Indirect Employer." In Meyer W. Bucovetsky, ed., *Studies in Public Employment and Compensation in Canada*. Montreal: Butterworths for Institute for Research on Public Policy.

Cannon, James B. 1989. "Directions in Canadian Regional Policy." *Canadian Geographer*, 33, 230–9.

Courchene, Thomas J. 1984. "The Citizen and the State: A Market Perspective." In Lermer (1984).

Davis, J. Tait. 1980. "Some Implications of Recent Trends in the Provincial Distribution of Income and Industrial Product in Canada." *Canadian Geographer*, 24, 221–36.

– 1983. "Government-Directed Money Flows and the Discordance between Provincial Consumption and Expenditure, 1961–1979." *Canadian Geographer*, 26, 1–27.

Economic Council of Canada. 1986. *Minding the Public's Business*. Ottawa.

Foot, David K. 1978. *Public Sector Employment in Canada: Myths and Realities*. Toronto: Butterworths for Institute for Policy Research.

Gore, Charles. 1984. *Regions in Question: Space, Development Theory and Regional Policy*. London: Methuen.

Gwyn, Richard. 1985. *The 49th Paradox: Canada in North America*. Toronto: Totem Books.

Lermer, George, ed. 1984. *Probing Leviathan: An Investigation of Government in the Economy*. Vancouver: Fraser Institute.

Macdonald Commission. 1985. *Report of the Royal Commission on the Economic Union and Development Prospects fro Canada, Volume Three*. Ottawa.

Mathias, Phillip. 1971. *Forced Growth*. Toronto: J. Lewis & Samuel.

Morrill, R.L. 1970. *The Spatial Organisation of Society*. Belmont, Calif.: Wadsworth.

Norrie, Kenneth, Simeon, Richard, and Krasnick, Mark. 1986. *Federalism and the Economic Union in Canada*. Toronto: University of Toronto Press.

Prichard, J. Robert S., ed. 1983. *Crown Corporations in Canada*. Toronto: Butterworths and Company.

Simmons, James W. 1984. "Government and the Canadian Urban System: Income Tax, Transfer Payments and Employment." *Canadian Geographer*, 28, 18–45.

Simmons, James W., and Bourne L.S., 1982. "Urban/Regional Systems and the State." *Progress in Human Geography*, 6, 431–40.

Thoreau, Henry David. 1886. *Excursions in the Writings of Henry David Thoreau*, vol. 9. Riverside Edition.

CHAPTER TWENTY-ONE

Socio-spatial Restructuring of Canadian Labour Markets

TOD RUTHERFORD

Recent changes in the Canadian economy have transformed the composition and regional structure of the Canadian labour market. This restructuring is related to the history of Fordism in Canada, which is labelled "permeable Fordism" by Drache and Glasbeek (1992) and by Jenson and Mahon (1993) because of the level of foreign ownership, the dominance of international unions, the open, resource based economy, and sectoral and occupational labour-market segmentation (Donner 1991; Norcliffe 1994). These characteristics have been reinforced by Canada's federal structure which allowed for increased intervention by Ottawa in labour-market operations during the height of the Fordist period (1945–75), while industrial relations were primarily provincial responsibilities or decentralized to the firm level.

This first part of this chapter reviews current trends in the restructuring of the Canadian labour market. These trends are in large part a response to increased international competition through agreements such as the North American Free Trade Agreement (NAFTA). Economic activities oriented toward domestic markets, such as manufacturing in southern Ontario, are becoming subject to the more cyclical types of external markets historically characteristic of resource-based regions and towns. Thus the period since the late 1980s has featured an unprecedented level of plant closures and permanent lay-offs in the traditional manufacturing heartland of Canada. These pressures are being reflected in almost all sectors and regions in reorganization of work through introduction of innovations such as lean production and, with them, new forms of occupational and gender segmentation. Moreover, these changes have occurred in a context of increased female labour-market participation and part-time employment. Thus, despite increased unemployment overall, the Canadian labour market has been confronting skill shortages and mismatches that reflect inadequate investment in public and private training.

In the second part of this chapter the response of provincial governments to this restructuring is examined. Like other advanced capitalist nations, Canada seems to be shifting the emphasis of its social and labour market policies from its (Keynesian) welfare focus to one concerned also with flexibility and profitability. In contrast to the welfare state this is sometimes referred to as the "workfare state," or the Schumpeterian workfare state (sws) (see Jessop 1993; Peck 1994; Peck and Jones 1995). The pattern of policy change is complex, however, and Canada's federal structure has led to considerable overlapping of labour market programs. Nevertheless, current changes involve adoption of a supply-side "active" labour market strategy involving forms of decentralization to the provincial and local levels.

Creation of the Canadian Labour Force Development Board, with accompanying training and adjustment boards at the provincial and local levels "regulated by, but a distance from the state" (Dehli 1993, 88), represents adoption of the sws in Canada. Moreover, these boards attempt to develop a consensus, or corporatist strategy, with equal representation from business, labour, and social equity groups. I argue, however, that the transition to an active labour market strategy based on consensus is self-contradictory, since it is occurring in the context of cuts to spending on training and adjustment programs, increased restrictions on support for the unemployed, and the legacy of a weak training culture and adversarial industrial relations.

The final section of this chapter examines the implications for labour of the decentralization of training and apparent adoption of a corporatist strategy. Fordism brought extension of union rights and organization in resource and mass-production industries and the public sector from the mid-1960s (Palmer 1992). After the mid-1970s, however, the weakening of the Fordist wage bargain, which linked wages to productivity, increased attacks on the right to strike in the public sector, and the growth of the service sector, which has traditionally been non-union, have all contributed to the challenge to labour's position (Meltz and Verma 1993; Panitch and Swartz 1993; Wells 1993). Nevertheless, labour has also developed its own agenda on both work organization and labour market policy, which will help determine the form of the sws. The strengths and weaknesses of these labour strategies will be examined through a brief overview of the participation of Canadian Auto Workers (caw) and the United Steel Workers of America (uswa) in sector-based training initiatives.

SOCIO-SPATIAL STRUCTURE OF THE
FORDIST LABOUR MARKET, 1945–75

The current period of restructuring in the Canadian labour market must be seen in historical context.[1] The development of industries in the industrial heartland and the "new staples" in the resource periphery (both heavily dependent on u.s. branch plants and investment) in the inter-war years and after 1945 was associated with adoption of a distinct set of Fordist labour market and industrial relations practices. This postwar settlement, which included adoption of the welfare state and Keynesian demand management, had major implications for the development of unionism. The spread of industrial unionism in the mass-production and resource sectors in the 1940s and 1950s led to development of internal labour markets based on the seniority system and the linking of wages to productivity (Holmes 1989; Premier's Council 1990; Norcliffe 1994). As Norcliffe (1994) points out, labour markets in the industrial heartland were complex and characterized by a segmentation similar to that in the United States (Gordon, Edwards, and Reich 1982). Male-dominated primary labour markets included high-status public- and private-sector professional and managerial workers with institutional and educational qualifications and a subordinate primary market of predominantly unionized workers based on seniority.

Outside these primary labour markets was a series of secondary labour markets in which women and visible minorities were disproportionately employed in service and manufacturing industries characterized by lack of unions and poor wages and conditions. Women's participation in Fordism was characterized by effective gender segregation. Advances made by women during the Second World War in gaining access to formerly male occupations were reversed in the immediate postwar period. However, by the early 1950s female labour market participation had begun to increase – especially by suburban married women in manufacturing and expanding clerical occupations. While this was in part intended to finance increased family consumption, there were also a shrinking pool of single women workers (from 80 per cent of all female workers in 1941 to 34 per cent in 1971) due to extended education and higher marriages rates, which made married women increasingly attractive to employers (Palmer 1992). While female labour market participation effectively doubled to nearly 40 per cent between 1941 and 1971 women were largely confined to traditional occupations such as clerical, secretarial, and teaching and labour-intensive manufacturing industries such as clothing. In the resource periphery there were fewer employment opportunities than in the industrial core, and gender patterns of wage work varied between sectors and regions. In fishing communities, for example, males worked in the inshore fishery while most women worked in fish-processing plants; in forestry and mining women often augmented family earnings by increasing domestic subsistence activity (Luxton 1980; Connelly and Macdonald 1983).

The Fordist settlement in Canada clearly differed from that in western Europe and the United States. The Canadian commitment to Fordism proved notably weaker than that in western Europe (Drache and Glasbeek 1992; Jenson and Mahon 1993; also see Tickell and Peck 1992). The dependence of the Canadian economy on the export of resources that vary in a cyclical fashion made the commitment to full employment and centralized collective bargaining less tenable than in western Europe. Similarly, governments proved averse to corporatist – social democratic forms of intervention into industrial relations (Jenson and Mahon 1993). Unionization increased from 16.3 per cent of the non-agricultural workforce in 1940 to 30.3 per cent in 1950 and to 33.6 per cent in 1971, but gains in income were uneven. During the initial postwar resource boom most gains were made by workers in mining and construction (Gindin 1989; Drache and Glasbeek 1992). By the mid- to late 1960s expansion of the Canadian automobile industry under the Auto Pact brought major gains to workers in the auto industry; unionization increased in mass-production sectors and petroleum.

Fewer gains were made in labour-intensive sectors, such as textiles and shoes, produced for local and regional markets (Drache and Glasbeek 1992, 26). Consistent with increasing continental integration of the economy, "international unions" based in the United States dominated Canadian private-sector unionism and by 1966 represented nearly 70 per cent of all organized workers in Canada (Palmer 1992, 410).

In Canada, while it is evident that the Fordist postwar settlement was weaker than that in western Europe, workers did have certain advantages over their American counterparts. Despite considerable regional segmentation of labour markets during the Fordist period, employers never had the equivalent of the "right to work" laws enacted in southern U.S. states. Regional policies did not really address the underlying causes of uneven development, but they did narrow regional income differentials through transfer payments from have to have-not provinces. Higher unionization rates and other social reforms made labour better able to "level up" wages and reduce income differentials. Within the industrial heartland unionized workers' wages in such sectors as the auto industry were only 20 per cent higher than the average manufacturing wage, compared to an over 40 per cent gap in the United States (Gindin 1989, 65). Social wages, such as unemployment insurance, had a more redistributive effect in Canada than in the United States (Bakker 1991), and there was also considerable migration of male skilled workers from weaker, resource-oriented regions such as the Maritimes to industrial regions (Norcliffe 1994). Also there was considerable long-distance migration for both skilled and semi-skilled male manual workers within the resource periphery (Bradbury 1984; Storey and Shrimpton 1994).

LABOUR MARKET RESTRUCTURING IN CANADA, 1980–94

Despite the greater cyclical nature and openness of Fordism in Canada it represented a period of relative stability and one in which workers made real gains (Mahon 1991; Drache and Glasbeek 1992; Palmer 1992). By the mid-1970s this period ended, as the Canadian labour market became characterized by greater volatility, higher levels of unemployment, and restructuring within the resource periphery and increasingly in heartland regions (Canadian Labour Market and Productivity Centre [CLMPC] 1994a). Not only has the labour market become increasingly volatile, it has also become characterized by new forms of segmentation and skill mismatches within and between regions (Donner 1991; Norcliffe 1994). The secular trend of increasing female participation and shifts in employment share from goods-producing to service sectors has continued. As in other advanced industrial nations a "two-tier" labour market has emerged, though with a smaller core, while a greater share of employment is held by a part-time or "contingent" labour force made up disproportionately of women, minorities, and white males formerly employed in core sectors (Atkinson 1987; Christopherson and Noyelle 1992; Jenson and Mahon 1993).

There has been considerable restructuring of work itself. Since the 1970s – especially in core sectors – there has been extensive adoption of computerized new technology and flexible forms of production (see Holmes 1989; CLMPC 1992; Hayter and Barnes 1992; Norcliffe 1994b). Moreover, introduction of new technology has been accompanied by considerable reorganization of work and by the late 1980s nearly half of all establishments in Canada had adopted initiatives such as employee involvement, team work, and delegation of decision making and responsibility for quality control and production planning to the shop floor (Sharpe 1994). These changes have not been confined to manufacturing, and in public and private services –, there has been increasing adoption of computers and innovations in work organization for example, total quality management (TQM). Shifts have taken place in the site of work. Innovations such as faxes and modems have allowed professionals in core labour markets to reduce office-based work, while outsourcing in peripheral sectors such as clothing has led to increased homeworking – often by immigrant women under very exploitative conditions.

Before assessing the implications of the above changes, we should review overall trends in the labour market itself during this period. Despite increased turbulence and increasing structural unemployment, Canada recorded a significant increase in jobs, which grew in number by one-third (from 9.3 to 12.4 million) between 1975 and 1993. Canada's rapid increase in population and labour-force size has contributed to unemployment (Meltz and Verma 1993).

In terms of the changing structure of employment, the most striking trend is growth of services. Between 1976 and 1991 employment grew by more than three times, from 2.8 million to 8.9 million. Goods-producing sectors such as agriculture, forestry, mining, manufacturing, construction, and utilities remained at 3.4 million employed, but manufacturing lost 195,000 jobs between 1981 and 1991 – a trend that accelerated after 1989. The fastest growth in service employment was in the finance, insurance, and real estate (FIRE) sector – by over 27 per cent 1981–91. This trend is also evident in occupational shifts during this period, with the fastest rates of growth being in managerial/administrative occupations, which grew by 64 per cent (a gain of nearly 680,000 jobs). The largest declines in employment occurred in production occupations, such as machining, assembly, and processing, which lost over 200,000 jobs.

These changes can be better understood in reference to the two severe recessions of 1981–83 and 1990–92, which had significantly different effects on labour market structure. The 1981–83 episode appeared to be more severe (Table 1), but a growing amount of research suggests that the later recession has had a more profound impact on labour market patterns (Canadian Labour Force Development Board [CLFDB] 1993; Gertler 1994). The 1990–92 recession was longer (13 quarters, compared with nine) and permanent lay-offs were concentrated in manufacturing and construction rather than in construction and primary resource sectors. Part-time employment continued to grow throughout the recession and for most of the recovery period of 1992–94 – in total, 15 per cent 1989–94 – but full-time employment fell by 5.6 per cent, from 10.7 million jobs in 1990 to 10.1 million in mid-1992, though by the end of 1994 full-time employment had recovered to its pre-recession peak.

In manufacturing, the 1990–92 recession, in contrast to previous recessions, precipitated a very high proportion (71 per cent) of permanent layoffs and a very high level

Table 1: Comparison of the 1981-82 and 1990-92 recessions:
key indicators (taken at worst time)

Indicator	1981-82	1990-92
Employment decline (percentage points)	5.4	3.1
Unemployment rate increase	5.5	4.2
Maximum unemployment rate	12.7	11.5
Job loser rate	7.9	6.8
Hiring rate decline	18.0	15.0
Average duration of unemployment (weeks)	19.5	19.6
Duration of recession (quarters)	9.0	13.0

Source: CLFDB (1993).

of cumulative impact – 8.8 million person-months lost, compared to only 2.8 million in the earlier recession (CLFDB 1993, 20). There was a delay of almost two years before employment in manufacturing began to recover, and by the end of 1994 employment in the sector was still 8.4 per cent below its pre-recession level.

The manufacturing focus of the 1990–92 recession had significant implications for spatial patterns of employment change. Core regions had become more exposed to cyclical fluctuations and higher unemployment, formerly characteristics of the resource periphery (Norcliffe 1994). By 1994, 55 per cent of Canadian manufacturing production was being exported, compared to only 25 per cent in 1980 (Freeman 1994). Southern Ontario, with 40 per cent of Canada's employment, suffered 70 per cent of the employment loss in 1990–92.

Regional patterns of employment change between 1989 and 1994 were highly variable (see Table 2). In Atlantic Canada the severe crisis of the fishing industry has become a regional problem (CLFDB 1993), with the provinces most directly dependent on the fisheries – Newfoundland and Nova Scotia – suffering the most. By 1994 both had employment levels below those of 1989, and in Newfoundland levels were still 8 per cent lower (Sunter 1994). In contrast both Prince Edward Island and New Brunswick experienced small initial declines/stagnation followed by modest employment growth after 1993. In central Canada, Ontario experienced significantly greater employment declines than Quebec, and the latter has been quicker to recover from the recession, with 1994 employment levels only 1.4 per cent below their pre-recession peak. Montreal has suffered extensive job loss in the textile and clothing industries since at least the 1981–82 recession (Brodeur and Galarneau 1994, 44), and it lost jobs in other manufacturing activities, financial and commercial services, and personal services. However, it suffered only a 1 per cent loss in employment during the 1990–92 recession, compared to Toronto's 3 per cent. In Ontario, employment recovery has been largely outside Toronto.

In western Canada, the 1990–92 recession was less pronounced, but by mid-1994 employment in Manitoba and Saskatchewan had not yet returned to pre-recession levels. In Alberta and British Columbia the relative buoyancy of the resource sectors (oil

Table 2
Canadian and provincial unemployment rates, 1989–94

Province	1989	1990	1991	1992	1993	1994
Nfld	15.8	17.1	18.4	20.2	20.2	20.7
PEI	14.1	14.9	16.8	17.7	17.7	15.0
NS	9.9	10.5	12.0	13.1	14.6	12.3
NB	12.5	12.1	12.7	12.8	12.6	11.6
Que.	9.3	10.1	11.9	12.8	13.1	12.1
Ont.	5.1	6.3	9.6	10.8	10.6	8.4
Man.	7.5	7.2	8.8	9.6	9.2	7.9
Sask.	7.4	7.0	7.4	8.2	8.0	6.1
Alta	7.2	7.0	8.2	9.5	9.6	7.6
BC	7.2	7.0	8.2	9.5	9.6	7.6
Canada	9.1	8.3	9.9	10.4	9.7	9.4

Sources: Statistics Canada, *Perspectives*, Cat. No. 75-001E, various years.

and gas and pulp and paper) meant that employment levels rose to 4 and 12.9 per cent, respectively, above pre-1990 levels (Sunter 1994, 9). Growth in British Columbia, especially Vancouver, reflects the province's increasing integration into a Pacific/ East Asian economy (Barnes et al 1992; Hayter and Barnes 1992). Vancouver's role as a staple-oriented exporter has been augmented by its financial and producer services; between 1971 and 1991 the number of workers in financial and commercial occupations tripled, and by 1991 more of the working population was employed in services (nearly 80 per cent) in Vancouver than Montreal or Toronto, which had approximately 75 per cent each (Brodeur and Galarneau 1994, 44).

There was a significant unevenness in unemployment within regions. Despite the apparent buoyancy of the Vancouver region, continued restructuring of resource sectors – especially pulp and paper (Hayter and Barnes 1992; Holmes and Hayter 1994) and mining – meant that interior BC communities continued to suffer significant employment rationalization.[2]

Unevenness within regions was evident in southern Ontario, where, at the end of 1994, Toronto was failing to recover full-time employment at the rate of the rest of southern Ontario. In the period between the "bottoming out" of employment in early 1992 and the end of 1994, when over 540,000 jobs were created in the Canadian economy, there was not net increase in jobs in Toronto. This may be partly linked to the rationalization effects of the FTA. Not only were many branch plants closed that did not have any continental or global mandate, but those that did often streamlined their managerial/office functions (Gertler 1994, 8). This current round of restructuring has weakened Toronto's historic specialization in manufacturing and office (especially clerical) work, moving it toward the national average (Bromley 1994, 6).

Whatever the uneven sectoral or regional impact, the recessions of 1981–82 and 1990–92 served to augment labour market tendencies already present by the late 1970s. One of the most significant is increased use of part-time or temporary employment. Between 1976 and 1992 in all industries the share of part-time workers employed increased from 11.0 to 16.8 per cent,[3] in goods-producing industries from only 4.7 to 7.1 per cent, in services from 14.3 to 20.4 per cent, and in accommodation, food and beverage, the retail trade, recreation, and personal services, to over 30 per cent. This difference is attributable to redeployment of labour within large, unionized firms and to high and rising capital/labour ratios in manufacturing, which make labour turnover more expensive and use of part-timers less attractive to employers. There is some evidence, however, that manufacturing employers may be using laid-off workers for temporary work (Corman et al 1993; Norcliffe 1994).

The trend toward part-time employment reflects employers' efforts to reduce their overall wage bill and a change in use of labour time. In the mid-1970s more than 70 per cent of full-time employees worked a standard working week of 40 hours. However, between 1981 and 1992 the number if those working a "normal" range of hours (30–49 hours per week) fell from 73.4 per cent to 69.1 per cent as the proportion working less than 30 hours a week rose from 14.7 to 17.5 per cent and those working over 50 hours a week rose from 1.9 to 13.4 per cent. Indeed, one of the most notable features of the recovery from the 1990–92 recession in manufacturing in both Canada and the United States has been increased reliance by firms on overtime rather than on hiring of new workers (Rigdon 1994).[4]

Another continuing trend is the increase in the female participation rate, which doubled from 30 to 60 per cent between 1961 and 1991. Since the 1970s the most rapid growth has been among women with children under six whose rate increased from 52 per cent in 1981 to 68 per cent in 1990 (CLMPC 1994b). The highest rate of participation (78 per cent) is for women between the ages of 35 and 44 with children at home, which is the same as the overall rate for males. During the period 1960–89 participation rates for women grew fastest in the urbanized industrial core, where not only manufacturing but the bulk of service-sector growth was concentrated. Female participation rates declined slightly during the 1990–92 recession, with young females (those under 25) suffering the steepest fall. This is attributable to the concentration of women in the most vulnerable manufacturing sectors and to a sharp decline in clerical employment in many areas of Canada.

The "feminization" of the labour market is linked to the shift toward service and part-time employment patterns, with the proportion of employment by women in all industries rising to 45.3 per cent and in the service sector to 53.4 per cent. Similarly, women made up a disproportionate share of those in part-time work – 26 per cent (1992), compared to 9 per cent of males. While women still experience considerable occupational segregation (Table 3), with the majority being in low-paid and temporary employment, they have made considerable gains in managerial/administrative positions. By 1992, managerial and professional occupations were held almost equally by men and women, though most of the women were in health care and teaching or were in lower-status administrative positions (CLMPC 1994b, 20), and these gains are also reflected in the fact that the male/female gap in earnings had narrowed.

Table 3
Five leading occupations for females, 1992

Occupational category	Female share of jobs (%)	Proportion (%) of women employed
Clerical	80	28.6
Professional	56	22.6
Service	58	17.3
Managerial/administrative	41	12.2
Sales	45	9.8

Source: CLMPC (1994).

The 1990–92 recession restricted women's promotion in administrative/professional occupations (Davidson 1993), and though there is a growing trend toward two-income households (71 per cent of couples with children aged 18 or younger in 1990) women still bear the disproportionate share of domestic responsibilities. Progressive workplaces offer daycare, especially to career-path women, but only 12 per cent of working mothers with children have access to regulated daycare.[5]

While the overall trend is toward advancement of women in professional occupations continuation of gender "ghettos" indicates new forms of labour market segmentation and growing polarization of opportunities (McDowell 1991). During the 1980s opportunities opened up for women, but class divisions increased as well-paid manufacturing employment declined, and there was growth in services, professional and administrative positions in health care, business services, and finance, insurance, and real estate. Nevertheless, the number of low-paid insecure, and short-term jobs also increased (Economic Council of Canada 1990; Drache 1991; Norcliffe 1994). Some routine producer services have been established in New Brunswick, confirming the pattern that the best jobs in the service sector are much more likely to be found in core regions.

Even within the industrial core manu of those laid off from manufacturing have accepted jobs in lower-paying manufacturing or service sectors (CLFDB 1993; Jenson and Mahon 1993).[6] By the late 1980s it was estimated that in Ontario males aged 45–54 suffered a wage loss of 25 per cent for their first job obtained after lay-off, and many of the gains in earnings made by women relative to men reflect the decline in male earnings. Even within unionized manufacturing sectors it is apparent that there has been a systemic undermining of the pattern under Fordism of linking of wages to productivity and the cost of living (Drache and Glasbeek 1992), which contributed to the share of national income going to labour falling from 72 per cent in the early 1970s to 67.2 per cent in 1987.[7]

Regional labour markets in Canada continue to be balkanized, and are often associated with skill mismatches resulting from poor public and private training and inadequate adjustment programs. There are long-standing differences in regional unemployment rates, and the degree of mismatch in skills between the unemployed and unfilled jobs has been rising both between and within provinces. In 1993, when there were 1.6 million unemployed, the mismatch between vacant jobs and available skills ranged between 9 and 27 per cent of total unemployment (or between 144,000 and 432,000 jobs) (Stoffman 1993, 54). Canadian firms continue to suffer from chronic shortages in many skilled manual and technical positions (Premier's Council 1990; Study Group on Technicians and Technologists 1993), with some seeking skilled workers from overseas (Van Alphen 1994).

These mismatches and increased international competition have resulted in heightened media and public policy attention to training (Premier's Council 1990; Stoffman 1993; Crane 1994). Some surveys suggest that private-sector provision of training is extensive (CLMPC 1993a, b), with 70 per cent of firms providing some form of "structured" training, but others show, for example, that only 26.9 per cent have any formal training program (*Report on Business* 1994), and only 6.9 per cent of the Canadian workforce has had more than 35 hours of structured training over the past year (Crompton 1994). Canadian firms spend only $100 per employed worker on training (Premier's Council 1990), compared to $200 in the United States and over $400 in Germany. Furthermore, the bulk of what is spent on training goes to programs for managerial/salaried employees (Crompton, 1994).

Shortages of technical and research skills also reflect the branch plant structure of Canadian industry and low rates of adoption of new technology as compared with other industrial nations (Study Group on Technicians and Technologists 1993, 3–5). Even when Canadian-owned firms invest in new technology major problems result from lack of skilled labour and/or inadequate training (Gertler 1994). Certainly, until very recently Canadian training policy could be best described as voluntarist – that is, a largely laissez-faire system in which firms invested in training or poached needed skills from other firms according to market conditions (Muszynski and Wolfe 1989). For skilled production work exceptions have been larger, often U.S.-owned branch plants in the automobile industry, where internal labour markets have led to development of in-house training for skilled trades.

In some industries such as automotive and pulp and paper there is evidence of increased entry-level skills (especially basic education) as shop-floor employees become increasingly responsible for monitoring complex, capital-intensive production systems (Automotive Parts Human Resource Study 1991; Holmes and Hayter 1994; Norcliffe and Bates 1994). Consequently, in both manufacturing and higher-order public and private services managerial systems are moving away from either seniority or bureaucratically determined employee evaluation/remuneration schemes to ones that are more individualized and flexible.

In the last decade there have been substantial investments in training announced by large firms such as GM, Chrysler, and McDonnell Douglas, but Robertson (1992) points out that often only a small proportion of budgets for training is actually spent. There is conflicting evidence whether those investments are linked to the need for increased technical skills. Many new training initiatives focus on non-transferable "cultural skills" such as problem solving human relations, and team building, and many new jobs in services are low skilled (Beckerman, Davis, and Jacks on 1992). Some high entry-level requirements reflect only current high unemployment levels (Dehli 1993; Rutherford 1995).

"Flattening" of supervisory structures and increased delegation of decision-making responsibilities to core personnel (individuals or teams) have coincided with increased application of computer-based technologies in both manufacturing and services. While technical training is often basic (Robertson 1992) corporate norms and goals are also made part of employee training. When these efforts are devoted to a narrow range of firm-specific skills, training takes on perverse characteristics (Premier's Council 1990, 95). Similarly, two-track labour markets are developing within particular occupations because of increased demand for specialized skills in computer science, health care, and financial services rather than broader, generic ones (Gibb-Clark 1995).

TRAINING AND THE RESTRUCTURING OF POLICY

Increased structural unemployment, low levels of training, and investment, and skill mismatches have changed state labour market policy since the late 1980s. Some argue that this has transformed the Keynesian welfare state (KWS) into what Jessop (1993) terms the Schumpeterian workfare state (SWS). While the KWS promoted the social rights of citizens, the SWS shifts social and labour market policy toward increasing the productivity of labour. Thus training and skill enhancement for the unemployed become a central theme of public policy, and national responsibility is reduced in favour of regional and local governments. This is especially true for education and training, where increasing local and regional involvement is associated with policies involving networks of business, unions, and other social groups (Jessop 1993, 24).

The tendencies for Canada to make the transition to the SWS must be seen within the context of a federal system characterized, even under Fordism, by considerable decentralization. Historically, the federal government has been responsible for labour market policy, while provinces have had a mandate for education. In practice, during the postwar period federal, provincial, and municipal governments overlapped considerably in labour market intervention, and there were private training agencies as well (Muszynski and Wolfe 1989, 259). In 1993, the three levels of government offered over 300 separate programs. Public investment in skill development occurred mainly in secondary schools, which offered basic "vocational" education, and there were only limited initiatives to support apprenticeship. Public policy attempted to resolve national skill shortages by encouraging skilled European workers to immigrate to Canada (Olsen 1988; Muszynski and Wolfe 1989). Labour market policy provided "passive" income support through unemployment insurance (UI) (Premier's Council 1990, 192). Since unemployment was assumed to be cyclical, unemployed workers were not seen as requiring enhanced or new skills and short-term public works projects or "make work" programs were implemented during recessions (Olsen 1988).

Under Brian Mulroney's Conservative government (1984–93) federal labour market and training policy began to move away from government-directed training and short-term job creation. In 1985 Ottawa introduced the Canadian Jobs Strategy (CJS), which attempted to increase labour mobility through skill development (Muszynski and Wolfe 1989, 259). An important element of this shift was increased privatization of training away from recognized institutions such as community colleges and Canada Employment Centres to private "managing co-ordinators" (Premier Council 1990, 111). Community Industrial Training Committees (CITCS) were organized at the local level to administer federal programs. Though largely unsuccessful, they decentralized federal labour market policy.[9]

By the late 1980s the UI and welfare systems were coming under pressure as a result of increases in both cyclical and long-term unemployment. Between 1981 and 1991 repeat users (three or more claims) made up by far the greatest proportion of all claimants, increasing from under 600,000 to over one million, and by 1993 nearly 15 per cent of the Canadian population was receiving either UI or welfare support while in Newfoundland the figure had reached nearly 25 per cent (Cox 1994). The policy response at both federal and provincial levels has taken two forms and has focused primarily on the unemployed. First, reform of UI has been tied to increasing employee responsibility. In 1992, in line with attempts over the previous 10 years to tighten qualifications for UI and to lower benefits, the Conservative government disallowed eligibility for UI for those who "voluntarity quit their jobs without just cause, or who lose their jobs because of their own misconduct" (Fraser 1993). At the same time, courts have given employers more hiring and firing flexibility, arguing that "businesses need more leeway today to restructure themselves or reassign employees in the face of free trade and a global recession" (Gibb-Clark 1992). Grounds for dismissal have been expanded to include behavourial attributes of employers, which formerly would not have been considered just cause for firing.

Second, as part of this new emphasis on "supply-side" flexibility, since the mid-1980s UI has become increasingly focused on training. This has been accomplished by diverting funds from income support under UI Developmental Uses (UIDU). By 1993 60 per cent of training was funded from UDIU, compared to only 20 per cent in 1990 (Stephen 1993, 3). But during this time the federal government significantly reduced its total training budget. Through the latter half of the 1980s Ottawa's expenditures on training fell 27 per cent, from $2.2 billion in 1984–85 to $1.6 billion in 1988–89, while in Ontario they fell 32 per cent over same period (Premier's Council 1990, 111; McBride and Shields 1993). These cuts have hit women especially hard. Between 1987 and 1992 federal training expenditures for women fell 30 per cent. Although in 1992 women accounted for 60 per cent of all new entrants to the labour market they received only 34 per cent of the total training budget.

The shift of UI toward training has significant social and regional implications. A recent proposal by the Liberal government aims to differentiate between "frequent" users, who would get lower benefits and more active job-search assistance, and "occasional" users, who would receive full benefits (Cox 1994). This would have adverse effects on workers in Atlantic Canada, who depend heavily on UI to supplement seasonal work in fishing, forestry, tourism, and agriculture. Since the collapse of the fishing industry Ottawa has linked compensation payments to career counselling and training.

The linking of UI to training has been extended to other social benefits, especially welfare. In New Brunswick, which has the lowest level of welfare benefits in Canada, those on social assistance will be asked to sign an "agreement of understanding" to undertake training or to to stay in school (Sheppard 1994). Other provinces are following suit: for example, even the NDP social services minister in Ontario was quoted as saying that people have to be redirected "towards Workfare – whether they are single parents, single individuals or married" (Diebel 1994).

Whatever moves there are toward forms of workfare, existing training and adjustment programs have often been subject to a confusing number of overlapping levels of government regulations and agencies (Premier's Council 1990; CLFDB 1993). The lack of a "seamless system" for those laid off has been exacerbated by considerable variations among provinces in the minimum notice required for layoff and the fact the most existing training programs are of short duration and unlikely to lead to any real upgrading of skills (Premier's Council 1990, 186–7). This is especially a problem for the long-term unemployed, many of whom require substantial training. Unfortunately, programs such as Jobs Ontario, designed to return the longer-term unemployed to employment, can stigmatize welfare recipients (*Globe and Mail* 1994b). Even in low-unemployment regions such as southern Ontario many of those on social assistance may not be truly available for wage work – for example, in early 1994 in Kitchener-Waterloo less than 8,000 of the 40,000 people on social assistance were considered "employable" (Nunn and Strathdee 1994). Lack of opportunity remains the major cause of high levels of unemployment. Thus in March 1994 in London for more than 32,000 people in need of work there were only 2,129 advertised openings – one-third of these positions were part-time (Greenspon 1994).

Cutbacks and shifts in training expenditures have occurred at a time when the federal government is devolving its labour market responsibilities to the provinces. One of the last acts of the Conservative government in Ottawa was to give Quebec responsibility over all federal labour market training in the province (Delacourt 1993), and Ontario has sought to tie UI benefits to provincial job training (Walker 1994).

The most significant part of devolution has been development of local labour force development boards (Mahon 1991). In 1990 Ottawa established the quasi-autonomous Canadian Labour Force Development Board (CLFDB), consisting of representatives from business, labour, and social groups and serving as a consensus-building organization on training policy. While an important part of its mandate is to establish both local and sectoral training boards, its major role is to "provide an effective and authoritative forum for establishing broad standards of access to training while ensuring the quality of programs available" (Employment and Immigration Canada 1991).

This board, and corresponding provincial and sub-provincial boards, facilitate private-sector involvement in design and implementation of training and, the UI program is envisioned as part of an active strategy that includes reductions in "work disincentives" (Premier's Council 1990, 113). These organizations will reduce duplication between federal and provincial labour market policies by channeling funding (CLFDB 1991, 11). However, such boards are not to act as mere "transmission belts" for government but are to exercise considerable autonomy in development of skills training, which will be delivered in a decentralized manner (Employment and Immigration Canada 1991, 7).

By the beginning of 1995 only Ontario and Quebec had begun to establish local boards. To understand this delay and the problems confronted by the boards in achieving consensus, the Ontario example is instructive. The Ontario Training and Adjustment Board (OTAB), set up in 1993, has responsibility for 48 training and adjustment and access programs currently operated by federal and provincial ministries. OTAB is to coordinate existing federal and provincial programs in four broad categories of labour-force development – workplace and sectoral training, apprenticeship programs and related services, labour-force adjustment (programs and services for workers affected by significant change in their workplace), and labour-force entry/re-entry programs, to assist individuals enter the paid labour force for the first time or to re-enter after a prolonged absence. Though OTAB is an agency of the provincial government it will operate as an independent, self-governing organization and includes all the "stakeholders" involved in training – labour, business, and social equity groups.

There is an implicit assumption that OTAB will lead to a break from the adversarial practices of the past (OTAB 1992, 7) and that business will increase investment in training "comparable to or surpassing leading international levels" (OTAB 1992, 10). Moreover, there is a strong sense that the restructuring required for economic competitiveness and social equity are not mutually exclusive and that training can secure both objectives (OTAB 1992, 1).

Training and skill development will be devolved to 25 local boards. The devolution of training responsibilities to that level is driven by the assumption that it is the appropriate spatial scale for delivery of training. Consultation with community stakeholders and existing market patterns and training programs were used to define the geographic base of the boards. Like OTAB these boards will include local labour market partners and will identify labour-force needs and priorities, develop strategic and operational plans based on labour market information; determine the mix of funding, programs, and delivery mechanisms required to address local needs and priorities; and promote a "life-long learning and training culture" among employers and workers (Employment and Immigration Canada 1991). These boards are to work with and report to OTAB for developing training policies and program delivery. They are also to work with local (federally run) Canada Employment Centres (CECs) to provide guidance and other inputs to strategic planning.

While it is evident that the strategy of local boards is consistent with some elements of neo-conservative strategy (Peck 1992; Peck and Jones 1995), the program draws strongly on the system of corporatism found in Scandinavia and Germany (CLFDB 1991, 4). Nevertheless, there are differences: historically, the Swedish active labour market system has provided a high level of income support and training for the unemployed, which generosity is less discernible in the Canadian model, given substantial cutbacks of funding to training and adjustment programs (CLFDB 1994, 14).

The implementation of local boards in Ontario has occurred while there has been significant "downsizing" and employment loss,[10] but government initiatives have aimed more to encourage labour market flexibility and to discipline the unemployed rather than to establish a highly rated training regime in Canada. The CLFDB initiative assumes a general shift toward a high skill–high wage economy, but this does not fit the evidence of emerging two-tier labour markets (Mahon 1991, 77). There is also considerable "voluntarism" in current policies, with employers being encouraged to upgrade skills, but the means provided are modest (Mahon 1991, 88).

The establishment of OTAB and local boards has proved conflictual and time consuming. There has been debate over how strongly OTAB should insist that training be delivered by publicly funded and accountable institutions. Unions, for example, wanted to co-ordinate all training through community colleges so as to resist any privatization of program delivery, while women's and immigrant groups argued that community-based and non-profit institutions should also be allowed to play a role. There has also been conflict between labour and business over who will pay for training, with labour favouring a levy system or tax on employers. Whether training should be an agent of restructuring and whether export-oriented sectors should be a focus have also been contentious (Dehli 1993, 99).

At both provincial and local levels there has been considerable debate over representation of unorganized workers. Some communities lobbied against being included in the larger board regions, and Aboriginal groups chose to opt out of the local board structure altogether. Many business groups have no experience in dealing with organized labour as equal partners on training policy at the local level and find this process problematic. While the local boards were supposed to take only a year to set up after the process was initialed in early 1992 (De Ruyter 1992), the launching meetings of the first of 25 such bodies did not occur until late 1994. Finally, with the political climate shifting away from the consensus strategy of the NDP toward a more business-led one of the recently elected Conservatives, there must be some doubt whether the boards will maintain their present form.

There is concern that the entire strategy of local boards will reinforce balkanization of Canadian labour markets. Employers have complained about lack of a national employment database to facilitate better meeting of demand with supply (Galt 1994), and there is a view that "globalization" requires a more centralized approach to training and labour market policy (Delacourt 1993). CLFDB documents, support the need to coordinate local skill requirements with nationally agreed standards, but increased decentralization of labour market policy to the provinces invites the opposite.

THE RESPONSE OF LABOUR

All the factors outlined above make a transition to a high skill–high wage economy less probable. The increased disciplining of the unemployed and the direction of those workers dependent on welfare into some form of workfare suggest underdevelopment of the Schumpeterian or innovative aspects of the sws in Canada. Nevertheless, the labour movement is assuming an increasingly important role in formation of training policy at all levels. Within public and private sectors it has influenced training strategies in individual workplaces and at the sectoral level and it has helped secure adjustment assistance for laid-off/unemployed workers.

Unlike their u.s. counterparts, Canadian unions have maintained a significant share of the non-agricultural workforce in Canada (Drache and Glasbeek 1992; Meltz and Verma 1993).[11] In many industries unions have successfully defended the gains made under Fordism and developed a proactive strategy in new economic activities. During the 1980s the separation of the Canadian section of the u.s.-based United Auto Workers (uaw) to form the Canadian Autoworkers Union (caw) reflected different economic conditions in the Canadian automobile industry and the commitment of Canadian workers to defend the linkage of wages to productivity/cost of living while the UAW was making wage concessions (Gindin 1989; Holmes and Rusonick 1991). This has been part of a "Canadianization" of unions, and by the mid-1980s over 60 per cent of organized workers in Canada were in a Canadian-based union (Palmer 1992, 410). Canadian unions have been more successful in organizing full- and part-time workers in the service sector (Mackie 1992; Meltz and Verma 1993). Even older industrial unions such as the caw and the Steelworkers have organized such traditionally non-unionized occupations as private security guards and employees in senior citizen homes. This has been assisted by the presence of ndp governments in several provinces, which have made labour organizing easier and have banned use of replacement workers during strikes (Mackie 1992).

The restrictions placed on the public sector's right to strike and federal and provincial freezes on public-sector wages have increased tensions between public-sector unions and ndp governments and have exacerbated strains between public and private-sector unions (Panitch and Swartz 1993). Given significant differences in the relative competitive positions of different sectors and regions, the ability of unions to defend earlier gains has been weaker in resource industry, especially pulp and paper and steel, than in the automobile industry (Holmes and Hayter 1994).

None the less, labour has proved proactive and innovative in responding to changing employer strategies and developing initiatives. Unions have been critical of management-initiated methods of workplace reorganization, such as tqm and team working, and lean production, because of concerns over increased work intensity, undermining of union structures, and deskilling, but this has not precluded their participation in workplace restructuring (Robertson 1992; Wells 1993; clmpc 1994a). Indeed, surveys of employee participation programs show unionized workplaces more likely to have such programs, which are more likely to last than those in non-

Table 4
Training issues and collective agreements: percentage of contracts and employees, 1980–92

	Agreements			Employment		
	1980	1987	1992	1980	1987	1992
Training on the job	n.a.	40	43	n.a.	45	48
Training – outside courses	n.a.	26	31	n.a.	33	35
Apprenticeship	n.a.	33	33	n.a.	32	30
Educational leave – job related	37	28	15	n.a.	38	19

Source: Meltz and Verma (1993).

unionized workplaces (CLMPC 1994a, 11–75). Furthermore, unions have initiated their own programs, such as the reductions in work time at the "Big Three" auto manufacturers and Bell Canada implemented by the CAW and the Communications, Energy and Paperworkers Union (CEP) (CLMPC 1994a). In the same agreements the CAW addressed issues of equity in terms of hiring and apparenticeship in addition to protecting workers from sexual harassment and increasing funding for childcare.

All of these initiatives have enhanced the presence of training in collective agreements. Since union membership is concentrated in large firms, so is union influence on training; thus it is three times more likely in large firms that unions will directly provide training or that training will be provided through the collective agreement (CLMPC 1993, 35). Provisions in collective agreements covering both on-the-job training and training involving outside courses have expanded since the early 1980s, but agreements covering apprenticeship and job-related educational leave have shrunk (Table 4).

These date understate the extent to which training has become more important to union strategies. Unions are included in the CLFDB, and they have adopted a "social bargaining" position initially developed as a corporatist strategy by the federal and provincial governments in the late 1980s (Jenson and Mahon 1993). The funding offered by governments for participation in the development of a training policy was also used for research by unions, and the pace of change of technology in the workplace meant that unions went beyond resistance to managerial initiatives and developed alternative strategies. The CAW and the Steelworkers have carried out extensive research on restructuring (Robertson and Wareham 1987; USAW 1991; CLMPC 1992). Since 1989 the Ontario Federation of Labour has used Ontario government funding (Training and Adjustment Research Project, or TARP) to develop and disseminate its own views on training, job redesign, and adjustment (Jenson and Mahon 1993, 83). Furthermore, labour has argued strongly for the development of transferable skills, has criticized the dilution of skills in "multiskilling" initiatives, and has attempted to adapt team working in ways that maintain union shopfloor strength (Beckerman, Davis, and Jackson 1992).

To illustrate the strategic choices pursued by labour I outline two initiatives – Sectoral Training Bodies and Adjustment Committees. The CAW and the Steelworkers negotiated with employers to establish sectoral training bodies in the automotive, steel, and electronics industries. The automotive parts sectoral-training initiative, which is federally and provincially–funded, is one of the most developed in this respect. In late 1991, the recently established Automotive Parts Sectoral Training Council (APSTC) set up a sectoral training program, administered jointly by management and labour. The goal of this three-level program is to develop skills that are portable across the industry – most firm-specific training programs are not. Thus the CAW is an equal partner and has an equivalent say with participating firms in developing training programs and selecting trainers. These programs will lead to the granting of recognized Auto Parts Certificates of training in modern technology, work organization, mathematics, and workplace communication and involvement. Thus training has both technical and political aspects, the latter involving exercise of workplace power (Piker 1994).

The initiative of the Canadian Steel Trades Employment Congress (CSTEC) is focused on adjustment. CSTEC, a joint management–union body, was established in 1985 and initially concentrated on trade issues, but by 1988 it had developed an adjustment program to handle increased lay-offs in the steel industry. A skills-training program similar to the APSTC also provides information about new employment. It is funded by the steel industry and Employment and Immigration Canada. It assisted 13,500 displaced workers between 1988 and 1994 and by the latter date had a job placement rate of 80–90 per cent and a dropout rate of less than 2 per cent (CLMPC 1994a; *Globe and Mail* 1994a). Because CSTEC's clients are usually semi-skilled, with few portable skills, there is a need for substantial skill upgrading and not just counselling and enhancement of job-search skills.

The CSTEC program has several advantages. It offers a "single window" of peer counselling. Teams of labour and management are coordinated by a small, permanent staff trained to interview and counsel laid-off steel workers, often well before the lay-off occurs. The program has also been able to use section 26 of the UI program, which allows UI benefits to be paid while clients receive employment services and training. Another important feature of the program is its local focus, with 59 local adjustment committees, many of which are grouped into regional committees and centres. A computerized national job bank/database allows CSTEC to tailor training more closely to local skill demand. Finally, in contrast to federal Employment and Immigration Canada (EIC) programs, CSTEC provides continuous assistance through training programs of up to three years in length (compared to a year or less for EIC). Often CSTEC assistance continues for up to 18 months after a client has found a new job, rather than ending, as EIC does, as soon as one is obtained (CLFDB 1993).

CSTEC is more successful than conventional adjustment measures in assisting laid-off workers in obtaining new skills and employment, but it has several shortcomings. It is dependent on considerable federal government funding, having received $32.5 million from Ottawa between 1988 and 1994 and being scheduled to get another $14 million by 1996; but now financing is in question, potentially the high standards of this program are in jeopardy, and it is vulnerable to shifts in government policy. CSTEC has not been able to assist those on temporary lay-off, but its Skills Training Program has led to skill upgrading for approximately 40 per cent of workers in participating firms. CSTEC is perceived to focus on assisting workers in leaving the industry and thus more directly benefiting employers in other sectors (CLMPC 1992, 45).

CSTEC and APSTC reflect different attitudes, though both endorse a "social union-ism" position toward labour-management relations. They have, however, developed different interpretations of this philosophy. The Steelworkers have confronted a se-vere downsizing of the steel industry and faced continued reduction of production ca-pacity. Unlike the CAW, they have remained part of an American-based international union and have participated in negotiating an employee buy-out of Algoma Steel in Sault Ste Marie, Ontario. At Algoma, the Steelworkers have adopted innovative prac-tices such as self-directed work teams and greater worker involvement in strategic management. This has placed them at odds with the CAW, which, though willing to participate with industry in strategic policy making, is suspicious of union involve-ment in overall management (Holmes and Rusonick 1991, 26). Over most of the last decade the CAW has witnessed expansion in output and, to a lesser degree, in employ-ment. Consequently, in the sectoral training agreements outlined above, the CAW ar-ticulates its own union agenda as an integral part of the training program.

Despite different views on training by the CAW and the Steelworkers, their partici-pation in formation of federal and provincial labour policy and sectoral training and adjustment programs is part of a shift toward some form of SWS in Canada. Labour in Canada could become an important agent in skill and labour market policy formation, yet this should not be viewed as a form of coopting, nor does it undermine innovative capacity and skill formation. If capital, when left to its own devices, systematically underinvests in training (Streeck 1989; Kern and Sabel 1991; Peck 1992) the labour movement's pressure on employers to protect skill transferability and worker bargain-ing may help offset this tendency while enhancing the Schumpeterian or innovative capacity of current restructuring.

CONCLUSIONS

The current restructuring of labour markets in Canada has been accompanied by a transition from a Fordist to a post-Fordist labour market regime. "Permeable Ford-ism" in Canada has been more open to cyclical world markets (especially in resource-based regions) and its policy commitment to full employment weaker than western Europe's. Restructuring, however, has led to a decline in core employment, emer-gence of a two-tier labour market, new forms of gender and occupational segmenta-tion, and increasing balkanization or regional labour markets in Canada.

At the level of state policy there has been an apparent transition from the Keynesian welfare state associated with Fordism to the Schumpeterian workfare state. "Perme-able Fordism" in Canada was an institutional renegotiation of federalism (Jenson 1989), and its crisis has resulted in decentralization of labour market responsibilities from the federal to the provincial level and below. Furthermore, labour and other social groups have been included as "equal partners" in labour-force development boards.

The coherence of this apparently corporatist approach in Canada is in some doubt. Decentralization has been accompanied by increased disciplining of labour through the way UI benefits have been tied to training, and cuts in federal funds for training and adjustment have been made when the increased openness of the Canadian economy has resulted in higher levels of displacement and structural unemployment. This process is more likely to reinforce than to reduce the already considerable regional balkanization of labour markets in Canada. Finally, in some instances such as in Ontario, new institutional arrangements for training are based on an assumption of consensus between labour and capital as "labour market partners" that rests uneasily on a legacy of adversarialism. In this context of growing structural unemployment, deindustrialization, and regional balkanization, if there is an ongoing shift toward a Schumpeterian workfare state in Canada, there is a good chance that it will be more workfare than Schumpeterian.

NOTES

1 Fordist labour relations were developed during the nineteenth and early twentieth centuries in a series of resource labour markets concentrated in western Canada, Northern Ontario/Quebec, and the Maritimes (see Drache 1984; Palmer 1992). By contrast, in south-central Ontario domestic manufacturing grew behind National Policy tariffs and generated a labour market led primarily by male, skilled craftworkers. In Quebec a declining agricultural sector led to mass migration of rural producers to the mill towns of New England but also to industrial centres such as Montreal. By the early 1920s a significant spatial division of labour had developed within the industrial heartland, and it still persists, with southern Ontario focusing on higher-value-added, capital-intensive manufacturing such as steel and machinery while Quebec became characterized by low-value-added, labour-intensive sectors such as shoes, textiles, and clothing (Brodie 1990, 107).

2 Norcliffe (1994), drawing on Mackenzie (1987), examined the impact of this rationalization and closure on the Trail-Nelson area of British Columbia. Laid-off male workers remained in the region and became engaged in the informal economy while women developed such community services as childcare and other domestic services.

3 Between 1975 and 1993 the highest annual rates of growth in part-time work were in western Canada, especially Saskatchewan, Alberta, and British Columbia, which all averaged annual increases of over 5.0 per cent (Pold 1994). By 1993 Saskatchewan and Manitoba had the highest proportion of total employment in part-time work, with 29 and 27 per cent, respectively, while Newfoundland had the lowest, at 17 per cent. The high average in Saskatchewan and Manitoba is in part explained by the mix of industries there, with agriculture – the dominant sector in the prairie provinces – being an above-average employer of part-time workers.

4 This is evident in both white-collar and blue-collar occupations. By 1993, 25 per cent of male managers were working more than 50 hours a week – up 6.4 per cent from 1985

(Scrivner 1994). Similarly, overtime by blue-collar workers rose significantly from mid-1992 to early 1994, when manufacturing began to hire more full-time employees. However, concern over the stress caused by excessive overtime has been increasing (Abbate 1994; Scrivener 1994), while Ontario and federal task forces and unions have called for increased regulation of its use by employers.

5 There is also evidence that women's occupational/career mobility is still limited by their low representation in jobs that require more than two years of training (i.e., training provided by employers). There is, however, relatively little difference in educational attainment of men and women in the labour force (CLMPC 1994b).

6 Whether male or female, older workers displaced from these sectors are at some disadvantage in the labour market, since 40 per cent of them lack a high school diploma and have skills in areas of low demand (CLFDB 1993, 32).

7 These trends have meant that even those able to gain full-time employment may not be able to escape low-income status. In 1992, it was estimated that the heads of 200,000 families (in addition to 135,000 single individuals) working full-time had incomes below a low-income cut-off determined by Statistics Canada (Drohan 1994).

8 These trends have been intensified by the most recent recession: increased unemployment and lower-paid and part-time employment contributed to a decline in family incomes of 6.7 per cent since 1989 – the deepest decline in 40 years (Freeman 1994).

9 The CJS has been heavily criticized. While it was oriented toward training unemployed workers (for example, 68 per cent of federal training expenditures in Ontario went to them in 1987–88) for the most part it offered only entry-level, short-term training and did not upgrade effectively the skills of those still in employment or those made redundant (Premier Council 1990, 111). It has also been attacked for a being a de facto wage subsidy to employers, who in turn provided only job-specific, non-transferable training. Only two-thirds of the planned CITCs were actually established. The CJS drew limited involvement from local groups involved in training and had difficulty gaining any labour participation.

10 Dehli (1993) has pointed out that many of the firms that have pushed for a high-skill, "knowledge-based" strategy in Ontario are precisely those that are laying off workers in production or moving such work to developing countries.

11 Since the mid-1970s the union proportion in Canada has been 37 per cent, compared to a U.S. figure of 20 per cent which declined to 16 per cent by the early 1990s. However, several researchers (Troy 1992; Meltz and Verma 1993) point out that these Canadian figures tend to overstate the relative performance of Canadian unions. Canadian union performance has been enhanced by the high level of public-sector unionization (over 60 per cent), while the private-sector proportion is only 21 per cent (Meltz and Verma 1993, 19). Others have argued that the Canadian labour movement has been weakened by attacks on the right to strike of workers in the public sector, the restructuring strategies of firms such as contracting out, and the growth of traditionally non-unionized private services (Drache and Glasbeek 1992; Palmer 1992; Panitch and Swartz 1993).

12 An example of the problems of CSTEC's dependence on federal funding occurred in March 1992, when the government ruled that funds could not be used to place workers in university courses.

REFERENCES

Abbate, G. 1994. "Ottawa Urged to Cap Workweek, Paid Overtime." *Globe and Mail*, 17 Dec., B4.

Atkinson, J. 1987. "Flexibility or Fragmentation?: The UK Labour Market in the 1980s." *Labour and Society*, 12, no 1, Jan.

Automotive Parts Human Resources Study. 1991. *Report of the Automotive Human Resource Study.* Toronto: Government of Ontario.

Bakker, I. 1991. "Canada's Social Wage in an Open Economy: 1970–1983." In D. Drache and M. Gertler, eds., *The New Era of Global Competition*, 270–87. Montreal: McGill-Queen's University Press.

Barnes, T., Edgington, D., Denike, K., and McGee, T. 1992. "Vancouver, the Province, and the Pacific Rim." In G. Wynne and T. Oke, eds., *Vancouver and Its Region*, 171–99. Vancouver: University of British Columbia Press.

Beckerman, A., Davis, J., and Jackson, N., eds. 1992. *Training For What? Labour Perspectives on Skill Training.* Toronto: Our Schools/Our Selves Education Foundation.

Bradbury, J. 1984. "Industrial Cycles and the Mining Sector in Canada." *International Journal of Urban and Regional Research*, 8, 311-31.

Brodeur, M., and Galarneau, D. 1994. "Three Large Urban Areas in Transition." *Perspectives*, Statistics Canada, winter, 37–45.

Brodie, J. 1990. *The Political Economy of Canadian Regionalism.* Toronto: Harcourt Brace Jovanovich.

Bromley, I. 1994. *Economic Change in the Toronto Region: Growth, Recession and Recovery.* Toronto: Metro Economic Development Division.

Canadian Labour Force Development Board (CLFDB). 1991. *A Proposal to Establish Local Labour Force Development Boards.* Ottawa: Canadian Labour Force Development Board.

– 1993. *Report of the Task Force on Labour Adjustment to the Canadian Labour Force Development Board.* Ottawa: Canadian Labour Force Development Board.

– 1994. *Annual Report 1993–1994.* Ottawa: Canadian Labour Force Development Board.

Canadian Labour Market and Productivity Centre (CLMPC) 1992. *Human Resource Study of the Canadian Steel Industry.* Ottawa: Canadian Labour Market and Productivity Centre and the Canadian Steel Trades Employment Congress.

– 1993a. *National Training Survey: 1991 National Training Centre.* Ottawa: Canadian Labour Market and Productivity Centre.

– 1993b. *National Training Survey: 1991 National Training Centre: Commentary.* Ottawa: Canadian Labour Market and Productivity Centre.

– 1994a. *Adjustment and Transition Issues.* Ottawa: Canadian Labour Market and Productivity Centre, March.

– 1994b. *Women and Economic Restructuring*, Ottawa: Committee on Women and Economic Restructuring, Canadian Labour Market and Productivity Centre, March.

Christopherson, S., and Noyelle, T. 1992. "The US Path towards Flexibility and Productivity: The Re-making of the US Labour Market in the 1980s." In H. Ernste and V. Meier, eds., *Regional Development and Contemporary Industrial Response: Extending Flexible Specialisation*, 163–78. London: Belhaven.

Connelly, P., and MacDonald, M. 1983. "Women's Work: Domestic and Wage Labour in a Nova Scotia Community." *Studies in Political Economy: A Socialist Review*, 10, winter, 45–72.

Corman, J., Luxton, M., Livingstone, D., and Seccombe, W. 1993. *Recasting Steel Labour: The Stelco Story*. Halifax: Fernwood.

Cox, K. 1994. "Maritime Seasonal Workers Fear Cut in Benefits." *Globe and Mail*, 6 Oct., A10.

Crane, D. 1994. "High-tech Heartland Hums Along." *Toronto Star*, 2 March, A19.

Crompton, S. 1994. "Employer-supported Training – It Varies by Occupation." *Perspectives*, Statistics Canada, spring, 9–25.

Davidson, J. 1993. "Equality Slowdown Denounced." *Globe and Mail*, 9 Dec.

Dehli, K. 1993. "Subject to the New Global Economy: Power and Positioning in Ontario Labour Market Policy Formation." *Studies in Political Economy*, 41, summer, 83–110.

Delacourt, S. 1993. "PM Faces Skill Test in Quebec." *Globe and Mail*, 3 Aug. A3.

DeRuyter, R. 1992. "Grassroots Process Delays Ontario Training System Overhaul." *Kitchener-Waterloo Record*, 3 Nov.

– 1993. "Unions Won't Give up Their 'Right to Fight.'" *Kitchener-Waterloo Record*, 26 Oct.

Diebel, L. 1994. "Bill Clinton's War on America's Poor." *Toronto Star*, 6 March, E1.

Donner, A. 1991. "Recession, Recovery and Redistribution: The Three R's of Canadian State Macro-policy in the 1980s." In D. Drache and M. Gertler, eds., *The New Era of Global Competition*, 26–50. Montreal: McGill-Queen's University Press.

Drache, D. 1984. "The Formation and Fragmentation of the Canadian Working Class: 1820–1920." *Studies in Political Economy*, 15, fall, 43–89.

– 1991. "The Systematic Search for Flexibility." In D. Drache and M. Gertler, eds., *The New Era of Global Competition*, 251–69. Montreal: McGill-Queen's University Press.

Drache, D. and Glasbeek, H. 1992. *The Changing Workplace: Reshaping Canada's Industrial Relations System*. Toronto: Lorimer.

Drohan, M. 1994. "The Rich Get Richer, the Poor Get Pummelled." *Globe and Mail*, 17 Dec., F2.

Economic Council of Canada. 1990. *Good Jobs, Bad Jobs: Employment in the Service Economy*. Ottawa: Economic Council of Canada.

Employment and Immigration Canada. 1991. *Local Boards: A Partnership for Training*. Ottawa.

Fraser, G. 1993. "Widening Split along Class Lines Revealed by Proposed UI Changes." *Globe and Mail*, 2 March.

Freeman, A. 1993. "New Brunswick Hits Books after Years of Hard Knocks." *Globe and Mail*, 16 Jan., B8.

– 1994. "Family Income Falls for Fourth Year." *Globe and Mail*, 22 Dec.

Galt, V. 1994. "Task Force Wants Jobs Data Base." *Globe and Mail*, 10 May, A1.

Gertler, M. 1992. "Flexibility Revisited: Districts, Nation-States, and the Forces of Production." *Transactions of the Institute of British Geographers*, 17, 259–87.

– 1994. "The Economic Context of Restructuring in the Toronto Region." Paper, presented at the Kyoto University–University of Toronto Symposium for Mutual Understanding of Japan and Canada, 15–16 Feb.

Gibb-Clark, 1992. "Hiring, Firing Flexibility Urged." *Globe and Mail*, 20 Nov.

– 1995. "Job Market Demanding Specific Skills." *Globe and Mail*, 3 Jan., B1.

Gindin, S. 1989. "Breaking Away: The Formation of the Canadian Auto Workers." *Studies in Political Economy*, 29, summer, 63–89.

Globe and Mail. 1994a. "Ontario Tackles the Long Term Unemployed." *Globe and Mail*, 30 Sept.

– 1994b. "Safety Net Bounces Workers Back." *Globe and Mail*, 29 Sept.

Gordon, D., Edwards, R., and Reich, M. 1982. *Segmented Work, Divided Workers: The Historical Transformation of Labour in the United States*. Cambridge: Cambridge University Press.

Greenspon, E. 1994. "Safety Net Snares Many in Web of Disillusionment." *Globe and Mail*, 28 Sept., A1.

Hayter, R., and Barnes, T. 1992. " 'The Little Town that Did': Flexible Accumulation and Community Response in Chemainus, British Columbia." *Regional Studies*, 26, 647–63.

Holmes, J. 1989. "New Production Technologies Labour and the North American Auto Industry." In G.J.R. Linge and G.A. van der Knaap, eds., *Labour, Environment and Industrial Change*, 87–106. Andover, Hants: Routledge, Chapman and Hall.

Holmes, J., and Hayter, R. 1994. *Recent Restructuring in the Canadian Pulp and Paper Industry*, Discussion Paper No. 26, Simon Fraser University.

Holmes, J., and Rusonick, A. 1991. "The Break-up of an International Labour Union: Uneven Development in the North American Auto Industry and the Schism in the UAW." *Environment and Planning A*, 23, 9–35.

Jenson, J. 1989. " 'Different' but not 'Exceptional': Canada's Permeable Fordism." *Canadian Review of Sociology and Anthropology*, 26, 69–74.

Jenson, J. and Mahon, R., eds. 1993. *The Challenge of Restructuring: North American Labor Movements Respond*. Philadelphia: Temple University Press.

Jessop, B. 1993. "Towards a Schumpeterian Workfare State? Preliminary Remarks on Post-Fordist Political Economy." *Studies in Political Economy*, 40, 7–39.

Kern, H., and Sabel, C. 1991. "Trade Unions and Decentralized Production: A Sketch of Strategic Problems in the West German Labor Movement." *Politics and Society*, 19, 373–402.

Little, B. 1994. "Toronto Struggles to Recover." *Globe and Mail*, 14 Nov. B1.

Lovering, J. 1990. "A Perfunctory Sort of Post-Fordism: Economic Restructuring and Labour Market Segmentation in Britain in the 1980s." *Work, Employment and Society*, special issue, May 9–28.

Luxton, M. 1980 *More than a Labour of Love: Three Generations of Women's Work in the Home*. Toronto: Women's Press.

McBride, S., and Shields, J. 1993. *Dismantling a Nation: Canada and the New World Order*. Halifax: Fernwood Publishing.

McDowell, L. 1991. "Life without Father and Ford: The New Gender Order of Post-Fordism." *Transactions of the Institute of British Geographers*, 16–4, 400–19.

Mackenzie, S. 1987 "Neglected Spaces in Peripheral Spaces: Women and the Creation of the New Economic Centre." *Cahiers du géographie du Québec*, 31, 247–60.

Mackie, R. "Unions Target Retail Chains for Ontario Membership Drive." *Globe and Mail*, 6 June, B1.

Mahon, R. 1991. "Adjusting to Win? The New Tory Training Initiative." In K. Graham, ed., *How Ottawa Spends, 1990–91: Tracking the Second Agenda*, 73–112. Ottawa: Carleton University Press.

Meltz, N. and Verma, A. 1993. "Developments in Industrial Relations and Human Resource Practices in Canada: An Update from the 1980s." In T. Kochan, R. Locke, and M. Piore, eds., *Employment Relations in a Changing World Economy*. Cambridge, Mass.: MIT Press.

Muszynski, L., and Wolfe, D. 1989. "New Technology and Training: Lessons from Abroad." *Canadian Public Policy*, 15, 245–64.

Myles, J. 1991. "Post Industrialism and the Service Economy." In D. Drache and M. Gertler, eds., *The New Era of Global Competition*, 351–66. Montreal: McGill-Queen's University Press.

Norcliffe, G. 1994. "Regional Labour Market Adjustments in a Period of Structural Transformation: An Assessment of the Canada Case." *Canadian Geographer*, 38 no. 1, 2–17.

Nunn, T., Strathdee, M. 1994. "Region's Unemployment Rate Lowest in Canada, But Numbers Could be Skewed." *Kitchener-Waterloo Record*, 8 Jan.

Olsen, G., ed. 1988. *Industrial Change and Labour Adjustment in Sweden and Canada*. Toronto: Garamond.

Ontario Training and Adjustment Board (OTAB). 1992. *Skills to Meet the Challenge: A Training Partnership for Ontario*. Toronto: Government of Ontario.

Palmer, B. 1992. *Working Class Experience: The Rise and Reconstitution of the Canadian Working Class*. Toronto: Butterworths.

– 1994. *Capitalism Comes to the Backcountry: The Goodyear Invasion of Napanee*. Toronto: Between the Lines.

Panitch, L., and Swartz, D. 1993. *The Assault on Trade Union Freedoms: From Wage Controls to Social Contract*. Toronto: Garamond Press.

Peck, J. 1992. "Labor and Agglomeration: Control and Flexibility in Local Labour Markets." *Economic Geography*, 68, 325–47.

– 1994. "Regulating Labour: The Social Regulation and Reproduction of Local Labour Markets." In A. Amin and N. Thrift, eds., *Globalization, Institutions, and Regional Development in Europe*, 147–76. Oxford: Oxford University Press.

Peck, J., and Haughton, G. 1991. "Youth Training and the Local Reconstruction of Skill: Evidence from the Engineering Industry of North West England, 1981–88." *Environment and Planning A*; 23, 813–32.

Peck, J., and Jones, M. 1995. Training and Enterprise Councils: Schumpeterian Workfare State, or What? Mimeo, School of Geography, University of Manchester.

Piker, J. 1994. *Auto Parts Certificate: An Innovative Approach to Training*. Markham, Ont.: Automotive Parts Sectoral Training Council.

Pold, H. 1994. "Jobs! Jobs!, Jobs!" *Perspectives*, autumn, 14–17.

Premier's Council. 1990. *People and Skills in the New Global Economy*. Toronto: Government of Ontario.

Rees, G., Winckler, V., and Williamson, H. 1989. "The 'New Vocationalism': Further Education and Local Labour Markets." *Journal of Education Policy*, 4, 227–44.

Report on Business. 1994 "Undercrowding: Classrooms in the Workplace," *Globe and Mail*, Nov. 8.

Rigdon, J. 1994. "Workers Getting Sick and Tired of Overtime." *Globe and Mail*, 7 Oct., B4.

Robertson, D. 1992. "Corporate Training Syndrome: What We Have Is Not Enough, More Would Be Too Much." In A. Beckerman, J. Davis, and N. Jackson, eds. *Training for What? Labour Perspectives on Skill Training.* Toronto: Our Schools/Our Selves Education Foundation.

Robertson, D. and Wareham, J. 1987. *Technological Change in the Auto Industry.* Willowdale, Ont.: CAW Technology Project.

Rutherford, T. 1995. " 'Control the Ones You Can' – Restructuring, Recruitment and Training in Kitchener-Waterloo Manufacturing, 1987–1992." *Canadian Geographer,* 39, no. 1, 30–45.

Scrivner, L. 1994. "Why We All Spend More Time at the Office." *Toronto Star,* 11 Dec., A1.

Senker, P. 1986. "The Technical and Vocational Education Initiative and Economic Performance in the United Kingdom: An Initial Assessment." *Journal of Education Policy,* 1 No. 4, 293–303.

Sharpe, A. 1994. *Developments in Work Re-organization in Canada: An Overview.* Ottawa: Canadian Labour Market and Productivity Centre, March.

Sheppard, R. 1994. "The New Deal: No Training, No Welfare." *Globe and Mail,* 13 Dec., A23.

Statistics Canada. 1992. *Labour Force Annual Averages,* Cat. no. 71-220. Ottawa.

Stephen, J. 1993. " 'What's Training Got to Do with It?' Training and Unemployed Workers." Paper presented at the CRWS Training and Education Workshop, York University, Toronto, June.

Stoffman, D. 1993. "The Skills Squeeze: Working-Class Heroes." *Globe and Mail, Report on Business,* Dec. 52–9.

Streeck, W. 1989. "Skills and the Limits of Neo-liberalism: The Enterprise of the Future as a Place of Learning." *Work, Employment and Society,* 3 no. 1, 89–104.

Storey, K., and Shrimpton, M. 1994. "From New Town to No Town: Implications of Increasing Use of Commuting by the Canadian Mining Industry." In J. Andrey and, J.G. Nelson, eds., *Public Issues A: Geographical Issues,* 87–110. University of Waterloo, Department of Geography, Geography and Public Issues Series 2.

Study Group on Technicians and Technologists. 1993. *Tapping Our Potential; Technicians and Technologists of Tommorrow.* Ottawa: Canadian Labour Market and Productivity Centre.

Sunter, D. 1994. "The Labour Market: Mid-year Review." *Perspectives,* Statistics Canada, summer, 2–11.

Sunter, D., and Morrissette, R. 1994. "The Hours People Work," *Perspectives,* Statistics Canada, autumn, 8–13.

Tickell, A., and Peck, J. 1992. "Accumulation, Regulation and the Geographies of Post-Fordism: Missing Links in Regulationist Research." *Progress in Human Research,* 16 no. 2, 190–218.

United Steel Workers of America (USWA). 1992. *Empowering Workers in the Global Economy: A Labour Agenda for the 1990s.* Toronto: USWA.

Van Alphen, T. 1994. "Magna Looks Overseas in Bid to Find Five Skilled Workers." *Toronto Star,* 10 April, A4.

Walker, W. 1994. "Province's Job Training May Be Tied to UI: Rae." *Toronto Star,* 6 Feb., A3

Wells, D.M. 1993. "Recent Innovations in Labour-Management Relations: The Risks and Prospects for Labour in Canada and the United States." In J. Jenson and R. Mahon, eds. *The Challenge of Restructuring: North American Labor Movements Respond.* Philadelphia: Temple University Press.

CHAPTER TWENTY-TWO

Technological Change and Innovation: Policy Issues

JOHN N.H. BRITTON, JAMES M. GILMOUR,
WILLIAM SMITH, AND GUY P.F. STEED

Recent and fundamental advances in science and technology (S&T) have weakened the long-standing relationship between national economic prosperity and natural resources. Now nations may prosper on the basis of their knowledge and intellectual capability in science and technology, particularly as these bear on management, organization, and production. The basis of trade between regions and nations and the wealth to be derived from it have also changed and increasingly reside in the quantity and richness of human skills and intellectual resources.

Canada is distinctive among members of the G7 in its continued reliance on its natural resources. In a recent analysis, Porter (1991) identifies five major clusters of natural resource/export activities, defined by the end-use of the products involved. Three of these are clusters of early-stage (processing) activities – metals, forest products, and petroleum/chemicals. Collectively these account for 62 per cent of Canada's exports. The only "final consumption" cluster, food/beverages, is also mainly composed of minimally processed products such as fish and grain. The fifth cluster includes automobiles, trucks and parts, aircraft and urban mass-transit equipment. In this case the market is a combination of final consumers and firms in the services.

The long-term goal of Canadian economic development policy has been to build on and diversify away from specialization in resource extraction and processing. The earlier chapters outline the many ways in which Canada has struggled to establish a modern economy in which secondary manufacturing and new service industries take their full place alongside the traditional resource sector. Some of these industries have shown themselves to be the innovative and production equals of foreign competitors. It remains true, however, that most of the export diversification over the past three decades is a direct result of the trade-enhancing effects of the Auto Pact.

In only two regional cases – southern Ontario and southern Quebec – has this diversification of the economy attained any significant scale, and any serious decline in Canada's natural resource industries would still pose a serious threat to these areas. The head office, finance, research, and marketing functions of many Canadian resource firms are centred in Toronto and, to a lesser extent, in Montreal, and the resource sector continues to be a source of demand for the financial, other-service, and manufacturing industries of these cities and their regions (see in this volume, Coffey, chap. 18; Semple, chap. 19).

Central to any assessment of Canada's industrial capabilities is its continued position as a heavy net importer of technology in the form of parts, patents, licences, specifications, and finished products. These imports are an established feature of the country's approach to maintaining its employment structure and standard of living. While Canada has a high standard of living in terms of average personal disposable income and social benefits, the growth in real income of Canadians has slowed considerably and Canada's competitiveness has slipped as unit labour costs have risen faster than those in other industrial countries. The 40 per cent differential in unit labour costs in manufacturing between Canada and the United States that has appeared since 1980 is attributed to slower productivity growth (40 per cent), higher rates of inflation (38 per cent), and faster real wage growth (16 per cent) (Economic Council of Canada 1992).

Part of the evidence about weak productivity growth is the lag in manufacturing productivity compared to the United States. Canada closed to within 24 per cent in 1980, but the gap widened to 45 per cent over the next decade, when Canada was also overtaken by other members of the G7 or lost ground to them and several newly industrializing countries. The origins of the productivity problem are complex, but they involve low levels of investment in machinery and equipment and in worker training (relative to GDP) throughout the 1980s, slow adoption of new technology in the workplace, and low levels of product innovation. As well, Canada remains constrained by the small number and size of its technology-intensive industries.

The task of catching up to some of the leading nations is urgent and is made the more pressing because the substantial and growing commitment to S&T by governments and firms in other industrialized countries is producing impressive growth in their economic capability. Meanwhile, Canada continues its long-run trend of high-technology trade deficits and relatively low expenditures on research and development (R&D). Solutions to its productivity problems involve large-scale investment in modern technology, knowledge and skills, and supportive physical capital. Canada does not lack scientific talent, but Canadian business is relatively weak in funding and performing R&D (see Britton, this volume, chap. 14) and similarly in turning the results of R&D into commercial gains. Given these characteristics of Canadian business, and the pattern of technological investment elsewhere, major changes in the strategic responses of many industrial firms are required to avert continued erosion of Canada's international economic position.

Revaluation of Resources and the Environment

While Canada is lagging behind the international leaders in applying new technology to industry, advances in S&T are reducing the economic value of its traditional resource base. Export competition is expanding faster than industrial demand for natural resource products, which are in addition experiencing a secular down-trend. In part, this decline in demand reflects the impact of technologies that substitute human-made for natural materials and also make more effective use of natural materials. These demand-reducing factors are reinforced by public concerns about environmental degradation and by resource depletion. Governments, individually and collectively, are setting tougher standards on emissions of toxic wastes, and though these standards may stimulate process innovation they frequently increase production costs.

Introduction of environmental controls into trading relationships has already been felt, as some European buyers demand chlorine- and chlorine dioxide–free pulp. There has also been talk in Europe of trade boycotts of Canadian forest products in view of Canada's reputedly poor forest management practices. Recycling is another non-tariff barrier to trade, now well established in some nations; for example, 50 per cent of used paper is recycled in Japan, but only 26 per cent in the United States and 11 per cent in Canada. Pressures for recycling, combined with the more general need to reconcile economic development with the environment, promise to force production changes throughout the Canadian economy.

Nowhere is this more evident than in Canadian agriculture. Until recently, Canadians used the need for economic competitiveness in international markets to deflect attention from environmental concerns. This is no longer possible. International pressures for new business practices are reinforcing Canadian domestic concerns about the environment. The U.S. Farm Bill of 1991 specifically supports sustainable agriculture. The European Union is integrating environmental objectives into its agricultural policies, and environmental practice will be a key factor in future trade negotiations. What is environmentally desirable is now also economically essential. Without an effective policy stance, Canada remains vulnerable on a broad front. About half of Canada's agricultural production is exported; the agriculture-food industry provides direct and indirect employment for 14 per cent of the labour force and generates as much as one-third of Canada's trade surplus in goods.

This level of exposure to global markets leaves agriculture-food vulnerable to sudden shifts in world trade conditions. On the demand side, shifts in taste and increased self-sufficiency in other markets for the crops that Canadian farmers used to provide reduce prices. On the supply side, environmental problems are reducing productivity and profit levels. Salinization on parts of the Canadian prairies has reduced crop yields by as much as 75 per cent; soil erosion is inadequately controlled and remains a major economic and environmental threat. The extended drought on the prairies in the 1980s provided a foretaste of the threat to production posed by global warming. Yields collapsed, as did export earnings; bankruptcies soared; and despite billions of dollars in aid to cushion the brunt of the disaster, serious depopulation resulted (Science Council of Canada 1992a). In short, the economic outlook for this sector is highly uncertain, and farmers face a complex task of adaptation to meet new circumstances.

The success of the Canadian agriculture-food industry is conditional on a shift in emphasis away from conventional measures of efficiency based solely on the ratio of crop yields to inputs of fertilizer, land, machinery, and the like. More complex goals and measures are needed that meet much broader environmental, economic, and social objectives.

Reorientation of the agriculture-food system to meet new market requirements and integrate economic and environmental strategies will require adoption of new technologies and management practices. The scientific and technological challenge is enormous. However, the major barrier to success is probably the enormous popular resistance to new products and processes, even to those that are more environmentally friendly than the ones already in use. Food irradiation is a useful example: though it would reduce use of preservatives and packaging it has generated more emotional than informed reactions. Advances in biotechnology research are also in the long term likely to be critical to the future of the sector. But the steps from R&D through innovation to implementation are slow and tortuous. They could be blocked altogether unless supported by the necessary institutional and policy reforms.

Problems in Canadian Industrial Innovation

The case of agriculture illustrates a general principle: if Canada is to improve its competitive position in any sector it must become more innovative. Over the last three decades our theoretical understanding of innovation has shifted from notions of technology-push and market-pull to more complex models that incorporate feedback from the market to the production, engineering, and design components of the process. From this base innovation is conceived of as a process that requires integration of a firm's R&D and manufacturing with the related activities of suppliers and the needs and product ideas of leading-edge customers. Most recent models focus on systems integration and networking (Freeman 1991) and these involve linkage of internal and external collaborators in the innovation process. Innovation should have primary place in corporate strategy: companies with aspirations to be competitive cannot ignore this principle.

Despite an improved understanding of the innovation process, there is a misleading perception among the general public that innovation follows a linear sequence from ideas that are initiated in scientific research, developed into technology, and commercialized after design, engineering, prototype production, and marketing. Combined with arguments about market failure because a firm performing R&D may not appropriate all the benefits that it generates, this view has justified the subsidy of expenditures by firms on R&D. It has failed, however, to provide deeper understanding of the (economic) factors that influence the willingness/ability of firms to do research (Rosenberg 1991) or to achieve commercial results. While greater R&D performance is essential to improved competitiveness, the factors that underlie Canada's limited industrial innovation are too complex to permit reliance on a single R&D target as a national innovation policy.

A more relevant goal is better management of those innovation processes that lead to new or redesigned and improved products, services, and processes across the broad frontier of established and new industries. To become more innovative, productive, and competitive, more Canadians, from the factory floor to the boardroom, in universities and colleges, and in federal and provincial bureaucracies, must develop an informed grasp of the nature of innovation; how it occurs – how it needs to be stimulated by corporate directors, owners, and managers, how it can be integrated into a comprehensive corporate strategy; and how it can be sustained with employees' involvement and supportive public policy.

The process of innovation is recognized internationally as highly complex, particularly as it reflects the peculiarities of each individual industrial, national and regional situation. Thus the speed and strength of the worldwide expansion in the role of human resources as a source of wealth are not inconsistent with the hope that the Canadian economy can continue to enjoy the benefits of its natural resources while developing competitive strength based on s&t, entrepreneurship, management, organization, and marketing. Such a composite strategy is plausible if generic technologies are used to transform production processes and if product development and redesign are strongly supported by business: the characteristics of the Canadian innovation system ensure that this will remain a major challenge.

The Current Pattern of Innovation

It is difficult to be optimistic. The dominant technological strategy of the leading firms in many industries is that of adopt/adapt (Science Council of Canada 1992b). This approach involves acquisition and minor modification of technology developed elsewhere and commits business to little expenditure on product development. Canadian firms generally undertake about one-third to one-half as much R&D (as a percentage of sales) as international competitors. Canada's telecommunications-equipment industry is the sole exception: Northern Telecom's ratio of R&D/sales is about twice that of AT&T. The pulp and paper industry is more representative of the general Canadian pattern: in Sweden, Finland, and Japan, firms spent between 0.8 and 1.0 per cent of sales on R&D, while the Canadian level is 0.3 per cent (Porter 1991). Moreover, in this industry, Canadian firms have been oriented to commodities – lumber, pulp, newsprint. They participate heavily in cooperative R&D and do not appear to consider technology a major source of competitive advantage.

Few firms in Canada employ a breakthrough strategy as the basis for competition because in order to attain some measure of technological lead – the goal of the strategy – firms must meet very high absolute thresholds of R&D expenditure and investment. A more feasible alternative (to adopt/adapt) for Canadian firms is incremental innovation, by which they continually build on their existing, successful technological competence: this is the favoured approach among the world's leading firms. Even this approach, however, demands that firms try to meet what may be high R&D thresholds, for such products as central-office switches (telecommunications), the auto parts that "Tier 1" suppliers contract to produce, reformulated gasoline standards by refineries (oil and gas), integrated circuit foundries (electronics), new drugs (pharmaceuticals), and new aircraft engines (Science Council of Canada 1992b). Few Canadian companies meet the R&D thresholds of this list; nevertheless, proportionally more small and medium sized firms in electronics, telecommunications, software, electric power, autoparts, and machinery pursue the incremental approach with products that have lower thresholds.

Canada needs more incremental innovators, but that will be a substantial task, as may be gauged from a more detailed profile of Canadian industrial innovation. The distinctive patterns of Canadian innovation have been recognized for some time, but they have been clarified recently by an analysis of interview data drawn from 25 Canadian industries that have, or once had, relatively high shares of world exports (Porter 1991).

Most of the nine resource-based industries identified by Porter derive their competitive advantage from relatively inexpensive inputs. Few have upgraded or widened their sources of advantage; they have not regarded investment in the latest process technology as a competitive necessity; and they have failed to upgrade to higher-value products and have not invested in skills development or used labour efficiently. Among the resource industries, nickel is the best known for the technological advantage that it has generated from its resource base. It is followed by aluminum, which now produces higher-value products, but mainly outside Canada.

Most non–resource-based industries have focused on risk-averse, inward-looking strategies to serve the domestic market. High tariffs left them with diversified product lines and dampened the stimulus to respond to changes in international markets or to new technology. Even export-oriented industries commonly look no further than the United States and have failed to benefit from the stimulus of a wider base of market competition and technological change.

Among 11 industries that are more innovation driven, consulting engineering, geophysical contracting, and ice skates have benefited from fierce domestic rivalry and achieved international success.

Canada has developed few relatively advanced industrial skills or technologies that are marketable in their own right or provide industry leadership on a world scale. Lack of sustained and sophisticated human and physical capital investment is the prime cause.

Often customers who generate demand for advanced products are overseas buyers. Domestic demand has rarely provided pressure for firms to upgrade their sources of advantage: few Canadian firms have generated sophisticated or technology-intensive home demand. The buyers of consulting engineering services are an exception.

Canada fails to generate competitive advantage from strong, supporting industries because these tend to be weakly developed. In part, this results from vertical integration of the leading firms that develop their own (capital and intermediate) inputs or source them from abroad. The consequence is limited development of those segments of the economy that might otherwise develop specialized skills and technology in machinery and advanced-components and producer services.

An Industrial Geographic Interpretation

Porter's study describes the core, competitive weakness of Canadian industry: it has never developed an innovation system in which there are sufficient creative and stimulating links (or effective competition) between firms to cause high rates of development, adoption, or modification of processes or technologies.

These problems have their origin in a set of characteristics common to all small countries. Small nations have scale disadvantages in mass-market innovation. Secondary manufacturing industries and some services therefore focus on custom design or non-standardized products. R&D is dominated by a few firms that are prepared to pay the entry costs for R&D and commercialization. The fortunes of innovative sectors tend to depend on the performance of a few firms and a few technologies. Small market size and low levels of R&D inhibit imitation as a form of innovation and thus slow technology diffusion (McFetridge 1991; Walsh 1987).

Compared with most industrial countries, Canada has a small manufacturing base, and both manufacturing and service sectors are dominated by small-scale production. Larger firms are disproportionally foreign owned, mainly from the United States. Covering nearly 50 per cent of all manufacturing activity, foreign ownership reaches even higher levels for technology-intensive industries (see Britton, this volume, chap. 14). Though the markets of these firms are mainly domestic, strategic decisions, especially those related to innovation, are closely controlled by the global strategies of their parent corporations. Thus these firms are less responsive to Canadian innovation policies.

The small scale of Canada's machinery sector is central to an explanation of its low rate of industrial innovation. Canada has very few internationally competitive machinery firms. This situation adversely affects access to and control over fast-changing process technologies and hinders the capacity for process innovation and upgrading. The trade deficit in machinery has increased by 31 per cent since 1978, a consequence of the size of the industry and a reflection of other features of industrial management and structure (such as foreign ownership). It also slows innovation diffusion among Canadian firms. Producers normally work more closely and easily with domestic equipment firms than with foreign-based ones and with local firms rather than those in more distant regions (see Gertler, this volume, chap. 15).

Theory and analysis in economic geography suggest that the most advantageous locations for manufacturing have a strong spatial concentration of firms that require and supply industrial services, equipment, and components (agglomeration economies). Both small and large firms benefit from improved access to inputs, flexible supply systems, and savings in communication and delivery times. Though Canada has a number of small industrial regions, especially in southern Ontario, only Toronto and Montreal provide the full range of benefits associated with major industrial agglomerations. These two places have sufficient scale and variety of industrial and other producer services and component suppliers, as well as infrastructural facilities such as international air service, metropolitan amenities, and research universities. Yet small and medium-sized firms in these locations also suffer from the problems of contact and linkage identified for the economy as a whole (Britton 1993).

There has been a widespread tendency to overlook any structural and behavioural weaknesses in Canada's major industrial locations, especially Toronto, in favour of policies to develop less-favoured regions. However, now that the long postwar boom has been replaced by episodes of stagnation, slow growth, and structural change, the problems of flawed and incomplete innovation systems in Canada's industrial regions are of greater significance and can no longer be ignored.

Some commentators argue that Canada's problems in industrial innovation will be reduced by the North American free trade agreements. From this perspective, the Canadian innovation system is open and the level of economic integration with the United States reflects geographical proximity, closely linked capital markets, mutual investment in one another's economy, and relatively unrestrained flow of goods. This is often the perspective of the larger firms that view all of North America as the innovation system in which they find their immediate competitors, key customers, and suppliers (Rugman 1992). But the management of most (smaller) Canadian-owned enterprises probably assumes a narrower frame of reference and conceives of Canada alone as home base and market. These two views may not be incompatible: if the strategies of large firms lead them into forming alliances and joint-ventures and seeking subcontracting arrangements and other links with firms in Canadian as well as U.S. cities, Canadian industrial regions could gain from more effective production systems and their closer connection with U.S. production points. This network view of production suggests a complex pattern of flows far removed from the unidirectional stream of components and subassemblies from the United States that has been the pattern of American branch plants.

Strategies to increase industrial innovation will differ somewhat according to the size of firms. A successful strategy must assist the large firms that dominate high-value product areas and technologies. Because Canada has few of these firms it is sorely pressed to build effective production systems. Business networks that stimulate innovation rely on the demand links of these "flagship" companies to align and harmonize the strategies of firms that make up each product system. In this way internal Canadian interdependence may contribute to external competitiveness (D'Cruz and Rugman 1992).

Threshold firms are also a vital innovative force in the economy (Steed 1982; Premier's Council, Ontario, 1988). These are young Canadian, technology-based companies of intermediate size that have been successful in highly specialized, international markets. Their technical leadership relies on global marketing to provide ideas to drive design and R&D. But small size and lack of capital restrict growth. Firms in this group had small beginnings but achieved fast growth despite management frailties and financial constraints (Macdonald 1991).

Some small and medium-sized enterprises (SMEs) have proved themselves capable of valuable product innovations, and some are noted as fast adopters of advanced, flexible design and production equipment. Such companies are a prime resource for production systems. The majority of small and medium-sized firms, however, are not innovative; they acquire new techniques slowly and take only halting steps toward re-design of their products. They have neither the financial resources of larger firms nor easy access to information that could enable them to be more productive. The federal government and the provinces themselves need to recognize the policy initiatives for innovation and technology diffusion that have had success in other industrial countries. Formation of the Canadian Technology Network is a step in the right direction, providing quick access to technology information across the country.

LOOKING FORWARD

In recent years deregulation, privatization of crown corporations, and various trade agreements – the FTA, NAFTA, and GATT – have begun to transform the economic environment of Canada. The federal government has replicated some of the change occurring in other major industrial regions and countries and encouraged a more competitive environment. For a small economy, much of that competition must come from more open external trading relationships. This is doubly important in Canada, where federal-provincial negotiations are only slowly reducing economic balkanization among provinces.

Through NAFTA and GATT, Canada has exposed its economy to intensified international competition. Secondary manufacturers and service firms that have grown to their present scale by focusing on the needs of the domestic market are now adjusting production and location to broader markets. Even Canadian firms in the auto industry, operating under an earlier set of trading arrangements, are developing new strategies to suit an increasingly competitive North American market.

Competition, however, is only one part of the solution. The resource industries have relied on export markets from their inception but have for the most part failed to adapt well to international competition (this volume, Barnes, chap. 3; Hayter, chap. 6; Wallace, chap. 7; Hayter 1987; Porter 1991). The competition encountered as part of the relatively "open" trading that they enjoyed in the past has proved an inadequate stimulus to innovation, and some of the resource industries are now struggling. They have uncertain futures as a result of inadequate Canadian corporate action, the greater success of competitors, and changes in the rules of international trade.

A major problem for Canada's firms in both the resource sector and in secondary manufacturing is how to increase their value added. Under NAFTA, firms have greater locational choice within North America. This leverage may give Canadian labour more reason to join cause with those firms that are innovative, progressive in their labour relations, and advanced in their marketing and other business strategies. Increasingly, too, there is a clearer fit between some corporate priorities and public goals; for example, the developmental value of purchases by foreign companies from Canadian sources, as a consequence of their establishment of world product mandates, is identified both as a policy objective (Government of Ontario 1992) and as a corporate goal (Porter 1991). Most small Canadian businesses could be helped immeasurably by this type of supportive purchasing practice. The corollary is that the information resources of larger firms and flexibility of small firms can work to boost the productivity of the larger businesses.

In recent years there have been small advances in innovation policies. One has been the faster and more effective dissemination of technology and scientific knowledge from government laboratories to business (Science Council of Canada 1991). Also notable are federal and provincial initiatives, including Ontario's Centres of Excellence program, which support collaborative (university-industry-government) precompetitive R&D. In this same vein, there is greater public support for alliances between SMEs, networks to provide access to a variety of technical information, and agencies to provide technology and to undertake brokerage functions between small firms (Science Council of Canada and Canadian Advanced Technology Association 1990; Porter 1991; Government of Ontario 1992). One major concern is how to improve trade associations to stimulate the awareness, knowledge, and competence of small firms.

There is, however, still some room for greater action along these lines and for new public and private initiatives to help small firms acquire information about new technology and implement new methods of work (Britton 1989; 1991). For too long Canada has ignored the substantial evidence that its industries need better extension or advisory services to increase the rate at which firms install best practices and new process technology, to increase the access of small businesses to sources of technical information, and to reduce bottlenecks in the functioning of capital markets (such as venture capital).

General networking services could help increase the level of design inputs and re-engineering and speed up adoption and adaptation of new product concepts and production techniques. Such improvements could assist Canadian firms of all sizes to keep up with their competitors. Faster adoption and more effective use of flexible technology are needed to produce high-quality, highly customized products. By taking less time to develop production systems and products that use less material and effort – "agile and lean systems" (Womack, Jones, and Roos 1990) – more Canadian companies could compete with firms in other industrial economies.

One potential stumbling block is satisfying the substantial demand that the continual advances in production methods make for worker training, whether in-house or within the formal education system. Fortunately, one of Canada's assets is its relatively well educated and experienced industrial labour force. More firms, in a knowledged-based, advanced economy, will require a skilled labour force if products (and processes) are to be developed and continuously improved. Firms need to maintain certain core competence so that the design process is fully integrated with other company functions – production, marketing, engineering, and R&D. Large and small firms alike can create production systems in which use of teamwork within firms is extended to similar interfirm relationships, leading to long-term agreements with a few preferred suppliers and customers (Tidd 1991). Successful development of these systems will require progressive linking of individual firms in a variety of informal and contractual ways.

REFERENCES

Britton, John N.H. 1989. "A Policy Perspective on Incremental Innovation in Small and Medium Sized Enterprises." *Journal of Entrepreneurship and Regional Development*, 1, 179–90.
– 1991. "Reconsidering Innovation Policy for Small and Medium Sized Enterprises: The Canadian Case." *Environment and Planning C*, 9, 189–206.
– 1993. "A Regional Industrial Perspective on Canada under Free Trade." *International Journal of Urban and Regional Research*, 17, 559–77.
D'Cruz, Joseph R., and Rugman, Alan M. 1992. *New Compacts for Canadian Competitiveness*. Toronto: Kodak Canada Inc.
Economic Council of Canada. 1992. *Pulling Together: Productivity, Innovation and Trade*. Ottawa: Minister of Supply and Services.
Freeman, Christopher. 1991. "Networks of Innovators: A Synthesis of Research Issues." *Research Policy*, 20, 499–514.
Government of Ontario. Ministry of Industry, Trade and Technology, Ontario. 1992. *An Industrial Policy Framework for Ontario*. Toronto: Queen's Printer for Ontario.
Hayter, R. 1987. *Technology and the Canadian Forest-Product Industries*. Ottawa: Science Council of Canada.
Macdonald, Mary. 1991. "Creating Threshold Technology Companies in Canada: The Role for Venture Capital." Discussion Paper, Science Council of Canada, Ottawa.
McFetridge, Donald G., ed. 1991. *Foreign Investment, Technology and Economic Growth*. Investment Canada Research Series. Calgary: University of Calgary Press.
Porter, Michael. 1991. *Canada at the Crossroads*. Ottawa: Business Council on National Issues and Minister of Supply and Services.
Premier's Council, Ontario. 1988. *Competing in the Global Economy*. 3 vols. Toronto: Queen's Printer for Ontario.
Rosenberg, Nathan. 1991. "Critical Issues in Science Policy Research." *Science and Public Policy*, 18, 335–46.
Rugman, Alan M. 1992. "Porter Takes a Wrong Turn." *Business Quarterly*, Winter, 59–64.
Science Council of Canada. 1991. *Reaching for Tomorrow – Science and Technology Policy in Canada, 1991*. Ottawa: Minister of Supply and Services.
– 1992a. *Sustainable Agriculture: The Research Challenge*. Ottawa: Minister of Supply and Services.
– 1992b. *Sectoral Technology Strategy Series No. 15: The Canadian Computer Software and Services Sector*. Ottawa: Minister of Supply and Services.
Science Council of Canada and Canadian Advanced Technology Association. 1990. *Firing up the Technology Engine*. Ottawa: Minister of Supply and Services.

Steed, Guy P.F. 1982. *Threshold Firms*. Background Study No. 48, Science Council of Canada, Ottawa.

Tidd, Joseph. 1991. *Flexible Manufacturing Technologies and International Competitiveness*. London: Pinter.

Walsh, Vivien. 1987. "Technology, Competitiveness and the Special Problem of Small Countries." *STI Review*, 2, 81–132.

Womack, J.P., Jones, D.T., and Roos, D. 1990. *The Machine That Changed the World*. New York: Rawson Associates.

CHAPTER TWENTY-THREE

Conclusion: Canada's Emerging Economic Geography

JOHN N.H. BRITTON

The focus of this book is on a short period in the evolution of Canada's economic geography. As a team of geographers, we have described and analysed the intriguing ways in which it has grown and been moulded by myriad influences, but we have done so only for recent decades. In a historical sense, the restricted time frame of our view must have limited our perception of change and influenced our results, but we have proceeded knowing that this is a period of considerable Canadian economic growth, the Canadian population of 18.2 million in 1961 having become half as large again by the census of 1991.

Despite this indication of dynamism, we have presented evidence of considerable stability in the geographic structure of the Canadian economy and its connections with foreign partners. There are good reasons for this situation. Stability results from both the inertia of sunk capital and the interdependence of activities in those locations where development is concentrated. In particular, as economies of agglomeration have emerged strongly to shape the Canadian urban system, so the backbone of Canada's economic geography has become established and has helped structure the location of much subsequent investment in infrastructure, production, and distribution. We have shown that the net locational effect of the various processes that operate within the urban or regional system has been to maintain the largest cities of 50 years ago as the major nodes of the contemporary economy. This is especially true for the location of finance and other services, manufacturing production, and the headquarters of corporations in all sectors. This is the case even despite significant changes in both the occupational structure and the sectoral distribution of employment.

Policies at each level of government also have had economic and spatially stabilizing effects. Agricultural income-support programs, especially supply-management schemes, for example, have tended to preserve the scale and location of rural production and the processing of foodstuffs; otherwise market forces would already have reduced activity of these industries in particular regions. Public policies such as "equalization" and "stabilization," too, have had the designed effect of dampening the impact of long-term declines (or persistent differences in regional income capacities) and shorter-term oscillations in regional economies. In doing so, they have added to spatial stability by reducing the incentives of the working-age population to migrate.

While social support programs and equalization of provincial ability to maintain public services generally function to stabilize the locational pattern of the economy, Canadian governments have tried by other means to engineer spatial economic shifts thought to be desirable from political or social standpoints. By working against market trends, regional development programs, for example, have sought to plant the seeds of economic renewal in low-income regions. They have not been particularly successful. Though this fact has been known for some time, successive federal governments have continued to steer development to nodes less favoured by market processes rather than speeding technological change in regions that have established investment, industrial skills, and necessary infrastructure. Though it is a simplistic comparison, federal support for regional agencies (including the Hibernia project) was $1.1 billion (1994–95) while the National Research Council (NRC) received only $446 million.

In many ways, the minimal success of locational policy supports the view that most such initiatives have been developed in ignorance of the interests of firms that have acted on their locational understanding of the advantages of agglomeration, particularly those that derive from close contact with suppliers, labour skills, and the market. Provincial governments, too, persist with protectionist arrangements, which, it is claimed, reciprocate restrictions on trade and employment. Canada's experience prompts the generalization that the policies directed to changing the market-based location of production generate inefficiencies.

Despite the degree of spatial stability, important changes in the organization and structure of economic activities have been initiated in recent years by firms, industries, and governments responding to new technologies, market opportunities, competition, and global patterns of economic change. The influence of these factors, however, has been complicated by the impact of Canadian monetary policy, which made the Canadian dollar worth more in relation to the U.S. dollar than it would otherwise have been in the late 1980s, and by the recessions of the early 1980s and 1990s, which have also stimulated particular forms of industrial restructuring. The contractions in demand, especially, have led to permanent plant shutdowns and unemployment, and changes in the structure of jobs and of the mix of skills required by firms have become evident. Unlike the situation in downturns in the business cycle over the past 40 years, skilled blue-collar workers have been laid off permanently, and the rate of early retirements facilitated by employers (and organized labour) has risen sharply. In many surviving plants, production has been reorganized so that inventories of parts and sub-assemblies are at a minimum (using just-in-time principles), and less labour is required on production lines, thus ensuring the permanently smaller relative size of the manufacturing labour force.

The structural changes in employment, reflecting new work processes, organizational structures, and technologies ushered in during the early 1980s, but especially during the early 1990s, have no strong parallels in Canada's recent industrial experience. Historically, however, major economic downturns have been associated with significant technological shifts in industrial economies. This was especially true for the "depressions" of the 1890s and the 1930s, and the recurring bouts of international recession at this late stage in the twentieth century have stimulated broadly based speculation that a comparably influential bundle of technologies represented by electronics, computers, and telecommunications networks – information technology (IT) – will be the leading origin of the next phase of growth. Already we have seen dramatic expansion in use of micro-processors in a wide range of applications and penetration of networked desk-top computers in all manner of workplaces. Adoption of reprogrammable – flexible – machines has allowed productivity gains to be made by small and medium-sized manufacturing firms and by larger enterprises and has allowed them to undertake product innovation to meet the demands of their market niches.

Canada is not noted internationally, however, as a producer of machine designs and equipment, though it does have some internationally competitive companies. By contrast, the telecommunications industry does promise to contribute to the production of new technology, and new telecommunications products have been adopted quite quickly in Canada by those parts of the service sector that provide data processing, radio and TV broadcasting, and other telecommunications products, such as value-added network services and access to proprietary databases. Use of IT has begun to affect Canada's economic geography; for example, it permits the devolution of routine data entry, data processing, and warehousing to lower-order centres within the urban system, thus loosening traditional hold of the metropolitan suburbs on these jobs. For regions with long-term job deficits, such as Atlantic Canada, IT has been generating modest job growth, especially in New Brunswick, which has an extensive fibre-optic network and has encouraged the location of firms in some of these activities. Innovations in marketing and retailing have also generated spread effects down the urban hierarchy in response to market processes driven by innovative firms and consumers: computerized inventory, sales, and ordering systems permit this type of locational flexibility.

Despite these shifts it would be inaccurate to portray IT as an exclusively decentralizing force and one that will thus strongly reshape the spatial structure of the economy in that direction. In the new information-based activities, high-order jobs have a strong locational focus on large urban centres, and thus the new activities, including those in the producer services, add a significant dimension to economies of agglomeration. Nevertheless, the availability – and mobility – of well-educated workers throughout the urban system have enabled some new activities in services (computer services and software design) and manufacturing (electronics) to establish toeholds outside heartland cities. Higher-order, information-based jobs are expanding in both the industries of the "new economy" and in Canada's older activities, which are also sources of demand for new services. This diversity of the industrial "impact zone" leaves no site of production or distribution immune to the restructuring of jobs and ensures that gains in productivity and flexibility may be experienced in all locations. Consistent with this, more of the head offices of resource companies, for example, are now found in the cities of western Canada – an arrangement assisted by telecommunications services.

Trade liberalization reinforces a core characteristic of the Canadian economy – its openness to the world. Several authors in this book argue, however, that in the absence of other concerted trade initiatives NAFTA is likely further to reduce the geographic breadth of Canada's trading pattern, since its most obvious effect will be even closer ties to the U.S. economy. This implies an expansion in the range of connections between Canadian cities and their U.S. neighbours. Two-way flows of retail shopping and radio/television programming and advertising are long-term influences that have, on the margin, reduced the importance of Canada's main centres within the domestic economy, in favour of a closer web of ties by cities along the border.

Changes in manufacturing trade have already begun, exports (mainly to the United States) have doubled their share of production (1980–93) to more than 48 per cent, and domestic output has fallen from 73 to 53 per cent of the Canadian market. Under liberalized trade, this implicit development of specialized flows or goods and services within North America is expected to characterize all aspects of the economy. It is not likely that this will reduce the national role of Canada's major metropolitan centres. Nevertheless, the functions that they exercise are bound to become more specialized as regional flows across the border compete with the current national pattern of connections.

Theoretically, one corollary of freer trade is that secondary manufacturing and tradeable services in Canada may become less locationally centralized than hitherto. This is a major change, since aspirations for industrial development in many parts of Canada have foundered on the domestic economic logic of spatial centralization and agglomeration. In the future, transborder regional markets have a better chance to sustain economic diversification, and neighbouring U.S. regions, rather than central Canada, may be viewed as the locations of immediate competitors and of markets to be acquired. The potential for industrial growth outside the Canadian heartland depends (as it does in central Canada) on the gains that can be developed through product specialization of individual companies realizing (regionally) defined market advantages. Whether the resolution of North American production patterns means a wider pattern of Canadian industrialization or not, the spatial model of Canadian industrialization has become less related to the logic of Ontario's central location within the Canadian urban system. Certainly greater national specialization is expected under NAFTA, and promises to increase the size and variety of flows of industrial goods and services across the border. Since this will lead to greater spatial interdependencies with the United States, it is likely to weaken relationships between Canadian regions and cities.

As firms rationalize their operations across North America, individual firms and industries in Canada are changing. Until very recently, many service and manufacturing industries, especially food processing, furniture, and domestic appliances, have functioned primarily in the national market (one-tenth the size of the u.s. market); now they are being integrated into the North American organization of production. Inadequate adjustment prior to the FTA has left some industries decimated. The furniture industry, which is localized in central Canada, is a good example of negative trade impact. As Canada's tariff advantage of 15 per cent has been eliminated, a trade surplus of $400 million with the United States has been converted to a trade deficit of $300 million, with the u.s. share of the Canadian market now being 30 per cent (compared with its former 10 per cent) and about 30 plants have closed. The high costs associated with small-scale production and wider product range in Canada left producers vulnerable to u.s. plants, which have longer production runs in low-wage locations. The question for Canada is whether furniture is representative of a class of secondary manufacturing industries that can navigate the shift to new competitive ground, a process requiring greater design inputs, flexible production, and a higher proportion of skilled workers.

A number of Canadian high-value–added products, such as electronics, software, communications equipment, and machinery, have always been exposed to competition from many parts of the globe. For these and a wide range of higher-value goods and services for industrial and consumer markets, which are increasingly available from a global array of production points, distance has had little influence on the geography of markets. Now, however, Canadian manufacturing industries such as low-price furniture and durable consumer goods (white goods), hitherto protected by distance and residual tariffs, have been exposed to competition from low-cost points in the United States and Mexico. Some u.s. firms have begun to reorganize their Canadian operations – for example, through purchase of the publicly held shares (a minority) in their corporation's Canadian company in order to create a wholly owned subsidiary[1] (Saunders 1994). Some foreign companies, as a result, have reduced the stature of their Canadian head offices and reduced manufacturing operations to warehouse and sales functions (such as Union Carbide), while others have rationalized production so that the Canadian plant supplies the North American market with a subset of the former product range.[2] In this context, plants in Canada are focusing more strongly on product development from a local base. Some Canadian companies have moved their head offices, the most notable being the Toronto–Buffalo transfer of the former Massey-Ferguson (now Varity). Others have formed alliances with u.s. firms (such as CN North America, with J.B. Hunt) or made major investments (BCE, in Jones Intercable).

Some Canadian firms are choosing a strategy that requires them to be offensive innovators, but this is a significant departure from the experience of the majority of Canadian companies in secondary manufacturing and the services that previously have not sought North American market positions. Such a strategy is implemented fully by only a small proportion of Canadian firms. There is therefore a disjuncture between the efficiency and product innovation achieved by most Canadian businesses and the level needed for Canadian firms to obtain advantage from trade liberalization.

The extended logic of Canada's participation in the FTA and NAFTA is that increased competition should also stimulate Canadian firms to adopt advanced technology, become more efficient, and create new products. But there is no inevitability in this rationale, and its logic is challenged by evidence supplied by firms in such industries as forest products, which rely on export markets but innovate at a low level, despite opportunities to add more value in production.

Much of Canadian industry is noted for its slow adoption of technology compared with international leaders, and even Canadian firms making the necessary capital expenditures on new equipment do not necessarily make comparable expenditures on training workers. As outlined in this volume, this is both a private-sector failing and a public-sector problem; until very recently government policies have been inadequate to the task of helping firms adjust to cyclical, competitive, and technological change. This failure reflects the conservative political-economic style of federal policy 1984–93 but also the difficulty experienced by provincial governments in taking relevant action. In the light of this constellation of problems, there is a pressing need for an effective response to the many industrial firms that lack the knowledge to organize cogent responses to emerging market conditions.

A new combination of public policy efforts is necessary. Previously, as we have outlined, investment in national network infrastructure – railways, interprovincial highways, airlines, and communications – overcame distance, connected regions, increased export earnings, and added to the occupational and industrial depth of the country's metropolitan centres. Equally important policies have built human capital through educational systems, developed industrial skills, supported creative businesses, and encouraged application of science and technology.

Now, policies for Canadian development should emphasize two elements. First, continued development of the advanced national telecommunications network – the "information highway" – is required. This initiative has started with existing fibre-optic cable systems, which the telephone companies developed, while strategic deregulation of the telecommunications service industry would permit advantage to be taken of the technical convergence of the networks of cable-TV companies and telephone companies. Early users of high-quality network connections have been larger businesses, governments, and smaller businesses willing to pay for access to a high-capacity, long-distance network. Research and educational users require facilities superior to those currently existing. Eventually, the effect of improved network connections on consumer behaviour is expected to expand the range of public and private services provided electronically, using text, voice, and video, reduce prices, and improve the quality of these services.

To complete the policy updating, new initiatives to develop human capital are required that will involve enhanced training in the use of IT in workplaces, information on technology acquisition, advice on the penetration of new market locations, and

reduction of impediments in other aspects of innovation. More rapid industrial innovation by Canadian business enterprises, including faster, successful use of new technology, is vital if the current standard of living in Canada is to be maintained and if
the spatial fabric of the Canadian economy is to be sustained.

NOTES

1 For example, Campbell Soup, CIL, GEC, Goodyear, Indal, Inglis, Kelsey-Hayes, Nabisco,
   Redpath, Teledyne, Union Carbide, and Westinghouse (Saunders 1994).
2 For example, Heinz (*Economist*, 15 Jan. 1994, 68).

REFERENCE

Saunders, John. 1994. *Globe and Mail*, 18 Jan.

Index

advanced manufacturing technologies, 281–3, 285–8; automation in clothing, 217, 224–5; Italy, 285; mining and processing technologies, 128, 129, 133–4. *See also* computer-based technologies; Germany; Sweden
agglomeration economies, 16, 270, 346, 378, 439–40, 446. *See also* Montreal; Toronto
agriculture, 98–9; capitalization, 156, 158; comparative advantage, 155–6, 158–9, 162; environmental limitations, 156–9, 166–7; farm incomes, 155–6, 158–9, 162; farm size, 156, 158; labour, 156, 158; marketing boards, 159, 163–6; mechanization, 156, 158; problems, 435–6; production, 156; productivity, 158; regional patterns of production, 156–9, 167; self-sufficiency, 159–60; subsidies, 99, 155, 162, 165, 167, 445; supply management, 159, 164, 166. *See also* land
airlines, 304
Alberta, 54, 55; capital stock in, 11, 275–6; forest harvest, 104–5; forest planting, 105; Heritage Savings Trust, 55; resources, 7; staples, 101–3; transfer-payment contributions, 377

Alcan, 125
Athabaska Tar Sands, 276
Atlantic Canada, 53, 54, 55; development initiatives, 376–7; employment growth, 184, 185, 186, 187, 191; forest harvest, 104–5; forest planting, 105; production, 12; public-sector jobs, 377; staples, 101–3; trade, 30–1; transfer payments (in), 297, 377; unemployment, 175–6; worker mobility, 188
Australia, 124, 127, 129, 131
Auto Pact, 52, 53, 172, 195–6, 230, 232–6, 249, 250n2, 250n3, 250n5, 307
automatic vehicle-monitoring system (AVMS), 311–12
automobile industry; assembly, 231, 233–5, 243, 245, 250n1; capital investment, 238, 242; competitive advantage, 243, 247; future, 245–50; integration with Mexican auto industry, 246–7, 252n19, 252n20; integration with U.S. auto industry, 232–3; Japanese production system, 238–40, 251n7; Japanese transplants, 231–2, 237, 240–2, 245–6; location, 8, 172–3, 231, 241, 243, 247, 252n21; parts, 231, 233–5, 243–9, 250n1, 251n14; R&D and training, 232, 248–9; size and structure, 230–1;

technological renewal, 230, 237–42. *See also* Auto Pact

balkanization, 401–3, 441, 446
base metals, 124, 130, 134
branch plants, 195, 196, 213. *See also* foreign subsidiaries
Bretton Woods, 54, 85
British Columbia, 49, 54, 55, 57–60, 377; capital stock, 276; employment growth, 185, 186, 187; fishing industry, 58; forest harvest, 104–5; forest industries, 57–60; forest planting, 105; manufacturing investment, 277; public-sector jobs, 377; resources, 7; staples, 57–60, 101–3; transfer-payment contributions, 377; unemployment, 176
Bronfman family, 130
business strategy, 331, 334. *See also* corporate strategy

Calgary: corporate head offices, 296; high-technology industry, 267–8
Canada Lands, 144, 145, 152n
Canada-U.S. Automotive Products Trade Agreement (1965). *See* Auto Pact
Canada-U.S. Free Trade Agreement. *See* Free Trade Agreement (Canada-U.S.)
Canada Wheat Board, 162–3

Due.